Understanding and Negotiating EPC Contracts

Volume 2

Any project that involves an EPC contract is also likely to involve a number of other complicated contracts. The challenge of the parties to an EPC contract is not to try to eliminate risk but rather to put into place a narrative structure that enables the parties to predict the contractual result that would obtain if a risk materializes. If the EPC contract does not allow the parties to determine the consequences of an unanticipated situation, they will have to look to an expert, mediator, tribunal or court to impart guidance or pass judgment.

The sample forms of contract contained in Volume 2 of *Understanding and Negotiating EPC Contracts* are intended to serve as a guide to demonstrate how risks and responsibilities can be allocated among project sponsors, EPC contractors and the various other parties that may be involved in a project.

Collectively the sample forms in this volume offer an extraordinary resource that provides the benefit of lessons learned and a priceless insight into any project being undertaken that can help assure the resilience of any EPC project.

Howard M. Steinberg is of counsel and a retired partner in the law firm Shearman & Sterling LLP and has more than 25 years of legal experience in the infrastructure sector in more than 100 countries. He is named a leading lawyer in project finance by Chambers & Partners and IFLR 1000. He represents sponsors, off-takers, miners, suppliers, engineering firms, consultants, financial advisors, contractors, operators, underwriters, lenders, export credit agencies and multi-lateral institutions in the development, acquisition, restructuring and privatization of projects around the world. He holds bachelor's and business degrees from Columbia University and law degree from New York University.

Understanding and Negotiating EPC Contracts

Volume 2

Annotated Sample Contract Forms

Howard M. Steinberg

LONDON AND NEW YORK

First published in paperback 2024

First published 2017
by Routledge
4 Park Square, Milton Park, Abingdon, Oxon OX14 4RN

and by Routledge
605 Third Avenue, New York, NY 10158

Routledge is an imprint of the Taylor & Francis Group, an informa business

Publisher's Note
The publisher has gone to great lengths to ensure the quality of this reprint but points
out that some imperfections in the original copies may be apparent.

British Library Cataloguing in Publication Data
A catalogue record for this book is available from the British Library

Library of Congress Cataloging in Publication Data
A catalog record for this book is available from the British Library

ISBN: 978-1-4724-2378-8 (hbk)
ISBN: 978-1-03-283710-9 (pbk)
ISBN: 978-1-315-54929-3 (ebk)

DOI: 10.4324/9781315549293

Typeset in Baskerville
by Out of House Publishing

Contents

Volume 2

Volume I

Reader's Note

Liquefied natural gas (LNG) is shipped around the globe these days as readily as wood and coal were shipped centuries ago. Only massive infrastructure and planning have made this LNG energy trade possible. This "virtual" gas pipeline connecting our continents is a testament not only to engineering prowess but also legal coordination of contracts that cut across jurisdictions and industries. Projects are only as good as the contracts that hold them together. Fear of legal and contractual uncertainty is far more likely to sink a feasible project than is a technical challenge or lack of capital.

When the unexpected strikes—and it will—contractual protection is the best insurance. Years of watching the unexpected occur has taught me one thing—never stop the search for drafting clarity. Nothing is more disappointing than reading a provision of a signed contract that is not clear once an unanticipated contingency has arisen. A vague contractual provision is as useless to a project participant as if it were a hieroglyph.

Ambiguity prevents prediction. Without prediction come judges, juries and arbitral panels. Not a good commercial result, even for the most technologically solid of projects.

Can problems always be avoided? No. But the following text goes a long way to help minimize problems that need not arise if the proper analysis has been made.

Gabriel Touchard
Senior Legal Counsel
Global Gas and LNG
Engie
Paris, France

Foreword

Several years ago I was being considered by a prospective client to serve as its counsel on a large construction project involving an engineering, procurement and construction (EPC) contract. Toward the end of the meeting the prospective client reviewed our discussions and noted that I seemed to have a good grasp of what they needed from their counsel—an understanding of what would be involved in the project's development and subsequent financing, and that I came highly recommended from their biggest competitor. Despite all that, the prospective client went on to say that this would be a very important project for their company and that they had not worked with me in the past; how could they be assured that I would not make a mistake? I responded that I had already made all the mistakes that could be made so they need not worry about that. Years later, they remain a client.

The objective of this Volume II is to try to assist in avoiding unnecessary mistakes. In some sense, mistakes are just another name for experience. Experience is always good. In this volume, experience is used as the basis to suggest how to implement the concepts and concerns that were outlined in Volume I. Lack of focus and clarity in a contract can lead to great disappointment for all parties involved in a project.

In composing the contractual provisions of an EPC contract, each provision must try to address the situations that could arise within its ambit. I have found the first six questions of any news reporter—who, what, when, why, where and how?—to be the most effective way to understand, analyze, draft and agree upon any contractual terms. Once there is tentative agreement on a contractual term, the next step necessary to gauge the provision's robustness is to try to anticipate the project's future at two different stages—construction and then operation. The parties should hypothesize an event that would trigger the operation of the contractual provision in question and posit that such event has just occurred and that the affected party has assembled its relevant work team in a conference room in order to determine what the EPC contract requires be done and also try to identify what might be the various courses of action. This "academic" exercise (as non-academics say) should expose any weaknesses that may exist in the formulation of the provision and any further textual elaboration of the contractual provision that could be helpful.

Risk is sometimes defined as "decision-making under uncertainty." The challenge of the parties to an EPC contract is not to try to eliminate risk but rather to put into place a narrative structure that enables the parties to predict what contractual result would be obtained if a risk materializes. If the EPC contract does not allow the parties to determine the consequences of an unanticipated situation, they will have to look to an expert, mediator, tribunal or court to impart guidance or pass judgment.

Keen foresight and rigorous scrutiny are the parties' only hope for a sturdy contractual structure. Once the ink on the signature lines of the EPC contract is dry, time will reveal whether the parties have anchored their EPC contract to a bedrock of jurisprudence or mired it in legal quicksand.

About the Author

Howard M. Steinberg is a retired partner and now "Of Counsel" to the project development and finance group of the international law firm of Shearman & Sterling LLP. He concentrates on transactions involving the energy sector and focuses on the power industry in particular. He has represented sponsors, offtakers, miners, fuel suppliers, contractors, operators, underwriters and lenders in the development, acquisition, restructuring and privatization of infrastructure projects around the world. He co-authored a chapter on Brazil in the book *The Principles of Project Finance* and has written articles in periodicals such as *Power*, *Astronomy*, the Institute for Energy Law's *Energy Law Advisor* and *Asia Law & Practice*. He is listed as a leading lawyer in project finance by Chambers & Partners and IFLR 1000.

Mr. Steinberg is a member of the New York State Bar, a member of the Advisory Board of the Institute for Energy Law and holds a J.D. from New York University School of Law, an M.B.A. from Columbia Business School and an A.B. from Columbia University's College, where he majored in philosophy and economics and graduated *magna cum laude* and was inducted into *Phi Beta Kappa*.

Introduction

Any project that involves an EPC contract is also likely to involve a number of other complicated contracts. For both commercial and competitive reasons, the project's sponsor (and its financiers if the sponsor is financing its project) may be the only party (or parties) that will see (and accept) the terms and conditions of all of these contracts. Any inconsistencies, overlap, interstices or ambiguities between or within contracts can only be the augur for delay, unanticipated spending and legal proceedings.

If a project sponsor has not cemented together an interlocking rebar of project contracts before it gives notice to proceed with the work to the EPC contractor, the project sponsor may have unwittingly created fundamental faults that can have the potential to rip apart the project's technical feasibility and financial promise. Figure 2.3 of Volume I outlines what contracts might be employed to carry out a typical project. This Volume II focuses its attention on the key construction arrangements of a large project—the EPC contract. The EPC contract is the pedestal that must support the weight of all other project contracts. Unless the project sponsor and its advisors analyze what contractual burden the EPC contract must bear (and whether some of this mass is "dynamic" load that may shift as a result of economic and technical forces that will be placed on the project), permanent instability of a project's economics may be the result. A project sponsor cannot eliminate (or usually even identify) all the risks that are present in its project. In the face of the unknown, the project sponsor must foster candid dialogue, unswerving trust and expert analysis among the project's participants (and also those whom the realization of the project may concern). Success of a project can never be assured, but without the foregoing efforts, failure is. The purpose of this volume is not to unlock the secrets to a project's success, but to expose the concomitants of failure. The following sample contracts should be used as a guide to expose potential weakness and suggest methods for reinforcement. The rest can only be left to providence and ingenuity.

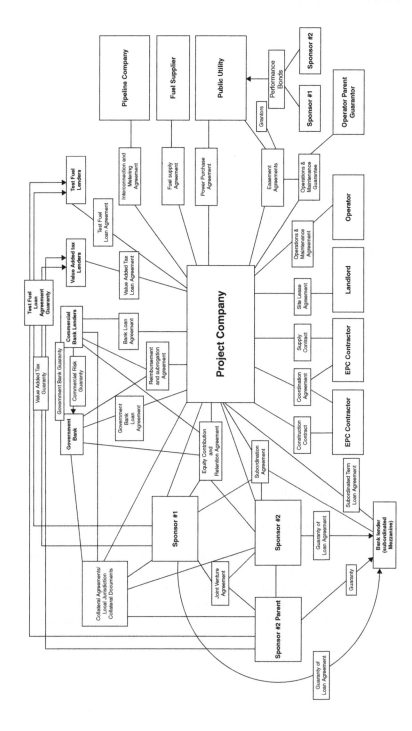

Typical contractual structure

How to Use These Sample Contract Forms

The sample forms of contract contained in this Volume II are intended to serve as a guide to demonstrate how risks and responsibilities can be allocated among project sponsors, EPC contractors and the various other parties that may be involved in a project. Clauses, provisions and phrases that are to be used in the alternative are framed by vector symbols ("< >") and the undesired formulation should be deleted (and the desired option maintained). Provisions that are more favorable to the EPC contractor and consequently shift risk to the project sponsor (and consequently to the lenders if the project sponsor is financing the project) have been placed in set notation brackets ("{ }") so that the parties can evaluate what balance of risk sharing is appropriate for the project in question. Clauses, provisions and phrases bounded by double vector symbols ("<< >>") contain suggested or typical terms but these suggested terms must be evaluated for the particular circumstances in question and modified or replaced as necessary in order to reflect any particular project's commercial context. Clauses, provisions and phrases that may not be applicable to all types of projects have been placed in square brackets ("[ ]") and may be retained as appropriate. Words in italics should be replaced with the proper reference for the project in question.

DISCLAIMER: The sample forms furnished in this book are by way of example only and are not to be construed as legal advice and readers must consult their own legal counsel to ensure that their commercial objectives will be advanced and their interests protected.

Part A

Sample Form of Letter of Intent for Fixed Price, Lump Sum, Turnkey, Engineering, Procurement and Construction Contract

[Binding][Non-Binding]¹ Letter of Intent for Turnkey Engineering, Procurement and Construction Contract

This <Binding> <Non-Binding> Letter of Intent is made by and between <<*insert name of Contractor*>> (the "<u>Contractor</u>") and <<*insert name of Owner*>> (the "<u>Owner</u>").

WHEREAS, the Owner is evaluating the development of a <<*insert type of facility*>> to be located <<*insert location*>> as further described in Exhibit A (the "<u>Facility</u>");

WHEREAS, the Owner and the Contractor have <begun> <concluded> discussions concerning the general terms and conditions of a definitive engineering, procurement and construction contract under which the Contractor <would> <will> design, engineer, procure, construct, commission and test the Facility on a fixed price, turnkey, date certain basis (the "<u>EPC Contract</u>");

WHEREAS, the terms below enumerated are <some of the>² <all the material> terms to be contained in the EPC Contract; and

WHEREAS, the parties wish to <negotiate> <memorialize their agreement> on <an exclusive> <a non-exclusive> basis regarding the EPC Contract [until the expiration of this Letter of Intent as set forth herein];

NOW, THEREFORE, <for good and valuable consideration>³ <since no consideration has been given or received>⁴ it is <agreed that> <under discussion whether> the EPC Contract will contain the following terms which [do not]⁵ constitute all the material terms of the EPC Contract.

1. <u>Relationship Among Owner, Contractor and Subcontractors</u>

The Contractor may enter into subcontracts ("<u>Subcontracts</u>") with vendors and subcontractors (collectively, "<u>Subcontractors</u>") with respect to any portion the Work (defined in clause 2 below) provided that the Owner will have the right to approve (in advance) all Subcontractors furnishing major or critical items of Work and their respective Subcontracts. The Contractor will not be relieved from any obligation under the EPC Contract by virtue of entering into any Subcontract or the Owner's approval or disapproval of any Subcontractor or Subcontract. The Contractor will assure that each Subcontractor will comply with all relevant provisions of the EPC Contract and will waive in its Subcontract any rights that it may have against the Owner.

2. **Parties' Responsibilities**

The Owner will provide the site, [certain equipment as listed in Exhibit B,] all permits to be obtained by it as listed in the EPC Contract and all relevant information within its possession needed by the Contractor to procure all other permits required for the Facility. The Contractor will provide all equipment, materials, permits, labor and services required for the design, engineering, procurement, construction, commissioning and testing of the Facility that are not required to be provided by the Owner (the "Work"). The Work will also incorporate any equipment and materials provided and services performed by the Contractor or any Subcontractor prior to the date on which the Owner gives the Contractor notice to proceed with the Work under the EPC Contract (the "Notice to Proceed"). The Owner will have the right to review and comment on drawings and specifications, and the Contractor will resolve any such comments to the Owner's satisfaction. No review or comment by the Owner, or the Owner's failure to conduct any review, will relieve the Contractor from any of its obligations under the EPC Contract.

3. **Change Orders, Adjustments and** *Force Majeure*

The Owner may order a change, suspension or termination of the Work at any time. Adjustments will be made to the Contract Price (defined in clause 4 below) and the Deadlines (defined in clause 7 below) only for certain limited circumstances (such as *force majeure*) but any such circumstances may not be asserted as an excuse for performance to the extent that the circumstances, among other things, were within the asserting party's control, did not materially affect the asserting party's performance, or could have been avoided or overcome by the asserting party; provided that the Contractor will [not] be entitled to an adjustment in respect of any site condition [, subsurface, latent or otherwise].

4. **Contract Price and Payment**

<The Owner will pay and the Contractor will accept a lump sum amount of <<*insert currency and amount*>> as full consideration for the Contractor's performance of the Work which <includes> <excludes> all taxes in respect of the Work (the "Contract Price").> <The Contractor will provide the Owner with an indicative price for the EPC Contract on or before <<*insert date*>>. [The parties will work together using an "open book", joint approach in developing a final price for the EPC Contract (the "Contract Price").] {Once the Contract Price is agreed upon by the parties, the Owner will no longer be entitled, except as specified in the EPC Contract, to further pricing information}. [The Owner agrees not to share this information with any other party without the written consent of the Contractor.]> Not more than monthly, the Contractor may apply for payment on the basis of, and provide substantiating documentation for, the actual work performed in accordance with a milestone payment schedule to be incorporated into the EPC Contract. The Owner will be entitled to withhold from each payment retainage equal to <<*insert number*>> percent of such payment [unless the Contractor delivers to the Owner a letter of credit equal to such retainage,] which will be paid to the Contractor upon Final Acceptance (defined in clause 6 below) after deducting the

cost of any unfinished items of Work. The Owner may also withhold any payment as may be necessary to protect the Owner from loss attributable to the Contractor.

5.   **Title and Risk of Loss; Insurance**

(a) The Contractor will warrant that title to each item of Work incorporated into the Facility will vest in the Owner free and clear of any liens or claims. The Contractor will bear the risk of loss for all Work and have care, custody and control of the Facility prior to the date on which Substantial Completion (defined in clause 6 below) is achieved.

(b) The <Contractor> <Owner> will maintain and pay for[, and will require all of its Subcontractors to maintain and pay for,] customary insurance, including builders all-risk, delayed start up, general and automobile liability, marine transportation, workers' compensation, employer's liability and excess liability coverages. Such insurance will have deductibles and limitations of liability, and be provided by insurance companies, satisfactory to the Owner, and will name the <Owner> <Contractor> as an additional insured.

6.   **Testing, Provisional Acceptance and Final Acceptance**

"Provisional Acceptance" will occur when, among other things, the Owner has issued a certificate of mechanical completion of the Facility, all of the performance tests of the Facility have been completed and each of the minimum performance levels of the Facility specified in Exhibit C (the "Minimum Performance Levels") has been achieved. "Final Acceptance" will occur when, among other things, the Owner has issued a certificate of Provisional Acceptance, the Facility has successfully passed all of its reliability tests (or, if only the Minimum Performance Levels have been achieved, the Contractor has paid all Buy Down Amounts (defined in clause 7 below)) and the Contractor has otherwise completed all Work and other obligations (including punch list items) as required under the EPC Contract (other than warranty obligations). The Owner may refuse to accept the Facility if the Contractor does not achieve each of the Minimum Performance Levels and otherwise meet all the requirements of Provisional Acceptance.

7.   **Warranties; Performance and Liquidated Damages**

(a) The Contractor will warrant, among other things, that:
 (i) it and its Subcontractors are capable of performing the Work in accordance with prudent industry practices, all applicable laws and permits and all requirements of the EPC Contract and the equipment manufacturers;
 (ii) each component of the Work will be new, free of defects and deficiencies and in conformance with the requirements of the EPC Contract;
 (iii) the Facility and each part thereof will be fit for its intended use, capable of operating in accordance with the requirements of all applicable laws and permits, the EPC Contract and all other project agreements;
 (iv) for a period of at least <<*insert number*>> days after Provisional Acceptance occurs, it will repair or replace any portion of the Facility which is not in conformance with all of the foregoing;

    (v) Provisional Acceptance will occur within <<*insert number*>> days after issuance of the Notice to Proceed and Final Acceptance within <<*insert number*>> days after Provisional Acceptance (collectively, the "Deadlines"); and

    (vi) it will pay to the Owner liquidated damages of <<*insert currency and amount*>> per day for each day following the guaranteed Provisional Acceptance date until Provisional Acceptance [and <<*insert currency and amount*>> per day per unit for each day the Facility is shut down as a result of the Contractor's breach of its warranty].

(b) If the Contractor fails to achieve the performance levels specified in Exhibit C (the "Performance Levels") by the guaranteed Provisional Acceptance date but the Contractor has achieved all of the Minimum Performance Levels, the Contractor will pay the Company the liquidated damages amounts listed in Exhibit C (the "Buy Down Amount") in respect of such failure.

(c) If all the Minimum Performance Levels have been achieved, the Contractor's obligation to pay liquidated damages will be limited to <<*insert percentage*>> percent of the Contract Price. The Contractor's liability to the Owner in respect of the EPC Contract and the Work (including liquidated damages) will not exceed <<*insert percentage*>> percent of the Contract Price except that the Contractor's liability for the indemnities referred to in clause 10 shall not be limited. Irrespective of the payment of liquidated damages or limitations on liability, the Contractor must achieve Final Acceptance.

8.   Representations and Warranties

The EPC Contract will contain customary representations and warranties by each party.

9.   Default, Termination and Suspension

(a) The EPC Contract will contain a list of customary events of default regarding the EPC Contractor, including defaults for non-payment of liquidated damages within <<*insert number*>> days after the due date and failure of the Contractor to achieve Provisional Acceptance within <<*insert number*>> days of the warranted Provisional Acceptance date, the occurrence and continuance of which will give rise to the Owner's right to terminate the EPC Contract, suspend payment, complete the Facility itself or appoint a replacement contractor to do so and/or exercise any other remedy the Owner may have under law. The Owner may terminate the EPC Contract for the Owner's convenience and pay the Contractor (as its full compensation) for all Work performed but remaining unpaid, plus all accumulated and unpaid retainage. The Owner may also suspend the Work for its convenience and the Contractor will thereafter resume performance of the Work after being directed by the Owner to do so. No remedy arising under the EPC Contract shall be exclusive of any other remedy arising under law except the (i) remedy of collecting liquidated damages payments for delay in completion of the Facility or the Facility's failure to perform at the Performance Levels once the Minimum Performance Levels have been achieved and (ii) remedy relating to warranty repair Work.

(b) It will only be an Owner event of default under the EPC Contract if the Owner (i) fails to pay Contractor any amount due to the Contractor <<*insert number*>> days after such payment is due or (ii) has suspended work for a continuous period of <<*insert number*>> consecutive days in which case the Contractor may stop the Work or, after an additional <<*insert number*>> days, terminate the EPC Contract and recover the amounts specified in paragraph (a) above. [Absent active interference in the Work by the Owner, a delay caused by the Owner shall not be an event of default or entitle the Contractor to damages or any adjustment to the contract terms except as specifically provided in the EPC Contract.]

10. **Indemnities**

The Contractor will indemnify, defend and hold harmless the Owner, the financing parties and their respective officers, consultants, agents, employees, shareholders, directors, successors and assigns and any one acting on their behalf (each, an "Indemnitee") against all losses, damages, expenses or liabilities resulting from, among other things, any injury or death, property damage caused by the Contractor, any breach of any of the Contractor's obligations under the EPC Contract and the Contractor's violation of any law.[6] The Contractor will also indemnify each Indemnitee for all losses arising from claims that Contractor or any Subcontractor has violated any patent or other intellectual property rights of any third party.

11. **Dispute Resolution**

An attempt will first be made to settle all disputes arising under this Letter of Intent and the EPC Contract through discussions between the <<presidents>> of each party. If the dispute is not settled through these discussions within <<*insert number*>> days, either party <may refer the dispute to arbitration to be exclusively settled in accordance with the arbitration rules of<<*insert governing rules*>> by the <<*insert arbitral body*>> seated in <<*insert location*>> > <may proceed with legal proceedings [exclusively] in <<*insert jurisdiction*>> > [and no arbitrations or mediations will be held].

12. **Credit Support**

(a) Concurrently with the execution and delivery of the EPC Contract, the Contractor will deliver an irrevocable Letter of Credit in form and substance and executed by a commercial bank acceptable to the Owner in a stated amount equal to <<*insert amount and currency*>> against which Owner may draw if the Contractor breaches any of its obligations under the EPC Contract and a guarantee executed by Contractor's ultimate parent entity of all the Contractor's obligations under the EPC Contract in form and substance satisfactory to the Owner.

{(b) Concurrently with the execution and delivery of the EPC Contract, the Owner will deliver to the Contractor an irrevocable Letter of Credit in form and substance and issued by a commercial bank acceptable to the Contractor in a stated amount equal to <<*insert amount and currency*>> against which Contractor may draw if the Owner breaches any of its obligations under the

EPC Contract, and a guarantee executed by the Owner's ultimate parent entity of all the Owner's obligations under the EPC Contract in form and substance satisfactory to the Contractor.}

13. <u>Assignment; Financing Parties' Cure Rights</u>

The Contractor may not assign its rights under the EPC Contract without the prior written consent of the Owner. The Owner may assign its rights under the EPC Contract to any financing party or to an affiliate without the Contractor's consent. The Contractor may not terminate the EPC Contract on account of an Owner default without providing the financing parties notice and an opportunity to cure for at least <<*insert number*>> days after the expiration of any cure period available to the Owner.

14. <u>Governing Law</u>

The EPC Contract and this Letter of Intent will be governed by the law of <<*insert jurisdiction*>>.

15. <u>General Provisions</u>

(a) [Until this Letter of Intent terminates, the <Contractor> <neither party> <will> <not> enter into any Letter of Intent, understanding, arrangement, agreement or contract or conduct negotiations with any party other than the other party concerning any engineering, procurement and construction arrangement for a facility similar to the Facility and to be located within <<*insert distance*>> of the Facility.] The Contractor represents to the Owner that the signing and performance of this Letter of Intent do not and will not violate any agreement, contract or other document to which the Contractor is a party or otherwise interfere with the rights of any other party with which the Contractor has dealt.

(b) <Each party agrees to hold in confidence this Letter of Intent and all proprietary information relating to the Facility disclosed or to be disclosed to it by the other party. This obligation does not apply to information that is in the public domain, was already possessed by the recipient party at the time of disclosure, was obtained by a third party entitled to disclose such information, is legally required to be disclosed or is disclosed to potential financing parties, equity investors, consultants, contractors, subcontractors or advisors whose involvement in the Facility can be reasonably expected.> <The Owner may use and share any information it receives from the Contractor with any person or entity at <any time> <once this Letter of Intent terminates.> >[7]

(c) [Except for the obligation to negotiate in good faith and the obligations set forth in paragraphs (a) and (b) of this clause 15, this Letter of Intent does not constitute or create any legal obligation or liability on the part of either party, and by signing this Letter of Intent, neither party obligates itself to enter into the EPC Contract. Signing of the EPC Contract remains subject to the satisfactory negotiation of all terms and conditions of the EPC Contract.][8]

(d) This Letter of Intent will automatically terminate upon the signing of the EPC Contract and will thereupon be superseded by the EPC Contract. [If the

EPC Contract is not signed by the parties by <<*insert date*>>, <either party> <the Owner> may, upon written notice, terminate this Letter of Intent, and neither party will have any liability to the other party by virtue of any such termination.]

(e) In connection with this Letter of Intent, each party agrees that it will not take any action or allow its respective agents or representatives to take any action which is prohibited by law applicable to the Owner.

(f) The parties agree that no services or Work is being performed under this Letter of Intent and that any services or Work to be performed by the Contractor shall be performed under the terms of a separate written agreement between the parties and therefore no payment obligations of the Owner for Work or services shall arise hereunder or by reason hereof.[9]

(g) The Contractor agrees, at its own expense, to assist the Owner in the Owner's negotiations with other parties involved in the Owner's business concerning the Facility.

[{(h) The Owner agrees that the Contractor may offer its services as a contractor directly to the <<*insert name of governmental authority and/or other bidders*>> in competition with the Owner provided that the Contractor agrees that it will not make any such offer on terms more favorable than those offered to the Owner hereunder.}][10]

(i) This Letter of Intent may be executed in two or more counterparts (and by facsimile), each of which shall be deemed an original, but all of which together shall constitute one and the same instrument.

The parties have caused this Letter of Intent to be signed on the date set forth below.

Date: <<*insert date*>>
<<*Insert name of Owner*>>   <<*Insert name of Contractor*>>

By:_____   By:_____
Name: Name:
Title: Title:

Exhibit A

Description of Facility

Exhibit B

Owner Supplied Equipment

Exhibit C

Minimum Performance Levels, Performance Levels and Buy Down Amounts

Notes

1 The Owner must consider whether it desires to use the Letter of Intent as an agreement to bind the Contractor to the terms of the EPC Contract or use the Letter of Intent as the basis for beginning discussions concerning the EPC Contract's possible terms.

2 The phrase "some of" should be deleted if the parties wish the Letter of Intent to be binding because courts will rule that a contract has been entered into once all its material terms have been agreed upon.

3 Only include if Letter of Intent is to be binding.

4 Only include if Letter of Intent is to be non-binding.

5 Use only if Letter of Intent is to be non-binding.

6 The Contractor may request a reciprocal indemnity from the Owner, which is typical.

7 The owner must determine whether or not it wants to maintain the flexibility to use information (especially pricing) that it receives from the EPC Contractor with other bidders.

8 Only include if Letter of Intent is to be non-binding.

9 Any work to be performed by the Contractor should be carried out under a separate services agreement.

10 This may be necessary if the project sponsor is a governmental entity and has not determined whether or not it desires to grant a concession for its project or build the project itself.

Part B

Sample Form of Limited Notice to Proceed Agreement

Limited Notice to Proceed Agreement[1]

This Limited Notice to Proceed Agreement (this "<u>Agreement</u>") is made this <<*insert date*>>, by and between <<*insert name of Owner*>> (the "<u>Owner</u>") and <<*insert name of Contractor*>> (the "<u>Contractor</u>").

WHEREAS, the Owner and the Contractor [have executed a Letter of Intent dated <<*insert date*>> (the "<u>Letter of Intent</u>") and] are negotiating [{and expect to enter into}] a turnkey engineering, procurement and construction contract (the "<u>EPC Contract</u>") regarding the facility described in Exhibit A (the "<u>Facility</u>");

WHEREAS, in order to support the substantial completion of the Facility by <<*insert date*>> (the "<u>Deadline</u>"), this Agreement is being executed prior to any execution of the EPC Contract that may occur because the Contractor has represented that the work [specified herein] must commence by <<*insert date*>> in order for the Contractor to meet the Deadline; and

WHEREAS, time is of the essence.

NOW, THEREFORE, it is agreed as follows:

1. **<u>Scope of LNTP Work</u>**

   (a) Upon receipt of an executed LNTP Notice in substantially the form of Exhibit B from the Owner, the Contractor shall proceed with any and all work (but only that work) necessary to enable the Contractor to meet the Deadline (the "<u>LNTP Work</u>").

   (b) The LNTP Work may include, among other things, the items described below.

   [(i) **Project Administration**
       (a) Identify, assign, and begin orientation of Contractor's project personnel.
       (b) Identify Contractor's office around which work will be centered.
       (c) Set up and initialize engineering, construction, and project accounting, cost control, and administration procedures.
       (d) Establish procedures for submittal and handling of drawings and data.
       (e) Kick off meeting with Owner's project management team.

   (ii) **Project Planning and Control**
       (a) Conduct constructability and schedule reviews.
       (b) Initiate preparation of the following:
           (i) Project instruction manual.
           (ii) Project design manual.
           (iii) Project procurement procedures.

   (iv) Quality assurance procedures.
   (v) Project construction plan.
   (vi) Permitting work.
   (vii) Site union labor agreement.
(c) Project scheduling:
   (i) Three-month look-ahead detailed bar chart.
   (ii) Summary level project bar chart.
   (iii) Initiation of project CPM.

(iii) **Design Engineering**
(a) Review existing soil data and determine the need for additional soils investigation work.
(b) Initiate detailed site arrangement including provisions for location of all structures, equipment, piping, permanent access routes, construction facilities, and provisions for the movement of materials and construction access.
(c) Initiate construction drawings for site preparation, grading, drainage, and fencing.
(d) Initiate engineering to support underground utility design.
(e) Finalize design concepts and criteria, including the following:
   (i) Process:
       Complete piping and instrumentation and equipment data sheets.
   (ii) Mechanical:
       Hold order definition meeting and manage the purchase order for major equipment.
       Issue inquiry requisitions for equipment.
   (iii) Civil, Architectural and Structural:
       Prepare scopes of work for site preparation and underground systems and building contracts.Commence design of foundations for major equipment.
       Prepare contract for water supply.
   (iv) Piping:
       Set up the 3D model and commence modeling equipment and alloy and large diameter piping.
       Prepare line lists.
       Extract material takeoffs for long lead material.
       Prepare design of underground piping.
   (v) Electrical Engineering:
       Issue inquiry requisitions for equipment.
   (vi) Controls Engineering:
       Prepare purchase requisition for direct control system ("DCS").
       Commence design and specification for in-line valves.
       Commence DCS configuration.
       Commence design for mechanical systems.
       Commence instrument database.
   (vii) Electrical and Instruments design:
       Complete one-line diagrams, hazardous classifications, grounding, underground, aboveground power and lighting.

Commence cable trays, instrument locations, connection diagrams, and buildings.

Commence material control.

(f) Initiate engineering design as necessary in each discipline to support the early procurements identified below.

(g) Finalize heat balance and thermal design.

(iv) Procurement

(a) Develop project specific procurement terms and conditions.

(b) Establish the procurement plan and identify any required advance procurements based on current market conditions.

(c) Initiate preparation of early procurement specifications.

(d) Commence procurement activities, including order placement if required to maintain the schedule.

(e) Review drawings and documents for Owner supplied equipment.

(f) Conduct initial coordination meetings with Owner and Owner's other contractors.

(v) Construction

(a) Commence subcontracting activities and initiate preparations and arrangements for mobilization, construction facilities, construction utilities, and construction equipment.

(b) Initiate pre-mobilization site and local area visits, inspections, and introductions.

(c) Hold kick-off meeting with site contractors.

(vi) **Process Engineering**

(a) Finalize Facility design.

(b) Validate process model.]

2. **LNTP Payment**

(a) In respect of the LNTP Work, the Owner shall pay the Contractor in accordance with the following schedule:

<<insert number>> days after LNTP Notice<<insert amount and currency>>
<<insert number>> days after LNTP Notice<<insert amount and currency>>

The Owner shall pay such amounts to the Contractor within <<insert number>> days after an invoice therefor has been delivered by the Contractor.[2] Upon payment hereunder by the Owner, all LNTP Work performed hereunder shall become the property of the Owner for its use [in any regard] [{in connection with the project}].

(b) If the Contractor believes that the Contractor must enter into any subcontract in order to meet the Deadline, the Contractor will explain its rationale therefor in detain in writing and will seek to obtain the Owner's advance written consent. If the Owner consents, the Owner shall thereafter pay any amount due under such subcontract within five (5) business days after receipt of the Contractor's invoice relating thereto. If the Owner does not give its prompt written consent to the Contractor's entering into such subcontract, the Deadline shall be extended appropriately to account for such failure or delay of the Owner. In connection with requesting any such consent from the Owner, the

Contractor shall furnish the Owner with a copy of the unpriced award letter (and the termination schedule therefor) proposed to be executed with such subcontractor which must provide that such subcontract can be assumed by the Owner from the Contractor at the Owner's option if the Owner terminates this Agreement. As of the date hereof, the Contractor anticipates that the only subcontracts that must be executed before <<*insert date*>> are the subcontracts for <<*insert categories of work or equipment order and/or supply*>>.

(c) All payments to the Contractor for LNTP Work shall be credited toward and reduce on a <<*insert currency*>> for <<*insert currency*>> basis the initial payment under the EPC Contract.

3. **Limitation of Liability**

(a) The Owner and the Contractor intend that <when> <if> the EPC Contract is executed, it will supersede this Agreement.[3]

(b) IN NO EVENT SHALL EITHER PARTY BE LIABLE TO THE OTHER PARTY FOR ANY INDIRECT OR CONSEQUENTIAL DAMAGES OR LOSS OF PROFIT.

(c) [{If the EPC Contract is not executed, the Owner shall defend, hold harmless and indemnify the Contractor against any claims, costs and liability arising out of the Owner's use of the LNTP Work.}][4]

4. **Termination Provisions**

(a) This Agreement shall terminate upon the earlier of (i) {<<*insert date*>>}, (ii) the execution of the EPC Contract, or (iii) termination hereof as provided below. The Owner may terminate this Agreement at any time for any reason in its sole discretion by giving the Contractor written notice thereof, which termination shall be effective upon the giving of such notice. If the Owner terminates this Agreement, upon receiving such notice of termination, the Contractor shall stop performing the LNTP Work and shall cancel (as quickly as possible) all commitments to any subcontractors (unless the Owner specifies otherwise) and shall use all reasonable efforts to minimize cancellation charges and other costs and expenses associated with the termination of this Agreement. In the event of a termination (other than by reason of the execution of the EPC Contract), the Contractor will be compensated for its costs incurred for LNTP Work completed through the termination date as provided in the next sentence, but under no circumstances shall the Owner be liable to the Contractor for work not yet performed by the Contractor as of the date of termination hereof. If for any reason the EPC Contract is not executed, the Contractor shall be entitled to payment of (i) all its reasonable and documented direct costs through the date of termination under this Agreement including the cancellation costs of subcontracts, plus (ii) <<*insert number*>> percent (<<*insert number*>>) of such costs described in clause (i), all less the amounts the Owner has previously paid to the Contractor hereunder but in no event shall the Contractor be entitled to more than <<*insert amount and currency*>> under this Article 4 plus the termination of charges of any subcontracts approved in writing in advance by the Owner in accordance with Article 2.

5. **Miscellaneous**

[Except with respect to the Letter of Intent,] [This][this] Agreement constitutes the entire agreement between the parties hereto and supersedes any oral or written representations, understandings, proposals, or communications heretofore entered into by or on account of the parties [with respect to the subject matter hereof][5] and may not be changed, modified, or amended except in writing signed by the parties hereto.

This LNTP Agreement may be executed [by facsimile exchange] in two or more counterparts, each of which shall be deemed an original, but all of which together shall constitute one and the same instrument. The Owner may [{not}] assign this Agreement without the Contractor's consent.

6. **Deadlines**

The Contractor agrees that it will achieve the Deadline if the Owner complies with this Agreement and the Contractor is directed to commence all work called for in the EPC Contract by *<<insert date>>* [(or *<<insert date>>* if the Owner has elected the Extension Option)].

[7. **Extension Option**

If the Owner delivers an executed Extension Option Notice in the form of Exhibit C to the Contractor by *<<insert date>>* (the "Extension Option"), the amount of *<<insert amount and currency>>* referred to in Article 4 shall be deemed to be *<<insert amount and currency>>*.]

8. **Warranty**

The Contractor warrants it shall perform the LNTP Work in accordance with the standards of care and diligence normally practiced by reputable firms performing services of a similar nature. If, during the *<<insert number>>* day period following the completion of the LNTP Work it is shown that the Contractor has failed to meet these standards [{and the Owner has notified the Contractor in writing of such within such period}], the Contractor shall perform such corrective services as may be necessary to remedy such failure at its sole expense [{and if the Contractor does so promptly it shall have no further liability to Owner with respect thereto}].

9. **No Solicitation of Employees**

Neither party shall, during the term of this Agreement or for a period of *<<insert number>>* days thereafter, directly or indirectly for itself or on behalf of, or in conjunction with, any other person, partnership, corporation, business or organization, solicit, hire, contract with or engage the employment of an employee of the other with whom that party or its personnel have had contact during the course of the LNTP Work under this Agreement, unless that party has obtained the written consent of the other to such hiring and that party pays to the other a fee to be mutually agreed upon.

10. **Governing Law**

This Agreement shall be governed by the laws of <<*insert jurisdiction*>>.

11. **Dispute Resolution**

An attempt will first be made to settle all disputes arising hereunder through discussions between the <<presidents>> of each party. If the dispute is not settled through these discussions within <<*insert number*>> days, either party may <refer the dispute to arbitration to be exclusively settled in accordance with the arbitration rules of <<*insert name of rules to be applied*>> by the <<*insert tribunal*>> seated in <<*insert location*>> > <proceed with legal proceedings>.

[12. **EPC Contract**

Neither party shall be bound to enter into the EPC Contract by reason of its execution of this Agreement.][6]

IN WITNESS WHEREOF, the parties have caused this Agreement to be executed as of the date first above written.

<<*insert name of Owner*>>	<<*insert name of Contractor*>>
By:_____	By:_____
Name:_____	Name:_____
Title:_____	Title:_____

EXHIBIT A

DESCRIPTION OF THE FACILITY

EXHIBIT B

LNTP NOTICE

The undersigned hereby directs Contractor to commence the LNTP Work.

<<*insert name of Owner*>>
By:_____

Name:_____

Date:_____

EXHIBIT C

EXTENSION OPTION NOTICE

1.   The Owner hereby elects to have the Contractor continue the LNTP Work.
2.   The Owner will continue to pay the Contractor for the LNTP Work according to the following payment schedule.
     <<insert number>> days after LNTP Notice: *<<insert amount and currency>>*
     <<insert number>> days after LNTP Notice: *<<insert amount and currency>>*
     <<insert number>> days after LNTP Notice: *<<insert amount and currency>>*
     <<insert number>> days after LNTP Notice: *<<insert amount and currency>>*
     <<insert number>> days after LNTP Notice: *<<insert amount and currency>>*
3.   The Owner understands that during *<<insert relevant number of days or relevant period>>*, the Contractor must award subcontracts for each of *<<insert categories of work and/or equipment>>* by *<<insert date>>* and if such are not awarded, an extension of the Deadline will be necessary.

<<insert name of Owner>>
By: _____
Name: _____
Date: _____

Notes

1   Typically, construction services will not be carried out under an agreement like this because engineering and procurement work generally must begin before construction begins. Thus, provisions covering *force majeure*, risk of loss, indemnification and insurance are probably unnecessary.
2   Since work at this point is organizational in nature, it is difficult to implement a "task based" payment structure but the Owner may wish to request progress reports from the EPC contractor.
3   Often, EPC contractors will try to "cap" their overall liability under this type of agreement.
4   Generally, EPC contractors will insist on this type of provision.
5   Include this language only if there is already another agreement between the parties.
6   The Contractor may want to consider deleting this clause in an attempt to compel the Owner to agree upon the terms of the EPC contract.

Sample Form of Fixed Price, Lump Sum, Turnkey, Engineering, Procurement and Construction Contract

FIXED PRICE, LUMP SUM, TURNKEY ENGINEERING, PROCUREMENT AND CONSTRUCTION CONTRACT[1]

dated as of <<*insert date*>>
by and <between> <among>[2]
<<*insert name of Sponsor*>>
and
<<*insert name of EPC Contractor*>>

This contract contains information that is confidential to <<*insert name of Sponsor*>>. Any such recipient's acceptance and/or use of this contract constitutes acknowledgment by such recipient that the recipient will not copy this contract or disclose it, in whole or in part, to any other person or entity without the express prior written permission of <<*insert name of Sponsor*>> and that the recipient will use this contract solely for the purpose of the evaluation for which it was furnished. This contract, and the information contained in it, remains the property of <<*insert name of Sponsor*>>, and the recipient shall return it to <<*insert name of Sponsor*>> when no longer required for the purpose for which it was furnished.

TABLE OF CONTENTS

EXHIBITS

This **FIXED PRICE, LUMPSUM, TURNKEY, ENGINEERING, PROCUREMENT AND CONSTRUCTION CONTRACT**, dated as of <<*insert date*>>, is by and <between> <among>:[3]

[_____], a [_____]
company (the "<u>Sponsor</u>");[4] and

[a consortium[5] consisting of:

[_____], a [_____] company ("_____"); and

[_____], a [_____]
company ([collectively, acting jointly and severally] "Contractor")].

WITNESSETH:

WHEREAS, the Sponsor desires to [finance,] build, own and operate [and has entered into the Offtake Agreement (as defined herein)] [which will <serve> <expand>] the [Facility][, the Existing Plant] (as defined herein);

[**WHEREAS**, the Sponsor has ordered various items pursuant to the Vendor Purchase Agreements (as defined herein);]

[**WHEREAS**, the Contractor participated in the negotiation of the Vendor Purchase Agreements and will assume the Sponsor's performance obligations thereunder;]

[**WHEREAS**, the Sponsor operates a simple cycle power plant in <<*insert location*>> which is being expanded to include another simple cycle unit and then convert the Existing Plant to combined cycle, and the Contractor and the Sponsor will perform baseline testing of the Existing Plant, at the Contractor's sole cost and expense, within <<*insert number*>> Days of the Contract Date (as herein defined) to establish its and each combustion turbine's baseload electrical output, exhaust mass flow, heat rate and noise and emissions levels in accordance with the Functional Specifications, and such information (the "<u>Baseline Existing Plant Levels</u>");]

[**WHEREAS**, the parties intend to address the Work to be performed as part of the combined cycle conversion so that each gas turbine unit of the Existing Plant can remain available to operate separately in simple cycle mode at its then existing performance capabilities once the Integrated (as herein defined) facility is available for operating in combined cycle mode in a manner so that Work will not negatively impact or jeopardize the operation of any other work heretofore installed (the "<u>Integration</u>" or to "<u>Integrate</u>") and so that the Integrated facility can achieve the Performance Guarantees and allow operation of the Integrated facility in all possible "one-on-one" and "two-on-one" [and "three-on-one"] [turbine generator] permutations (i.e. each or any two [or all three] [gas turbines] can operate to drive the [steam turbine[s]]);]

WHEREAS, pursuant to a [<<*Turnkey Engineering, Procurement and Construction Contract*>>] dated [<<*insert date*>>], Contractor built a <<*insert type of facility*>> facility in [<<*insert date*>>] for the Sponsor and the Sponsor now desires that the Contractor itself or through qualified Subcontractors (as herein defined), provide all the services necessary for the design, engineering, procurement, construction, start-up and testing on a fixed price, turnkey basis of:

<(a) the Phase II Expansion (as herein defined); and>

(b) the Phase III Expansion (as herein defined) in the case that the Sponsor has accepted a then valid Proposal (as defined below) within <<*insert number*>> Days after the Phase II Notice to Commence and within <<*insert number*>> Days after receiving the Contractor's written, detailed and firm pricing proposal (and proposed changes to this Contract concerning the following as they relate to Phase III: <<*list sections and exhibits to change*>> but without requiring any other changes to this Contract) (the "Proposal");

WHEREAS, the Contractor, itself or through qualified Subcontractors, desires to provide all the services necessary for the design, engineering, procurement, construction, commissioning and testing for the Facility [including Integration of the New Facility with the Existing Plant (as defined herein)][6] [{in accordance with the requirements of this Contract}][7] on a fixed price, lumpsum, date certain, complete and fully integrated "turnkey" basis;

NOW, THEREFORE, in consideration of the mutual promises contained herein, the Parties, intending to be legally bound, hereby agree as follows.

ARTICLE I. DEFINITIONS

All capitalized terms used in this Contract have the meanings given them below.

"**Acceptable Bondsman**" shall be any of <<*insert list of acceptable bonding companies*>>,[8] so long as such Person has at the time of issuance [and maintains] the Required Rating.

"**Acceptable Issuer**" shall be <<*insert list of acceptable letter of credit issuers*>>,[9] so long as such Person is an Eligible Bank that has at the time of issuance [and maintains] the Required Rating and offices in <<*Manhattan, London, Paris, Madrid, Hong Kong, São Paulo, or Frankfurt [insert additional or eliminate cities if desired]*>>.

"**Accounting Standards**" means generally accepted accounting principles of <<*name applicable accounting standards board*>>.

["**Additional Scope Option**" is defined in Article 7.10].

["**Additional Scope Option Facilities**" shall be those facilities, as further described in the Functional Specifications, to <<*describe additional items*>>].

"**Adjustment**" shall mean a Price Adjustment or a Time Adjustment.

"**Adjustment Claim**" shall mean a document, in substantially the form of Exhibit 1A, prepared by the Contractor as a proposal to the Sponsor for an Adjustment.

"**Advance Payment LC**" means a letter of credit issued by an Acceptable Issuer in the form of Exhibit 10.1.2.

"**Antiquity**" is any man-made object or structure, fossil or human remain which has a reasonable likelihood of being determined to require preservation by any Governmental Authority.

["**Annual Availability Guaranty**" is defined in Exhibit 9.2].

["**Arbitration**" is defined in Article 20.2].

"**As Built Drawings**" shall mean the final drawings and documents that have been certified by a duly licensed professional engineer as reflecting the actual Equipment installed and the actual Facility as constructed.

"**Asserting Party**" is defined in Article 11.1.1.

["**Assigned Agreements**" shall mean those agreements listed in Exhibit 1B].

["**Back-up Member(s)**" is defined in Article 20.3.1].

"**Baseline Existing Plant Levels**" is defined in the recitals hereof.

"**Baseline Level 3 Schedule**"[10] shall mean the expanded version of the Baseline Schedule, produced utilizing <<[*insert name*] software>>, reflecting all activities essential for performance of the Work, all Critical Path Items and the critical path to completion, prepared in accordance with the Functional Specifications and Article 5.6, in the form approved expressly in writing by the Sponsor.

["**Base Price**" is defined in Article 10.1.1].

"**Baseline Schedule**" is attached as Exhibit 1C. For the avoidance of doubt, the Baseline Schedule shall never change [and includes the Existing Plant Schedule].[11]

"**Bid Documents**" means <<*insert title of bid documents*>>.[12]

["**Board**" is defined in Article 20.2].

["**Board Costs**" is defined in Article 20.2.15].

"**Btu**" means British thermal unit.

"**Business Day**" shall mean any Day other than a Saturday or Sunday or a Day on which commercial banks are not authorized to transact business or are required to close in the << *insert location*>>.

["**[<Cable> <Pipeline>] Route**" means the route for the Facility as set forth in the Functional Specifications.][13]

"**Certificate of Facility Final Acceptance**" shall mean a certificate issued by the Sponsor to the Contractor in accordance with Article 8.7, evidencing the Sponsor's acknowledgment that Facility Final Acceptance has occurred.

"**Certificate of Facility Provisional Acceptance**" shall mean a certificate issued by the Sponsor to the Contractor in accordance with Article 8.6, evidencing the Sponsor's acknowledgment that Facility Provisional Acceptance has occurred.

"**Certificate of Facility Commissioning Completion**" shall mean a certificate issued by the Sponsor to the Contractor in accordance with Article 8.4, evidencing the Sponsor's acknowledgment that Facility Commissioning Completion has occurred.

"**Certificate of Unit Commissioning Completion**" shall mean a certificate issued by the Sponsor to the Contractor in accordance with Article 8.4, evidencing the Sponsor's acknowledgment that Unit Commissioning Completion has occurred.

"**Certificate of Unit Provisional Acceptance**" shall mean a certificate issued by the Sponsor to the Contractor in accordance with Article 8.5, evidencing the Sponsor's acknowledgment that Unit Provisional Acceptance has occurred.

["**Change in Control**" means that a person who did not have such power on the Contract Date becomes able, indirectly or directly, to direct the affairs of the Contractor, by means of voting, contract or otherwise].

"**Change in Law**" shall mean:

(a) the adoption, enactment or application to either Party, the Facility or the Work of any Law relating to the environment, taxes, customs or duties not existing or not applicable to such Party, the Facility or the Work on the Contract Date; or

(b) any change in any Law relating to the environment, taxes, customs or duties or the interpretation or application thereof by a Governmental Authority after the Contract Date.

For the avoidance of doubt, Change in Law does not include:

  (i) any Law or application thereof in existence [or proposed] on the Contract Date that, by its terms, became or will become effective or applicable to either Party, the Facility or the Work after the Contract Date;
 (ii) any change in a Permit directly or indirectly caused by the acts or omissions of any Contractor Person; or
(iii) any change in Contractor Taxes or withholding taxes or taxation levied on the income, payroll, employee count or revenue of any Contractor Person or any of their affiliates.

"**Change Order**"[14] shall mean a written order, issued pursuant to Article 19.4, authorizing a change in the Work and/or an adjustment to the Project Variables.

["**Chairman**" is defined in Section 20.2.1].

"**Claim**" shall mean any claim, action, suit or proceeding asserted by any Person.

"**Code**" shall mean any industrial, national, state, provincial, local or municipal law, rule, regulation, statute, ordinance, code, standard, interpretation or other requirement of any Governmental Authority, institute or organization pertaining or relating to the Facility or the Work including any Code or Standard mentioned in the Functional Specification or Exhibit 1D.[15]

["**Cold Weather Test**" is defined in Article 8.1].

"**Completion Guarantees**" shall mean the guarantees set forth in Article 9.1.

"**Computer Programs**" shall mean a sequence of instructions, data or equations in any form intended to cause a computer, a control data processor, a distributed control system or the like to perform any kind of operation. A computer program may at times be referred to as software or hardware, and the provisions herein applicable to Computer Programs shall be applicable to software and hardware.

"**Confidential Information**" shall have the meaning ascribed thereto in Article 21.3.1(a).

"**Consenting Party**" shall have the meaning ascribed thereto in Article 20.17(j)(iii).

["**Consortium Member**" shall mean <<*insert name*>> or <<*insert name*>>].[16]

"**Consumables**" shall mean all items consumed or needing regular periodic replacement during the operation and maintenance of the Facility, including, but not limited to, water treatment chemicals (such as ammonia, etc.), small tools, lubricants, rags, oils, filter media, additives, anti-corrosion devices, gases (such as CO_2, O_2, H_2, Halon, etc.) and other expendable materials, but not spare parts or Fuel.

"**Contract**" shall mean this Fixed Price, Lump Sum, Turnkey, Engineering, Procurement and Construction Contract between the Sponsor and the Contractor and all the Exhibits hereto.

"**Contract Date**" shall mean the date referred to in the Preamble of this Contract.

"**Contract Price**" shall have the meaning ascribed thereto in Article 10.1.1.

"**Contractor**" shall mean <each Consortium Member acting jointly and severally hereunder> <<*insert name*>>.

"**Contractor Document Deliverables**" shall have the meaning ascribed thereto in Article 14.5.

"**Contractor Event of Default**" shall have the meaning ascribed thereto in Article 17.1.

"**Contractor Land Use Agreements**" means any contractual rights obtained by the Contractor to use real property in connection with the construction of the Facility.

"**Contractor Permits**" shall mean all Permits necessary to be obtained for the Contractor to carry out the Work (except for the Sponsor Permits), including the Permits, if any, described in Exhibit 1E.

"**Contractor Person**" shall mean:

(a) the Contractor;
(b) any subsidiary, affiliate, agent, representative, employee, director, officer, successor or assignee of the Contractor; or
(c) any Subcontractor of any entity listed in clause (a) or clause (b) above.

["**Contractor Primary Work Area**" shall mean the portions of the <Site> <Route> on which the Contractor will construct:

(a) <the power island and the HRSG;
(b) the switchyard for the Facility;
(c) the gas pipeline for supplying fuel to the Facility;
(d) the tank for supplying the Facility with demineralized water; and
(e) the laydown area for Equipment for the Facility.>][17]

"**Contractor's Commissioning Manual**" is the commissioning manual for testing the Facility, the Facility Systems, all Equipment and subsystems for proper operation.

"**Contractor's Project Director**" shall have the meaning ascribed thereto in Article 5.10.

"**Contractor Taxes**" shall mean any and all taxes and duties (import or otherwise) connected with the performance of this Contract with respect to which, under applicable Law, <the Contractor> <any Consortium Member> is the taxable Person, tax payer, or tax collector, including taxes on any item or service that is part of the Work or the Facility, such as occupational, sales, works, use, value added (within and outside of <<*insert country*>>), excise, social security, health security premiums, unemployment, corporate or personal income and other taxes and import, customs or other duties, and stamp duties to be paid on bills of lading, bills of exchange and Subcontracts, etc., regardless of whether such tax or duty is normally included in the price of such item or service or is normally stated separately (including labor furnished by any Contractor Person) or otherwise arising out of the Contractor's performance of the Work, including any increases in any such taxes during the term of this Contract, and any duties, fees or royalties imposed with respect to any Equipment, labor or services.

"**Correction Curves**" shall mean those output and efficiency correction curves set forth in the Functional Specifications or, if not set forth therein, in any output correction curves supplied by a Vendor relating to the Equipment supplied by such Vendor.

["**Cost Plus Work Amount**" is defined in Article 4.2.8].

["**Counterclaim Response**" is defined in Article 20.7.3].

"**Critical Path Item**" shall be any activity included in the Baseline Schedule, the delay of which, if it arose on the Contract Date and not at the time of the event in question, would result in a delay in the Contractor's achieving the Guaranteed Facility Provisional Acceptance Date.[18]

"**Current Schedule**" shall mean a critical path method type schedule prepared and periodically updated in accordance with Article 5.6.6 and based upon the form set forth in Exhibit 5.6.7.[19]

["**Customary Utility Requirements**" shall mean the requirements, methods and procedures in effect on the Contract Date which are employed by the <Offtaker> <Thermal Customer> local telephone company, Wheeling Company, Fuel, Transporter, Sewage Company, Grid, water authority, Fuel Supplier and <<*insert others if any*>>[20] in connection with the tie-ins and interconnection facilities for utilities for the Facility].

"**Day**" shall mean a calendar day, including Saturdays, Sundays and holidays.

"**Deadlines**" shall mean the [Guaranteed Unit One Provisional Acceptance Date,] [Guaranteed Unit Two Provisional Acceptance Date,] the Guaranteed Facility Provisional Acceptance Date, [the Guaranteed Unit One Commissioning Completion Date,] [the Guaranteed Unit Two Commissioning Completion Date,] the Guaranteed Facility Commission Completion Date and the Guaranteed Facility Final Acceptance Date.

"**Delay Damages**" shall have the meaning ascribed thereto in Article 9.1.

"**Delay Damages Cap**" shall have the meaning ascribed thereto in Article 18.9.

"**Delayed Payment Rate**" shall mean:

(a) in the case of all payments to be made hereunder other than those expressly stated to be made in <<*insert currency*>>, <the <<*London Eurobor Interbank Offered Rate of*<<*bank name*>> for <<*thirty (30)*>> Day <<*insert currency*>> deposits plus> [<<*insert incremental rate*>>] percent per annum; and

(b) in the case of payments hereunder expressly stated to be made in <<*insert currency*>>, the rate per annum equal to the offered quotation which appears on the <<*insert currency*>> page of <<*insert publication*>> designated for the display of an average of the <<_____>> Interbank Offered Rate for deposits in <<*insert currency*>> plus <<*insert incremental rate*>> percent per annum.

["**Delivery Point**" means the <<*pothead/flange*>> at the <<*insert name*> <pipeline> <substation> to be modified by the Contractor.]

["**Differing Site Conditions**" is defined in Article 11.2.]

["**Dispatcher**" is defined in Article 5.1(jj).]

"**Dispute**" is defined in Article 20.1.

"**Dispute Notice**" is defined in Article 19.6.

["**Dollars**" and "**$**" shall mean the lawful currency of the United States of America.]

["**ECA**" shall mean any export credit agency or similar agency or corporation providing financing or insurance or guarantees to the Sponsor or any Financing Party in respect of the Facility or the Work.]

["**Electrical Interconnection Agreement**" means the interconnection agreement between the Sponsor and <<*Wheeling Company*>>.]

["**Electricity Wheeling Agreement**" is the agreement between the Sponsor and the Wheeling Company.]

"**Eligible Bank**" shall mean any bank or financial institution established under the laws of the United Kingdom, United States, or any State thereof <<*add any other financial institutions from acceptable jurisdictions*>>.

"**Emergency Entities**" shall have the meaning ascribed thereto in Article 5.4(c).

"**Environmental Baseline Report**" means the report prepared by <<*insert name of consultant*>> concerning environmental conditions of soil and ground water located adjacent, at or [under] [along] the <Site> <Route> (including the presence or absence of contaminants).[21]

"**Environmental Report**" shall mean the Environmental Impact Assessment Report, a copy of which is attached as Exhibit 1G.[22]

"**Environmental Standards**"[23] shall mean:

[(a) <<*add appropriate standards*>>; and]
[(b) any other environmental standard set forth in Exhibit 1H, the Environmental Baseline Report or in the Environmental Report.]

"**Equipment**" shall mean all of the materials, apparati, structures, supplies and other goods required to construct and equip the Facility including Computer Programs.

["**Escalator**" is defined in Article 10.1.1.]

["**Euro**" means the lawful currency of the member states of the European Union.]

"**Execution Plan**" is defined in Article 3.4.

["**Existing Plant**" is defined in Article 5.1(c).][24]

["**Existing Plant Outage**" shall mean an occurrence when the Existing Plant is required to be taken out of service or its output reduced, resulting in one or more Existing Plant Shutdown Hours, in order for the Contractor to complete any Work necessary to complete the Tie-Ins or to test or commission the Facility.][25]

["**Existing Plant Schedule**" shall mean the tie-in schedule according to which the Work will be tied in to the Existing Plant.][26]

["**Existing Plant Shutdown Hour**" shall mean any hour or part thereof in which the Existing Plant cannot be safely operated at full capacity in accordance with <Good> <{Prudent}> <Utility> <{Industry}> Practices.][27]

["**Expected Outage Duration**" shall have the meaning set forth in Article 6.10.][28]

"**Facility**" shall mean [<the Units, the common facilities, completed buildings and structures, Tie-ins, Interconnection Facilities, Equipment and related Site improvements required by this Contract relating to the production of electricity [and thermal energy], including fuel storage, as further described in the Functional Specifications> the <<__>> kilovolt <alternating> <direct> current owner and related facilities necessary to transmit <<*insert amount*>> MVA of electrical power from <<*name injection point*>> to the <<*name delivery point*>> as more fully described in the Functional Specifications <a liquefied natural gas unloading, storage and vaporization facility including <<*LNG tanks*>>, together with appurtenant facilities and equipment related thereto>].

"**Facility Buy Down Amounts**" are set forth in Exhibit 9.2 and with respect to any particular Performance Guarantee shall be calculated in accordance with Exhibit 9.2.

"**Facility Final Acceptance**" shall have the meaning ascribed thereto in Article 6.7.1.

"**Facility Final Acceptance Date**" shall mean the date on which the Sponsor issues the Certificate of Facility Final Acceptance.

"**Facility Minimum Performance Levels**" shall be met when the Facility is capable of achieving all of the performance levels set forth as minimum performance in Exhibit 9.2.

"**Facility Provisional Acceptance**" is defined in Article 6.6.1.

"**Facility Provisional Acceptance Date**" shall mean the date on which Facility Provisional Acceptance occurs.

"**Facility Systems**" shall mean the systems and Equipment that are required for the Facility or a Unit to perform in accordance with the Functional Specification.

["**FCPA**" is defined in Article 21.5.]

["**Final Price**" shall be the Base Price if such is in effect on the Work Commencement Date or any revised price if such is in effect on the Work Commencement Date as a result of the express application of a provision of this Contract on or before the Work Commencement Date.]

["**Financial Closing**" shall mean the date that all of the conditions precedent to each Financing Party's obligation to make its initial extension of credit have been satisfied or waived [{in writing}] by each such Financing Party.]

["**Financing Party**" shall mean any Person providing debt or equity financing or insurance (of any nature or against any peril, whether natural or not) or guarantees to the Sponsor or any of its Affiliates, including any trustee or agent representing any such Person.][29]

["**Financing Party's Engineer**" shall mean any engineer retained by a Financing Party to review the Work on behalf of such Financing Party.]

["**FEED Work**" shall have the meaning ascribed thereto in the definition of "Preliminary Services".]

"**Float**" means, at the time in question, the time the Work can be delayed without delaying the Guaranteed Provisional Acceptance.

"**Force Majeure Excused Event**"[30] shall mean any event or circumstance or combination of events and/or circumstances listed in the next sentence to the extent such is beyond the control of the Asserting Party and materially and adversely affects the performance by the Asserting Party of its obligations (other than payment obligations) under or pursuant to this Contract and which the Asserting Party could not have prevented or overcome, in whole or in part through the exercise of due diligence or <Good> <{Prudent}> <Utility> <{Industry}> Practices. Force Majeure Excused Events shall <{include}> <be limited to>[31] the events and circumstances in the following list:

(a) lightning, drought [{or insufficient <water> <steam> <or fuel> flow for Performance Testing}], fire, earthquake, volcanic eruption, landslide, flood, famine, plague, epidemic, or other natural disaster;

(b) explosion or chemical contamination occurring as a result of an event or circumstance described in clause (a) above;

(c) war, blockade, embargo, act of the public enemy, sabotage, acts of any Governmental Authority, or riot;

(d) storms {or storm warnings issued by [<<*insert name of weather service*>>]}, floods or rain provided that each of the foregoing is of a magnitude which has not occurred within the past <<*insert number*>> years or is a Named Storm, and provided further that neither the first <<*insert number* >> Days of any weather-related delay in any calendar month nor the first<<*insert number*>> Days of all weather-related delays shall be considered a Force Majeure Excused Event [{but once such occurs <all><<*insert number*>> of such non-eligible Days shall be eligible to be included as a Force Majeure Excused Event}];

(e) [{damage caused by a railroad carrier if such railroad carrier is the only railroad carrier operating the railroad line on which the damage occurred};][32]

(f) radioactive contamination;

(g) labor strikes, lockouts, go slows, work stoppages, boycotts, walkouts, labor disputes to the extent any of the foregoing such actions are national or regional in nature, but excluding labor shortages or the ability to obtain and/or retain qualified labor {unless caused by any of the foregoing circumstances}; and

(h) failure to issue, suspension, revocation or modification of a [Sponsor] Permit [if the Sponsor is the Asserting Party].

Force Majeure Excused Events shall not include any of the events in the previous two sentences to the extent that:

(a) such were caused or provoked by the Asserting Party (or any Contractor Person in the case that the Contractor is the Asserting Party);

(b) with respect to the <Site> <Route>, such result from weather conditions generally experienced in the local area;

(c) such result from a weather condition not expressly enumerated therein;

(d) with respect to the <Site> <Route>, an experienced contractor employing <Good> <{Prudent}> <Utility> <{Industry}> Practices could have taken reasonable precautions to prevent any of the forgoing circumstances;

(e) such arise as a result of any conditions of the <Site> <Route> contemplated by Article 11.2 but excluding Antiquities;

(f) such occurred outside the local area;

(g) such were caused by delays in customs clearances in any country;

(h) such do not result in a delay of Critical Path Items; or

(i) such result from waterways, roads, bridges or tunnels which are inadequate for transportation of Equipment.

Changes in market conditions and changes in the financial condition of any Contractor Person shall not be considered a Force Majeure Excused Event and failure to obtain any Permit shall not be a Force Majeure Excused Event [{unless the Contractor has made a timely and complete application therefor}]. No obligation to make payment shall be excused by a Force Majeure Excused Event. Force Majeure Events shall not include any natural physical condition of the surface or subsurface of the <Site> <Route> which influences the suitability of the <Site> <Route> for the Facility, including natural physical conditions which influence the ground's capability to retain or absorb water, support structures or resist load or which should be taken into account in determining foundation design, soil stability or methods of construction{; provided, that as an exception to the foregoing, Force Majeure Events shall include those natural physical conditions of the surface or subsurface of the <Site> <Route>, other than the water table and rock formations, that are not reasonably using <Good> <{Prudent}> <Utility> <{Industry}> Practices ascertainable from the geotechnical reports and data attached in Exhibit 1J and Exhibit 5.1(a) (iv), including sinkholes, underground cavities, underground aquifers or streams, and unusually soft stratum for the area}. Collapse of any Work (whether intended to be temporary or permanent) at any stage of construction or completion shall not be a Force Majeure Excused Event unless caused by a Named Storm or a Threshold Earthquake.

["**Fuel**" shall mean fuel meeting the specifications therefor set forth in the Functional Specification.]

["**Fuel Interconnection Agreement**" shall mean the interconnection agreement between Sponsor and the Fuel Transporter.]

["**Fuel Interconnection Facilities**" shall mean (i) the Fuel Interconnection Facilities and (ii) the interconnection facilities required in order to establish an interactive interface and interconnection between the Facility and [the Fuel Transporter] [the Oil storage tanks of the Sponsor and the system of the Fuel Transporter.]

["**Fuel Supplier**" is any Person that supplies Fuel for the Facility [or Existing Plant]]

["**Fuel Supply Agreement**" shall mean any agreement entered into by Sponsor and a Fuel Supplier.]

["**Fuel Transportation Agreement**" shall mean any agreement between Sponsor and a Fuel Transporter.]

["**Fuel Transporter**" is any Person that delivers to the Facility [and/or Existing Plant.]

"**Functional Specifications**" shall mean the design criteria, plant performance requirements and other requirements for the Work, which are contained in Exhibit 1K.[33]

"**General Warranties**" shall have the meaning ascribed thereto in Article 13.1.

"**<Good> <{Prudent}> <Utility> <{Industry}> Practices**" shall mean those practices, methods, equipment, specifications and standards of safety and performance, as the same may change from time to time, as are commonly used by professional construction and engineering firms performing engineering and construction services in *<<insert country>>* in the *<<insert name of industry>>* <utility>[34] industry for facilities of the type and size similar to the Facility that, in the exercise of reasonable judgment and in the light of the facts known at the time the decision was made, are considered good and safe <utility> <{industry}> practices in connection with the design, construction and use of *<<insert name of industry>>* and other equipment, facilities and improvements, with commensurate standards of safety, performance, dependability, efficiency and economy.[35]

"**Governmental Authority**" shall mean the state, governmental, municipal and union authorities of *<<{insert <country in which project will be located> <any country having jurisdiction over any Contractor Person, Sponsor Person, any of the Work or the Facility>}>>* or any ministry, department, subdivision (political or otherwise), municipality, instrumentality, agency, corporation or commission under the direct or indirect control thereof including any political subdivision or state.

["**Grid**" shall mean the electrical transmission system [owned and] operated by *<<insert name of operator>>.*]

"**Guaranteed Facility Commission Completion Date**" shall mean the date set forth in Article 9.1.1.

"**Guaranteed Facility Final Acceptance Date**" shall have the meaning ascribed thereto in Article 9.1.1.

"**Guaranteed Facility Provisional Acceptance Date**" shall mean the date set forth in Article 9.1.1.

["**Guaranteed Unit One Commissioning Completion Date**" shall mean the date set forth in Article 9.1.1.]

["**Guaranteed Unit One Provisional Acceptance Date**" shall mean the date set forth in Article 9.1.1.]

["**Guaranteed Unit Two Commissioning Completion Date**" shall mean the date set forth in Article 9.1.1.]

["**Guaranteed Unit Two Provisional Acceptance Date**" shall mean the date set forth in Article 9.1.1.]

"**Guarantees**" shall mean the Completion Guarantees and the Performance Guarantees.

"**Hazardous Material**" means any hazardous substance, pollutant, contaminant or substance that may be the subject of liability for costs of response or remediation under applicable Law or any Antiquity found.[36]

["**Hot Weather Test**" is defined in Article 8.1].

"**Impaired Performance Liquidated Damages**" shall mean the Facility Buy Down Amounts.

"**Insurance Proceeds**" shall have the meaning ascribed thereto in Article 15.15.

"**Insured Parties**" shall have the meaning ascribed thereto in Article 15.14.

["**Integration**," "**Integrate**" or "**Integrated**" is defined in the recitals hereof.]

["**Interconnection Facilities**" shall mean, collectively, the Site Electrical Interconnection Facilities and the Thermal Interconnection Facilities.][37]

["**Interconnection Requirements**" means the interconnection requirements of the [electrical grid operator/Sponsor].]

["**Kcal**" means kilocalorie].

"**Key Contractor Personnel**" are listed in Exhibit 5.10.

["**kJ**" means kilojoules.]

["**kW**" means kilowatts.]

["**kWh**" means kilowatt hours.]

"**Latent Defect**" shall mean a defect or deficiency in the Work {that was present during the Warranty Notification Period, but} which would not have been disclosed by [careful] visual examination.[38]

"**Latent Defects Liability Period**" shall mean in respect of the Work, including the Facility, but excluding the civil works, the period commencing on the expiration of the Warranty Notification Period and expiring <<*insert number*>> Days later, and in respect of the civil works, the period commencing on the expiration of the Warranty Notification Period and expiring <<*insert number*>> Days later.[39]

"**Law**" shall mean:

(a) any Governmental Authority's:
    (i)   constitution;
    (ii) charter;
    (iii) act;
    (iv) statute;
    (v)   law;
    (vi) ordinance;
    (vii) code;
    (viii) rule;

   (ix) regulation;

   (x) order; or

(b) any specified standards or objective criterion, Permit, or legislative or administrative action of any Governmental Authority;

(c) the Environmental Standards;

(d) Customary Utility Requirements;

(e) any final decree, judgment or order of a court existing by act of any Governmental Authority; and

(f) any Code.

["**Letter of Intent**" means the letter of intent between the Sponsor and the Contractor dated <<*insert date*>>.]

"**Lien**" shall mean any lien, Claim, charge, security interest, attachment or encumbrance placed on the Work or the Facility, including materialmen's, laborers', mechanics', subcontractors' and vendors' liens.

["**Limited NTP**" means a notice to proceed with only the Work expressly specified therein.][40]

"**Liquidated Damages**" shall mean the Delay Damages [, Shutdown Liquidated Damages] [, Loss of Availability Damages] and the Impaired Performance Damages.

"**Loss**" shall mean any loss, damage, fine, penalty, expense or liability (including court costs and attorneys' fees).

"**Major Subcontract**" is defined in Article 4.2.

"**Milestone Payment**" shall mean a payment to the Contractor by the Sponsor of a portion of the Contract Price for the Work completed pursuant to Article 8.3.

"**Milestone Payment Schedule**" shall mean the milestone payment schedule set forth in Exhibit 10.3.1A.

"**MPT**" is defined in Article 10.1.4.

["**MW**" refers to a megawatt of electrical capacity.]

"**Named Storm**" shall mean a storm, hurricane, typhoon, severe tropical cyclone, severe cyclonic storm, tropical cyclone, atmospheric disturbance, depression, or other weather phenomenon designated by the <<*United States National Weather Service and/or the United States National Hurricane Center or appropriate entity*>> to which a number or name has been assigned.

["**National Content**"] [means <<*insert applicable definition from concession agreement or bidding requirements*>>][is defined in Article 5.12.1].

["**National Content Actual Damages**" means the damages payable by Contractor to the Sponsor for Contractor's failure to meet the national content requirements set forth in Article 5.12.1, as more fully described in Exhibit 5.12.1.]

"**New Test Date**" is defined in Article 8.8.

"**Notice to Proceed**" shall mean the written notice from the Sponsor to the Contractor that directs the Contractor to commence the Work.[41]

["**O&M Agreement**" shall mean any agreement for the operation and maintenance of the Facility.]

["**Offtake Agreement**" shall mean the <*Offtake Agreement*> <*power purchase*> <*LNG sale/supply*> <*electric and thermal supply*> <<<*insert chemical name* >>> supply and <*refined product supply*> Agreement, dated <<*insert date*>>, between the Sponsor and the Offtaker.]

["**Offtaker**" shall mean <<*insert name of counterparty to the Offtake Agreement*>>.]

["**Offtaker Completion**" as such term is defined in <<*insert proper name of term from Offtake Agreement*>> in the Offtake Agreement, shall have occurred [as evidenced by the Offtaker's refraining from assessing liquidated damages with respect thereto, or a final determination under the dispute resolution procedure in the Offtake Agreement resolving that no liquidated damages are due from the Sponsor to the Offtaker in respect of the deadline to achieve such in the Offtake Agreement.]

["**Offtaker Tests**" shall be the tests for the Facility imposed by the Sponsor Information and Exhibit 8.1.]

["**Offtaker Facility Acceptance Date**" is the date that the Facility has been determined under the Offtake Agreement to be acceptable to the Offtaker.]

["**Operating Spare Parts**" shall mean Parts and Consumables required for the operation and maintenance of the Facility after Facility Provisional Acceptance including those set forth in Exhibit 5.29.1.]

["**Operator**" means the operator under the O&M Agreement.]

["**Outage Start Date**" shall have the meaning set forth in Article 5.10.]

"**Parent Guarantee**" shall have the meaning ascribed thereto in Article 10.17.[42]

"**Parent Guarantor**" shall mean [each of] <<*insert name*>> [and <<*insert name*>>].[43]

"**Parties**" shall mean the Sponsor and the Contractor.

"**Parts**" shall mean those items or assemblies, which are generally long lead time parts, the deployment of which will either be unique in the life of the Facility or typically be required infrequently and shall include all types of parts that in case of failure during the Warranty Notification Period shall be replaced by the Contractor. [Parts do include <<*insert categories*>>.]

"**Payment Application**" shall mean the Contractor's request for a Milestone Payment, in substantially the form of Exhibit 10.3.1B.

"**Performance Guarantees**" shall mean each of the guarantees set forth in Exhibit 9.2.1 [and Exhibit 9.2.2].

"**Performance LC**" shall have the meaning ascribed thereto in Article 10.16.

"**Performance Tests**" shall mean the tests conducted in accordance with the Functional Specification, Exhibit 8.1 and the Sponsor Information in order to determine whether the Performance Guarantees have been achieved.

"**Permit**" shall mean any permit, approval, license, consent, variance, notification or authorization required by any Governmental Authority in connection with the Work or the Facility or any agreement concerning the Work or <Site> <Route> by which the Sponsor is bound and a copy of which has been delivered to the Contractor [{prior to the Contract Date}].

"**Person**" shall mean any individual, corporation, partnership, joint venture, limited liability company, association, joint stock company, trust, unincorporated organization, Governmental Authority or other entity.

["**Plaintiff**" is defined in Article 20.7.]

"**Preliminary Services**" means any portion of the Work and all other services and work relating to the Facility which have been or will be performed by the Contractor or any Subcontractor or otherwise on behalf of the Contractor prior to the Work Commencement Date [including any so-called "front end engineering design" ("FEED Work")].

"**Price Adjustment**" shall mean an increase or decrease in the Contract Price agreed upon by the Parties and shall be the basis for adjusting the Contract Price.

"Progress Report" shall mean a progress report prepared by the Contractor and provided to the Sponsor in accordance with the form thereof contained in Exhibit 5.6.5, which shall include the latest Current Schedule as of the date of such Progress Report.[44]

[**"Project"** shall mean <<the entire integrated <<*insert type of facility*>> facility of which the Facility will <be part> <serve> >>.]

"Project Description" shall mean the overall project description as set forth in the Functional Specifications.

"Project Variables" shall mean the Contract Price, the Guarantees, the Project Schedule, the Functional Specification and all warranties and guarantees.

[**"Proposal"** is defined in the recitals hereof.]

"Provisional Performance Tests" shall have the meaning ascribed thereto in Article 8.

"Punch List" shall mean the list of items of uncompleted Work which should have been completed by such date.

"QA/QC Manager" shall have the meaning ascribed thereto in Article 6.4.

"Reference Conditions" shall have the meaning set forth in Exhibit 8.1.

"Remedial Plan" shall have the meaning set forth in Article 6.7.3.

"Replacement Contractor" shall mean any Person selected to replace the Contractor.

[**"Request for Review"** is defined in Article 20.7.3.]

"Required Rating" shall mean, with respect to (a) any Person that is not a bank, that such Person's non-credited enhanced long term senior unsecured debt is rated at least <<*insert desired rating*>> by <<[Moody's Investors Service]>> <and> <or> <<*insert desired rating*>> by <<[Standard and Poor's Ratings Services]>>; and (b) with respect to any Person that is a bank, that such bank has both a short-term deposit rating of at least <<*insert desired rating*>> by <<[Moody's Investors Service]>> <and> <or> <<*insert desired rating*>> by <<[Standard and Poor's Ratings Service]>> and a long-term deposit rating of at least <<*insert desired rating*>> by <<[Moody's Investors Service]>> <and> <or> <<*insert desired rating*>> by <<[Standard and Poor's Ratings Service]>>.

[**"Respondent"** is defined in Article 20.7.3.]

[**"Response"** is defined in Article 20.7.3.]

"Retainage" shall mean the funds withheld by Sponsor from the payment of the Contract Price or available under the Retainage LC, all of which may be applied, if necessary, by the Sponsor towards completion of the Work or in satisfaction of any of the Contractor's other obligations hereunder. With respect to any Milestone Payment, Retainage shall be <<*insert percentage*>> percent thereof.

"Retainage LC" shall be a letter of credit in the form of Exhibit 10.4.1.

[**"Rules"** is defined in Article 20.6.]

"Safety Coordinator" shall have the meaning ascribed thereto in Article 6.4.

[**"Seawater Extraction Agreement"** means the agreement with <<*insert name*>> by which the Sponsor may extract seawater from <<*insert location*>> for use in the Facility.]

[**"Seawater Intake/Outfall Connection Facilities"** means the seawater pumps and all auxiliary equipment necessary for the Facility to extract and convey screened and chlorinated seawater and waste discharge from and to the seawater intake/outfall facilities to be provided by the Sponsor under the terms of the Seawater Extraction Agreement.]

["**Seawater Intake/Outfall Facilities**" means the seawater intake, seawater pumping station, screens, chlorination plant, electrical substation and used water outfall together with all supporting facilities for the purposes of conveying seawater from the sea to the seawater intake pumps and conveying effluent produced in the desalination process to the sea.]

["**Seawater Supply Failure**" means an inability of the Sponsor to obtain sufficient supplies of seawater of adequate quality under and in accordance with the Seawater Extraction Agreement to enable the Contractor to perform its obligation hereunder to the extent that such inability is attributable to a breach of, or Force Majeure Excused Event affecting, <<*insert name*>>'s obligations under and in accordance with the Seawater Extraction Agreement.]

"**Security Coordinator**" shall have the meaning ascribed thereto in Article 5.4.

["**Senior Debt**" shall have the meaning ascribed thereto in Article 10.15.2.]

["**Sewage Company**" means <<*insert name*>>.]

"**Shareholder**" shall mean any direct or indirect owner of the Sponsor.

"**Shipment Notice**" shall have the meaning ascribed thereto in Article 5.12.

["**Shutdown Liquidated Damages**" is defined in Article 9.1.9.]

"**Site**" shall mean the real property on which the Facility shall be constructed, together with easements and rights of way relating thereto, all as described in the Functional Specifications.

"**Site Conditions**" shall have the meaning set forth in Article 11.2.[45]

["**Site Electrical Interconnection Facilities**" shall mean the interconnection facilities, protection devices, electric lines, facilities and other facilities located on the Site up to but not beyond the Site Electrical Interconnection Point which will be connected to <<*insert point*>> so that the Offtaker <<and/or the Wheeling Company>> can establish an interface between the Facility and the <<Offtaker's <<*insert name*>> facility>> <<Wheeling Company's transmission system>> in accordance with <<the Offtake Agreement, Electricity Wheeling Agreement and the Electrical Interconnection Agreement>>.]

["**Site Electrical Interconnection Point**" is the physical points at which the Site Electrical Interconnection Facilities are connected to <<the Grid>> as further defined in the Functional Specification.]

["**Site Lease**" means any and all leases and subleases relating to the <Site> <Route> [and the Facility].]

["**<Site> <Route> Assessment**" shall mean a written assessment of the geophysical condition of the <Site> <Route> <prepared in accordance with the protocol and meeting the scope set forth in> <is attached as> Exhibit 5.1(a)(iv)].[46]

"**Specifications and Drawings**" shall mean the specifications and drawings prepared by the Contractor with regard to the Work in the manner required by this Contract.

"**Specified Purpose**" is defined in Articles 14.5 and 21.7.

"**Sponsor**" is defined in the first sentence hereof.

"**Sponsor Event of Default**" shall have the meaning ascribed thereto in Article 17.5.

["**Sponsor Information**" shall mean [the excerpts from] <<*list project agreements relevant such as Thermal Energy Supply Agreement, Water Purchase Agreement, Ash Disposal Agreement, Water Discharge Agreement, O&M Agreement, Fuel Supply Agreements, Fuel Transportation Agreements, Fuel Interconnection Agreements, Electrical Interconnection*

Agreement, Electricity Wheeling Agreement, Offtake Agreement, Site Lease, Limestone Supply Agreement, Gas Turbine Purchase Agreement, Steam Turbine Purchase Agreement, Seawater Extraction Agent, Feedstock Supply Agreement, Sponsor Land Use Agreement>> all of which are set forth in Exhibit 1J.][47]

["**Sponsor Land Use Agreements**" mean any contractual rights obtained by the Sponsor to use real property [(including the Site Lease)] in connection with the construction of the Facility listed in the Functional Specifications.]

["**Sponsor Parent Guaranty**" is defined in Article 10.19.]

"**Sponsor Permit**" shall mean any Permit listed in Exhibit 7.1.1.

"**Sponsor Person**" shall be the Sponsor, the Sponsor's Engineer, [any Financing Party, the Financing Party's Engineer,] Sponsor's operators and contractors and any and all of the agents, representatives, affiliates, officers, directors, and employees of each of the aforementioned Persons.

["**Sponsor Risk Event**" shall mean:

(a) war; or
(b) sonic boom.]

"**Sponsor's Engineer**" shall mean any engineer employed or retained by the Sponsor to review the Work, the Facility or the design documents relating thereto.

"**Sponsor's Project Manager**" is defined in Article 7.3.

"**Sponsor's Responsibility Items**" is defined in Article 2.5.

["**Start Up Fuel**" shall mean fuel, meeting the specifications set forth in the Functional Specification.]

["**<< *insert name* >> Station Injection Point**" means the pothead to be constructed at the switchyard in the *<<insert name of station>>*.]

"**Subcontract**" shall mean any contract between the Contractor and any Subcontractor, or between any Subcontractor and any other Person, relating to the services, Equipment or other Work to be provided by such Subcontractor in respect of the Facility.

"**Subcontractor**" shall mean any other contractor, Vendor or supplier that contracts to perform services or provide materials or Equipment constituting part of the Work, whether under a Subcontract directly with the Contractor or a Subcontract with another Person.[48]

["**Substation**" is the electrical substation of *<<insert name>>* located at *<<insert location>>*.]

["**Technical Services Agreement**" shall mean the services agreement to be entered into by and between the Contractor and the Sponsor.]

["**Thermal Customer**" is *<<insert name>>*.]

"**Threshold Earthquake**" is defined in Article 11.2.

["**Tie-ins**" shall mean those pipelines, electrical lines, conveyors and other systems up to and including the respective termination points as described in the Functional Specifications required in order to establish an active interface between the *<<Existing Plant/Grid>>* and the pipelines and electrical lines and Facility.]

"**Time Adjustment**" shall mean an increase or decrease in time for the execution of the Work agreed upon by the Parties and shall be the basis for adjusting any Deadline affected thereby.

["**Unit**" shall mean either of Unit One or Unit Two as the case may be.]

["**Unit Commissioning Completion**" is defined in Article 8.4.1.]

["**Unit Commissioning Completion Date**" shall mean the date on which the Sponsor issues a Certificate of Unit Commissioning Completion for a Unit.]

["**Unit One**" shall mean that portion of the Facility which includes:<<

(a) the first CTG to be placed into service;
(b) the [boiler and] turbine generator unit which achieves Unit Commissioning Completion first;
(c) all other Equipment comprising such unit and all parts, components and accessories relating thereto and the common facilities; and
(d) all Work to be performed hereunder in connection with the items described in clauses (a), (b) and (c) above>>.]

["**Unit Performance Guarantees**" are each of the guarantees of performance for a Unit in Article 9 <<__>>.]

["**Unit Provisional Acceptance**" is defined in Article 8.6.1.]

["**Unit Provisional Acceptance Date**" shall mean the date on which the Sponsor issues a Certificate of Unit Provisional Acceptance for a Unit.]

["**Unit Two**" shall mean that portion of the Facility which includes:<<

(a) the CTG which is not part of Unit One;
(b) the boiler and turbine generator unit which is not part of Unit One;
(c) all other Equipment comprising such unit and all parts, components and accessories relating thereto; and
(d) all Work to be performed hereunder in connection with the items described in clauses (a), (b) and (c) above>>.]

["**Utility Tie-Ins**" shall mean those pipelines, electrical lines, conveyors and other systems to be located within or outside the Site, up to and including, the respective termination points as described in the Functional Specifications, required by the Facility in order to establish an active interface between the Facility and the pipelines, electrical lines and other facilities of Persons providing utilities and other services to the Facility.]

"**VAT**" shall mean any value added taxes on goods or services imposed by any Governmental Authority of <<*insert country*>>.

"**Vendor**" shall mean any supplier of any materials or Equipment for the Facility to any Contractor Person.

["**Vendor Purchase Agreements**" refers to <<*any and all Subcontracts for* [<<*insert type*>>] *Equipment*>>.]

"**Vitiation Loss**" shall mean all losses, costs, claims, demands and liability of whatsoever nature suffered or incurred by the Sponsor in respect of which (but for any misrepresentation, non-disclosure, or break of declaration, condition or express or implied warranty by or on behalf of the Contractor, or its Subcontractors, agents or employees) the Sponsor would have been entitled to claim compensation or be indemnified under any policy of insurance to be taken out by the Sponsor.

"**Warranty LC**" shall be a letter of credit in substantially the form of Exhibit 1L.

"**Warranty Notification Period**"[49] shall commence on Facility Provisional Acceptance [{or, if earlier, the date that the Facility is ready for [performance testing but the Sponsor or Offtaker has prevented such]}] and end <<*insert number*>>

Days thereafter or <<*insert number*>> Days in the case of latent defects or deficiencies; provided that in the case of any addition, repair or replacement to the Facility, the Warranty Notification Period shall be extended by <<*insert number*>> Days from the date of completion of such addition, repair or replacement [{but in all cases no more than <<*insert number*>> Days after the Facility Provisional Acceptance Date}].[50]

[{"**Warranty Stipulations**" shall mean that the Sponsor has:

(a) operated, inspected and maintained the Equipment generally in accordance with Contractor's instructions and manuals;
(b) operated the Equipment generally within the express design operating conditions if any described in the Functional Specifications; and
(c) generally used Fuel falling within the characteristics specified as safe in the Functional Specifications[{, except to the extent variations in Fuel properties will result in significant Equipment damage}].}]

["**Water Discharge Agreement**" means <<*insert title, date and parties to agreement*>>.]
["**Water Purchase Agreement**" means <<*insert title, date and parties to agreement*>>.]
["**Water Supplier**" means <<*insert name*>>.]
["**Wheeling Company**" is <<the owner of the Grid>>.]
"**Work**" shall have the meaning ascribed thereto in Article 5.1 and includes spare Parts.
"**Work Commencement Date**" shall mean the date specified in the Notice to Proceed as the date on which the Contractor is to begin the Work [{which date may not be later than <<*insert number*>> Days after the Contract Date}].
"<<*insert currency*>>" means the lawful currency of <<*insert country*>>.[51]

ARTICLE 2. INTENT

2.1 Intent

It is the intent of the Parties that the Contractor design, engineer, procure, construct, start up, test and put into operation a highly reliable <<*name type*>> facility. The Contractor shall perform all of the Work specified or [{reasonably}] necessitated or implied by this Contract. The Contractor's performance under this Contract shall include everything requisite and necessary to complete the entire Facility notwithstanding the fact that every item necessarily involved may not be specifically mentioned. Details and items not indicated by the Functional Specifications shall be adequately and properly performed by the Contractor at no extra cost if such details or items are necessary [{to carry out the intent of this Contract}{so that the Facility can operate in accordance with <Good> <{Prudent}> <Utility> <{Industry}> Practices}].[52] The intent of this Contract is to relieve the Sponsor of the necessity of engaging or supplying any labor, service or material to complete the Facility unless the labor, service or material is expressly stated in Article 5 of this Contract as being furnished by the Sponsor.[53] The Contractor acknowledges that by entering into this Contract, the Sponsor has placed great trust and confidence in the Contractor to design and construct the Facility in an economical and timely manner so that the Sponsor will receive a Facility that conforms fully with the intent of in this Contract. The Contractor accepts this position of trust and confidence and agrees that it shall utilize its best skill, efforts and judgment to comply with the provisions of this Contract. No prototypical materials or equipment shall be incorporated in the Work, but rather the Contractor shall only incorporate materials and equipment of well-demonstrated reliability. [Time is of the essence.][54]

2.2 Discrepancy in This Contract

Should the Contractor find any error, omission, inconsistency, ambiguity or other discrepancy in this Contract, the Contractor shall provide written notice to the Sponsor within <<*insert number*>> Days thereof.

2.3 Notice of Objection to Different Terms[55]

The Sponsor hereby gives notice of its objection to different or additional terms and conditions than those contained in this Contract. This Contract is expressly conditioned on the Contractor's assent to the terms and conditions stated herein.

2.4 Written Clarifications.

The Sponsor shall have the right to issue written clarifications should it deem necessary. Under no circumstances shall the Sponsor be liable for any oral clarifications, instructions, or interpretations. All determinations, decisions, instructions, judgments, interpretations or clarifications of the Sponsor regarding this Contract and the Work to be provided shall be final, binding and conclusive, unless determined to have been made in bad faith.

2.5 Interpretation of Functional Specifications

If:

(a) the Contractor has any doubt as to the meaning of any portion of the Functional Specifications;
(b) the Contractor believes that a conflict exists within this Contract; or
(c) a standard of material is not specified,

then the Contractor shall perform the Work in the best and most workmanlike manner, and the Contract Price shall be deemed to include the most prudent method, material, finish, system or similar item, unless such uncertainty is clarified by a Change Order. Contractor shall be solely responsible for requesting instructions or interpretations and shall be solely liable for any cost and expenses arising from its failure to do so. Contractor's failure to protest Sponsor's determinations, instructions, clarifications or decisions pursuant to Article 7.3.1 above in writing in detail within <<*insert number*>> Days thereof shall constitute a waiver by Contractor of all its rights to protest such.

2.6 Preliminary Services

2.6.1 Preliminary Services have been performed prior to the Contract Date [(including under the <<*insert name of agreement*>>)] and that various Preliminary Services may be performed after the Contract Date and before the Work Commencement Date. All of the Preliminary Services are hereby incorporated into the Work for all purposes, including the standards of the Contractor's performance, General Warranties, Performance Guarantees and Liquidated Damages. The Contractor shall be fully responsible for the performance or non-performance of the Preliminary Services as if the Preliminary Services were performed under this Contract following the Work Commencement Date.

2.6.2   [Before using the FEED Work, the Contractor shall satisfy itself as to its appropriateness and the Sponsor will not have any responsibility for any use by the Contractor of the FEED Work and the Contractor shall not be entitled to claim an Adjustment if the FEED Work provides inadequate for the Work.]

2.6.3   The Sponsor shall not be liable for any data or information contained in the Sponsor Information except as expressly noted which data and items Contractor will not investigate (the "Sponsor's Responsibility Items"). Before using any documents of the Sponsor [(including the Sponsor's FEED Work)] other than those in the Sponsor's Responsibility Items, the Contractor shall conduct its own review and analysis and its own studies, taking all measures necessary to verify such information or data to satisfy itself of the accuracy and completeness. Contractor shall have full responsibility for use by itself or by any Subcontractor of any data or information [(including the FEED Work)], not contained in the Sponsor Responsibility Items (including any conclusions drawn from it), and assumes the risk of accuracy and completeness thereof. The Contractor shall indemnify and hold the Sponsor harmless from any damage and any loss resulting from the use or reliance on the documents [(including the FEED Work)], other than those in the Sponsor Responsibility Items. The Parties agree that no Adjustment or other relief will be granted as a result of the Contractor having relied on data or information [(including the FEED Work)] not covered by the Sponsor's Responsibility Items.

2.6.4   The Sponsor shall be responsible for any errors in these specified or notified items of reference contained in the Sponsor's Responsibility Items, but the Contractor shall use reasonable efforts to verify their accuracy before they are used.

2.6.5   If the Contractor suffers delay in any Critical Path Item and/or incurs costs from executing work which was caused by an error in items contained in the Sponsor's Responsibility Items, and an experienced contractor using <Good> <{Prudent}> <Utility> <{Industry}> Practices could not have discovered such error and avoided this delay and/or cost, the Contractor shall give notice to the Sponsor and shall be entitled to an Adjustment under Article 19.

2.7 Escrow of Bid Documents

At the Sponsor's request, copies of all Contractor's bid documents, including but not limited to, all data, information, documents, references, calculations, unit costs, and backup sheets, shall be preserved and placed in escrow with such third party as may be designated by the Sponsor. In the event of any disputes between the Sponsor and the Contractor, said bid documents shall be released to the Sponsor upon the Sponsor's written request to said third party. The Sponsor shall have the right to require that the Contractor make the representation, supported by affidavit, that all such bid documentation has, in fact, been preserved and submitted to escrow as provided above. The Contractor's failure to preserve, submit, and provide such bid documentation, at the Sponsor's request, as provided herein, shall constitute a default and material breach of the Contract.

2.8 [Value Engineering]

During the performance of the Work, the Contractor may identify a more cost-effective means to achieve the Facility's performance objectives that would require a modification to the Functional Specifications. If so, the Contractor shall promptly notify the Sponsor and provide the Sponsor with a complete analysis of the proposed change, including installed and operating cost differences and project and payment schedule impacts. The Sponsor shall review the analysis of the proposed change and notify the Contractor of a decision within <<*insert number*>> Days after receipt of the Contractor's notice. If the change is accepted by the Sponsor (although the Sponsor shall have no obligation whatsoever to consider or accept such proposal), the Contractor shall receive a fee equal to <<*insert number*>> percent of the reduction of the Contract Price, and the parties shall execute and deliver a Change Order reflecting the effect of the proposed change on the Project Variables.

ARTICLE 3. COMMENCEMENT OF THE WORK[56]

3.1 Notice to Proceed[57]

The Notice to Proceed shall be a full release of the Contractor to commence all the Work, and the Contractor shall commence the Work <no later than <<*insert number*>> Days after the Work Commencement Date [{so long as the Contractor has been provided with evidence that Financial Closing has occurred}]>[58] <<on the Work Commencement Date>>. If the Work Commencement Date has not occurred within <<*insert number*>> Days after the Contract Date, either Party may, at any time before the Work Commencement Date occurs, upon written notice to the other Party, terminate this Contract without any liability hereunder except in the case that the <Site> <Route> Assessment has been performed and delivered to the Sponsor within the time period specified herein, the Sponsor shall pay Contractor pursuant to the last sentence of [Article 8.1.3] for the <Site> <Route> [Assessment]. The Contractor understands that under no circumstances ever shall the Sponsor be obligated to give Notice to Proceed or make any payment hereunder until Notice to Proceed has been given [{and Financial Closing has occurred}].

3.2 [Terminal Condition Subsequent]

If the Sponsor has not given Notice to Proceed by <<*insert date*>>, this Contract shall automatically terminate on such date without any action necessary by any Person, and thereupon, neither Party shall have any liability whatsoever of any nature to the other Party[, except such as may arise under Article 10].

3.3 [Condition Precedent]

This Contract shall be of no force or effect unless:

(a) [the Sponsor has received the executed originals the Parent Guarantee executed by the Parent Guarantor by <<*insert date*>>, and if it does receive such Parent Guarantee by such date, this Contract shall come into full force and effect automatically and without any further action necessary by any Person][59] [; and] [; or] [.]

(b) [the Sponsor has received approval of this Contract from <<*insert name of Governmental Authority and/or internal governing body*>>, and if it does receive such approval by <<*insert date*>>, this Contract shall come into full force and effect automatically and without any further action necessary for any Person.]

3.4 Execution Plan

No later than <<*insert number*>> Days after <the Contract Date> <Notice to Proceed has been given>, the Contractor shall, separate and apart from its obligation under Article 5.6, deliver to the Sponsor for its review and approval, a Facility-specific execution plan (the "Execution Plan"), that shall set forth the specific procedures, manuals and method statements for use in the performance of the Work. The Execution Plan proposed by the Contractor shall be consistent with the terms and conditions of this Contract and the contents of its exhibits and schedules. [{The Sponsor shall provide comment and shall accept or reject the Execution Plan within <<*insert number*>> Days of its receipt by the Sponsor. If the Sponsor neither accepts nor rejects the Execution Plan within such <<*insert number*>> Day period, then the same shall be deemed accepted.}]

3.5 [Selection of Certain Equipment][60]

The Contractor shall, no later than <<*insert number*>> Days from its receipt of Notice to Proceed, send written notice to Sponsor identifying which of the <<*insert name/ models of equipment*>> from the Vendors identified on Exhibit 4.2 it has selected for use in the Facility.

ARTICLE 4. RELATIONSHIP OF THE SPONSOR, THE CONTRACTOR AND SUBCONTRACTORS

4.1 Independent Contractor

The relationship of the Contractor to the Sponsor under this Contract is that of an independent contractor with respect to the performance of the Work. Neither the Contractor nor any Subcontractor, nor any of their employees, are servants, employers or agents of the Sponsor. No Subcontractor shall be deemed a third-party beneficiary of, or have any interest in, this Contract.

4.2 Subcontractors[61]

The Contractor may cause any of the services or Equipment included within the Work to be furnished by one or more Subcontractors pursuant to Subcontracts.

The Contractor shall provide to the Sponsor copies of each Subcontract [{(without price information)}] for the supply of any item [in excess of <<*insert currency and amount*>> [{and] listed on Exhibit 4.1.13}] (whether or not such Subcontract is let by the Contractor or any Subcontractor, each a "Major Subcontract") within <<*insert number*>> Days of its execution to allow the Sponsor to satisfy itself that the relevant provisions of this Contract have been fulfilled. Exhibit 4.2 contains a list of the Subcontractors that have been approved by the Sponsor with respect to supplying each item of Equipment or portion of the Work set forth opposite such Subcontractor's name in Exhibit 4.2. The Sponsor's prior written approval [{(which shall not be unreasonably withheld)}] shall be required for:

(a) any Person not listed on Exhibit 4.2 to become a Subcontractor;
(b) any Subcontractor listed on Exhibit 4.2 to supply any item of Equipment or Work[62] not listed next to its name on Exhibit 4.2; or
(c) any Major Subcontract to be amended, modified or terminated or any provision thereof to be waived by the Contractor, {if any of such actions would increase any of the Contractor's payments thereunder by more than <<*insert currency and amount in words and in numbers*>> or extend the original delivery schedule in such Major Subcontract by more than <<*insert number*>> Days}.

No failure of the Sponsor to give any such approval shall[{, if based on reasonable grounds}]:

(d) affect, increase or diminish the Sponsor's or the Contractor's obligations or rights under this Contract; or
(e) result in any Adjustment.

   Notwithstanding the Sponsor's review, approval or disapproval of any Subcontractor or Subcontract, the Contractor shall not be relieved from any obligation under this Contract.

4.2.1   The Contractor agrees that all Major Subcontracts will be supported by bonds, guarantees [{or other security instruments customary for industry practices [and reasonably}][63] acceptable to all Financing Parties [{should any Financing Party have a [{reasonable}] basis to believe that such Subcontractor may not be able to perform its material obligations under its Major Subcontract}]].

4.2.2 [64]   The Contractor shall use <{reasonable}> < its best> efforts to ensure that each [{Major}] Subcontract [{with a value in excess of <<*insert currency and amount in words and numerals*>>}] shall preserve and protect the rights of the Sponsor under this Contract with respect to the Work to be performed by such Subcontractor so that the subcontracting thereof shall not prejudice such rights and each such Subcontract shall contain the following provisions, in form and substance acceptable to the Sponsor:

(a) a provision that such Subcontractor shall comply with the provisions of this Contract insofar as they apply to the Work to be performed by such Subcontractor;

(b) a provision that the Subcontractor shall be bound by the dispute resolution procedures set forth in Article 20 and agrees to participate in those procedures if requested by the Sponsor;

(c) a provision that such Subcontractor shall have no rights against the Sponsor and shall not file any Lien against the Facility, any of the Sponsor's assets or any of the Work;

(d) a provision that if this Contract is terminated pursuant to Article 17, such Subcontract shall be automatically assigned, without further action, to the Sponsor (or, at the Sponsor's direction, any Replacement Contractor) except that the Sponsor may, at its option, refuse to accept any such assignment;

(e) a provision that the Sponsor is a third party beneficiary[65] thereof; and

(f) a provision that each such Subcontract may:

   (i) upon the Sponsor's request, be assigned by the Contractor to any Sponsor Person or Contractor of any Sponsor Person, or any of their respective successors, assignees or transferees without the consent of such Subcontractor; and

   (ii) not be assigned by such Subcontractor without the prior written consent of the Sponsor which may be denied for any reason whatsoever or no reason.

4.2.3   No Subcontract shall create any contractual relationship between any Subcontractor and any Sponsor Person, except for any contractual warranty rights created in favor of the Sponsor.[66] The Contractor shall be solely responsible for all Work (whether performed by the Contractor or a Subcontractor) and the engagement and management of Subcontractors in the performance of the Work and no failure of a Subcontractor's performance for any reason whatsoever shall be deemed a Force Majeure Excused Event. The Contractor shall properly supervise and coordinate the Work performed by Subcontractors so as to ensure that all Work performed and Equipment furnished by Subcontractors conforms to the provisions of this Contract.

4.2.4   The Sponsor shall have the right, upon written request, with respect to Work performance by any Subcontractor, to receive from Contractor a copy of such Subcontractor's equipment specifications, performance guarantee data and Subcontractor warranties, as well as a copy of all shop and field performance test reports and vendor field representative reports for any reason whatsoever.

4.2.5   [The Contractor shall cause its Subcontractors to utilize, in carrying out the Work, National Content of a level equal to or greater than <<*insert percentage*>> as elaborated in Article 5.12.1.]

4.2.6   The Contractor shall, and shall cause its Subcontractors, to give priority to {competent and qualified} local personnel materials, services and contractors [{provided that quality, delivery times, reliability, skills, experience and other terms are comparable to those offered by other reputable contractors}].

4.2.7   The Contractor shall cause its Subcontractors to comply with the Law, including the Law insofar as it relates to and regulates the Government's policy and requirements for local content and employment and training of its nationals.

[4.2.8   Sponsor shall deliver to Contractor the specifications for all <<*insert items*>> within <<*insert number*>> Days of the Work Commencement Date, as well as the names of the Subcontractors that the Sponsor proposes that Contractor requests bids therefor. Contractor shall solicit bids therefor within <<*insert number*>> Days after its receipt of such information from Sponsor and Contractor shall inform such bidders that they must respond with proposals within <<*insert number*>> Days. Contractor shall forward all proposals to Sponsor for Sponsor's selection of the relevant Subcontractor[, {which shall be notified to Contractor within <<*insert number*>> Days}]. After having received the offers the Contractor shall prepare a technical and commercial comparison of the offers within <<*insert number*>> days and present the same to the Sponsor [{The Contractor may reject any proposal by giving reasonable arguments therefor}]. The nomination of the Subcontractor by the CEL shall occur within 7 days of the presentation of the commercial and technical comparison to the CEL. The Contractor shall enter into the Subcontract with the relevant Subcontractor within 7 days of CEL's choice of the Subcontractor. Upon Sponsor's selection, the Contractor will enter into Subcontracts with the relevant Subcontractor in a form acceptable to Sponsor and Contractor will be responsible therefor for all matters pertaining to such Work and Sponsor shall not be responsible in any way hereunder except that Sponsor shall pay Contractor the invoices received by Contractor from such Subcontractors to the extent they are reflective of the pricing in the original proposal of such Subcontractor plus a margin of <<*insert number*>> percent thereon (the "Cost Plus Work Amount"). Any amendment to any such Subcontract shall be at Contractor's risk and responsibility provided that no change shall be made to the <<*specifications or warranty*>> in such Subcontract.]

[4.2.9   Contractor and Sponsor agree to cooperate in attempting to lower the cost to Contractor of certain Subcontracts by allowing Sponsor or its representatives to assist Contractor in Contractor's negotiation of certain Subcontracts with potential Subcontractors with which Owner (or its Owners or its or their affiliates) has prior commercial relationships. Contractor in good faith shall determine, in consultation with Sponsor, which of the Subcontracts may benefit from this approach and advise Sponsor of the list of potential Subcontractors and of the assumptions it has made as to the cost (or the basis for calculating the cost) of the relevant Subcontracts to Contractor. In any such circumstances, Contractor and Sponsor shall cooperate in such a manner as to ensure no delays in the Work arise. In any event, Contractor shall remain the counterparty to such Subcontracts and continue to be solely responsible for the performance by such Subcontractors under such Subcontracts. Any cost savings attributable to the foregoing assistance (calculated by the difference between the actual cost to Contractor of each of such Subcontracts and the assumptions provided by Contract to Sponsor in good faith) shall be shared on the basis of <<*insert number*>> percent to Contractor and <<*insert number*>> percent to Owner. Sponsor shall issue a Change Order reducing the Contract Price to reflect the cost savings accruing to Sponsor. The Contract Price shall be reduced only if and as such cost savings are achieved. [Each then unpaid Milestone Payment shall be adjusted by a percentage equal to the total savings divided by the number of Milestone Payments unpaid.].]

4.2.10   The Contractor will ensure that each [{Major}] Subcontract requires that the Subcontractor give the Sponsor notice of any default thereunder and the Contractor hereby grants the Sponsor the right to cure any Contractor default thereunder and charge to the Contractor any costs of the Sponsor incurred in connection therewith.[67]

4.2.11   If, at any time during the progress of the Work, the Sponsor determines in its sole judgment that any Subcontractor is incompetent or undesirable, the Sponsor shall notify the Contractor, and Contractor shall take immediate steps to terminate the Subcontractor's involvement with the Work. The rejection or approval by the Sponsor of any Subcontractor or the termination of a Subcontractor shall not relieve Contractor of any of its responsibilities under the Contract, nor be the basis for additional charges to the Sponsor or entitle the Contractor to an Adjustment.

ARTICLE 5. CONTRACTOR'S RESPONSIBILITIES

5.1 Work

The Contractor shall take all necessary action to perform its obligations hereunder[68] including engineering, designing, procuring, constructing, installing, commissioning and testing the Facility on a fixed price, turnkey basis, including all the items specified in, or [{reasonably}] implied by, this Contract (the "Work") and, toward this end, the Contractor, at its own expense, shall:

[(a) prior to the Work Commencement Date:
    [(i)   cooperate with the Financing Parties concerning the Sponsor's financing of the Facility;]
    [(ii) perform design work to the extent expressly specified in the Baseline Schedule;]
    [(iii) carefully review the Environmental Report and take into account the contents thereof in connection with the design and engineering of the Facility and the performance of all other portions of the Work;][69] [and]
    [(iv) select reputable experts to conduct, prepare and submit to the Sponsor within <<insert *number*>> Days] of the Contract Date, the <Site> <Route> Assessment,[70] which shall be based solely on the results of a thorough investigation of the <Site> <Route> [according to the scope set forth in Exhibit 5.1(a)(iv)] to ensure that the Facility's design is compatible with existing <Site> <Route> conditions and before any part of the Work on <Site> <Route> is commenced, all necessary levels, including ground water levels, shall be measured by the Contractor and reported to the Sponsor and the recorded levels shall be submitted by the Contractor to the Sponsor in the form of a drawing;]]
(b) execute the Work and structure its purchase orders (including provisions for cancellation and the longest available payment terms under the circumstances) in such a manner as to minimize the economic impact to the Sponsor if the Work is canceled after the Work Commencement Date;[71]
(c) [carry out the Work in a manner that will minimize any interference with the ongoing operations of the currently operating <<*insert type*>> facility located <adjacent to> <at> the Site (the "Existing Plant");]

(d) arrange for and provide at its expense all Consumables to be used in connection with the Facility until the Facility Provisional Acceptance Date except to the extent such is consumed in connection with the operation of the Units or the Facility by the Sponsor under Article 8.11;

(e) prepare on behalf of the Sponsor, on a timely basis sufficient to allow the Sponsor to meet all its notice obligations contained in the << [*Sponsor Information*] *insert relevant agreements*>>, all notices required to be given to any Person thereunder [including notices concerning the dates of testing the Facility and initial synchronization of the Units];

(f) [cooperate with the Sponsor [{in all reasonable respects}] concerning the Sponsor's application(s) for benefits under the <<*insert name of program* >> program including with regard to preparing and certifying documentation;]

(g) promptly provide all necessary technical support and prepare all documentation and application (including engineering and design information) related to the Work required by the Sponsor for the Sponsor to obtain the Sponsor Permits by the times contemplated in the Baseline Schedule;

(h) perform all Work continuously and diligently in accordance with the Baseline Level 3 Schedule in order to achieve the Completion Guarantees and provide any information the Sponsor [or any Financing Party] may [{reasonably}] request to verify progress of the Work and predict future progress of the Work;

(i) immediately notify the Sponsor in writing at any time that there is reason to believe that any Critical Path Item on the Baseline Level 3 Schedule (including the Completion Guarantees) will not be completed by the times predicted therein and specify in such notice the reasons therefor and the corrective action planned by the Contractor;

(j) if the Sponsor determines [{and is advised in a written report by a reputable advisor approved by the Contractor from the list set forth in Exhibit 5.1(j)}] that the Contractor is unlikely to maintain the progress that is necessary to achieve the Completion Guarantees, work such hours, including night shifts, weekends and Days, and furnish such additional personnel and/or construction equipment as may be necessary to return to, and thereafter maintain, the Baseline Level 3 Schedule;

(k) provide specifications, engineering and designs of the Facility, which shall be consistent with the design and equipment parameters set forth in this Contract, including the Functional Specifications, and construct the Facility in accordance with such specifications, engineering and design;

(l) design, construct, install and commission:
   (i) the Interconnection Facilities in accordance with this Contract [including those contained in the Sponsor Information] [and the Interconnection Requirements] [and Customary Utility Requirements];[72] and
   (ii) the Tie-ins and Utility Tie-Ins in accordance with the requirements of this Contract and the [Customary Utility Requirements] [requirements][73] of the Persons providing utilities and other services to the Facility;

(m) procure and supply all machinery, Equipment, supplies and services for the Facility and the engineering, design, construction, start up, testing and commissioning thereof, all in accordance with this Contract;

(n) provide in writing to the Sponsor not later than <<*insert number*>> Days after the Work Commencement Date [[Unit] [Facility] Commissioning Completion [of Unit One] a list of recommended spare Parts to be kept on Site for the Facility [for the <<*insert number*>> Day period commencing on the termination of the Warranty Notification Period];[74]

(o) provide all services of, and materials for, engineers, designers, schedulers, accountants, buyers, inspectors, estimators, supervisors, superintendents, foremen, skilled and unskilled laborers, Subcontractors and all other Persons required for the proper execution of the Work;

(p) provide all construction tools and equipment, other tools, office facilities, telecommunications and other items required to complete the Facility and achieve Facility Final Acceptance;

(q) clear, level and develop the Site and erect and construct the Facility thereon, including all civil works, foundations, structures, buildings and process systems of the Facility;

(r) cart away and dispose of all debris and fill at Contractor's own expense (except in the case of Hazardous Materials as provided in Article 11.3) in a manner that minimizes all fugitive dust emissions;

(s) transport to the Site, receive, unload and store at the Site, all Equipment and other components of the Work and obtain all necessary customs clearances for all Equipment;

(t) obtain and maintain all Contractor Permits;

(u) provide all necessary technical support and prepare all documentation (including engineering and design information) required by the Sponsor in order to obtain the Sponsor Permits so that all submissions the Sponsor makes to the relevant Governmental Authorities corroborate that the Facility will comply with all Environmental Standards [when operated at <loads> <levels> above <<*insert number*>>];

(v) [furnish, within <<*insert number*>> Days of the Contract Date, a noise map of the Site which is consistent with the noise levels guaranteed in the Performance Guarantees;]

(w) provide all services, personnel and materials <and make available up to a total of <<*insert number*>> <<*insert number*>> hour Days of group training for> <to> train the Sponsor's personnel and agents in the commissioning, testing, operation and maintenance of the Facility as further provided in [the Functional Specifications] [Exhibit 5.1(w)][75] and submit to the Sponsor, within <<*insert number*>> Days of the Work Commencement Date, the Contractor's schedule of dates for such training program;[76]

(x) maintain in good order at the Site all necessary documentation for the performance of the Work, including at least one (1) record copy of the design and engineering documents, project execution plan, construction procedures, quality assurance materials, drawings, specifications, product data, samples, modifications, marked currently to record changes made during construction, all of which will be available to the Sponsor for inspection and use at all times;

(y) [after Facility Provisional Acceptance, but prior to Facility Final Acceptance, if requested by the Sponsor, maintain qualified personnel on site <<*twenty-four (24)*>> hours per Day to support the Sponsor's operators regarding operation and maintenance of the Facility;]

(z) provide all Facility documentation, including system and operational manuals, As-Built Drawings and Computer Programs and any documentation set forth in <Exhibit 5.1(z)> <the Functional Specifications> by the <Facility> <Provisional Acceptance> <Final Acceptance> Date;[77]

(aa) coordinate with other contractors hired by the Sponsor [or <<*the Offtaker* [etc.]>];[78]

(bb) provide all other items and services that are specified in, or may be [{reasonably}] inferred from, this Contract or which may be reasonably required by <Good> <{Prudent}> <Utility> <{Industry}> Practices;

(cc) [except with respect to a Unit in operation under Article 8.13,] arrange for, and provide at its expense:

    (i) all [demineralized] [construction] [potable] water required by the Facility through the Facility Provisional Acceptance Date;

    (ii) all water required in connection with the Work or the Facility until the date [of first fire of Unit Two provided that after the first fire of Unit One, it shall not be required to provide or pay for any water (other than demineralized water) consumed by Unit One];

    (iii) all electricity (including backfeed electricity to start the Facility) required in connection with the Work or the Facility until the date of first fire [of Unit Two provided that after the first fire of Unit One, it shall not be required to provide or pay for any electricity consumed by Unit One];

(dd) ensure that all Work is completed in accordance with the Baseline Level 3 Schedule and so as to achieve all Performance Guarantees;

(ee) [pay to the Sponsor the Sponsor's actual and direct [burner tip] cost per <MMBtu> <ton> <gallon> <barrel> for Fuel used by the Facility before the Facility Provisional Acceptance Date in excess of <<*insert number*>> MMBtu and pay to the Sponsor the Sponsor's [actual [burner tip] cost for Startup Fuel used by the Facility before the Facility Provisional Acceptance Date in excess of <<*insert number*>> MMBtu <tonnes> <barrels> gallons;] and pay to the Sponsor <<*insert currency amount*>> per <<insert number>> <MMBtu>, <tonnes> <gallon> <barrel> for Fuel used by each Unit before the applicable Unit Provisional Acceptance Date in excess of <<*insert number*>> <MMBtus> <gallons> <tonnes> and pay to the Sponsor <<*insert currency amount*>> per <MMBtu> <ton> <gallon> <barrel> for Start Up Fuel used by each Unit before the applicable Unit Provisional Acceptance Date in excess of <<*insert number*>> <tonnes> <gallon> <barrel> <MMBtu>;][79]

(ff) [administer and perform all activities under the Vendor Purchase Agreements as the Sponsor's agent until such time [(if ever)] as the foregoing are assigned to the Contractor;]

(gg) [give the Sponsor at least << *insert number* >> Days' advance written notice of the date the Contractor is requesting an Existing Plant Outage to the Sponsor and the Contractor can agree upon a schedule therefor;]

(hh) take all reasonable steps to protect the environment within and near the <Site> <Route> and avoid damage or nuisance to persons or property resulting from pollution, noise or other causes arising out of the performance of the Work;

(ii) {use its commercially reasonable efforts to} maintain good relationships with local communities;

(jj) [coordinate with and observe the instructions of <<*insert name*>> (the "Dispatcher") [and the Fuel Transporter and Fuel Supplier] with respect to the synchronization and dispatch of the Units and the Facility;][80]

(kk) give the Sponsor <<*insert number*>> Business Days' prior written notice before the commencement of any Performance Tests or Provisional Performance Tests;[81]

(ll) [provide authorized representatives of the <<*Thermal Customer, Dispatcher, Fuel Supplier, Fuel Transporter, Wheeling Company and the Offtaker*>> with reasonable access to the Facility so that they can assess compliance with the Offtake Agreement [or <<*insert agreement title*>>, as the case may be];][82]

(mm) [attend monthly meetings with the <<Thermal Customer, Dispatcher, Fuel Supplier, Fuel Transporter, Wheeling Company and the Offtaker>> to discuss the progress of construction the Work;][83]

(nn) comply with <<*insert name*>> requirements for work on the <<*insert name*>> <<*substation*>>;

(oo) prepare and submit [in <English and> <<*insert language*>>] for the Sponsor's approval, at least <<*insert number*>> Days before the <Unit> <Facility> Commissioning Completion Date [of Unit One], the Contractor's Commissioning Manual;

(pp) appoint, by written notice to the Sponsor, an executive vice president of the Parent Guarantor who shall be the senior executive with ultimate responsibility for supervising the successful completion of the Work and such senior executive shall remain aware of the Contractor's obligations and performance and shall be available at reasonable times for consultation with the Sponsor;

(qq) promptly upon the Sponsor's request execute the Contractor Indemnity and Contractor Priority Agreement in both of the forms set forth in Exhibits 5.1(qq)(A) and 5.1(qq)(B), respectively;

(rr) cause all Major Subcontractors, promptly upon the Sponsor's request, to enter into a Subcontractor Priority Agreement in the form set forth in Exhibit 5.1(rr);

(ss) cooperate and coordinate with any other contractor of the Sponsor, so as to enable it to conduct its work, and in particular provide it with reasonable access to the <Site> <Route> so as to enable its use of site services and utilities; and

(tt) ensure that all trucks, vehicles, equipment, machinery, or the like provided by the Contractor shall be in safe operating condition and at all times shall be properly protected, maintained and safely operated.

5.2 Professional Design and Engineering and Construction Management[84]

The Contractor shall be solely responsible for the professional design and engineering and construction management of the Work, and shall provide to the Sponsor professional engineering certifications if required by any applicable Law. Any omissions from this Contract of design, services or equipment shall not relieve the Contractor from furnishing such design, services or equipment if such are required in order to furnish a complete, operable, safe and reliable Facility capable of performing as required hereunder and no change, addition or deletion in design, services or Equipment resulting from any such omission shall constitute a basis for an Adjustment. [No error, inconsistency, inadequacy or omission in the FEED Work, whether or not prepared by the Contractor, shall relieve the Contractor of any obligation hereunder nor entitle the Contractor to an Adjustment.]

5.3 Supervision[85]

The Contractor shall provide full time, on Site management and day-to-day supervision of the Work. In this regard, the Contractor shall:

(a) maintain an office on the Site, and shall at all times keep in such office a copy of this Contract and all Specifications and Drawings as certified for construction by a professional engineer if required by Applicable Law, including all revisions, supplements and modifications thereto;
(b) provide supervisory personnel to be on location whenever and wherever Work[86] is being performed and to ensure that all Work is being carried out in a safe and efficient manner and in accordance with this Contract;
(c) ensure that all Contractor's supervising personnel shall be proficient in English; and
(d) [provide the Sponsor with an office at the Site including:
    (i) <<insert number>> desks;
    (ii) <<insert number>> chairs;
    (iii) <<insert number>> phone and data lines; and
    (iv) a facsimile machine].

5.4 Health, Safety and Quality[87]

The Contractor shall be solely responsible for the health, safety and welfare of all Persons working at the Site or who enter the Site for any purpose until the Facility Provisional Acceptance Date including any Sponsor, Person and Financing Party.[88] The Contractor shall ensure that all Subcontractors, employees and agents comply with all applicable Laws relating to health and safety, as well as the Contractor's health and safety regulations, while they are on the Site. The Contractor shall provide security and surveillance services at the Site and be solely responsible for any theft, loss or vandalism that occurs at the Site and also any consequent delay as a result thereof. Contractor shall be responsible for establishing and maintaining security at the Site from and after the Work Commencement Date include fencing, lighting, guard service, controlled access and any other measures required to prevent vandalism, theft and danger to the Facility, the Site, equipment and personnel and no such occurrence shall be considered a Force Majeure Excused Event or entitle the Contractor to an Adjustment. Without limiting the generality of the foregoing, the Contractor shall:

(a) take all precautions to prevent injury to all Persons on the Site and arrange to have first aid administered to all Persons who are injured or become ill on the Site, and promptly report all accidents and injuries to the Sponsor;

(b) provide sufficient light for all Work that is to be performed at night;

(c) meet with representatives of the local fire department and nearest hospital (the "Emergency Entities") in order to become familiar with the Emergency Entities and to familiarize the Emergency Entities with the Facility and the Site (and access thereto) and coordinate emergency plans with the Emergency Entities; provided that, if an Emergency Entity is not available within a reasonable distance from the Site, the Contractor shall arrange to have fire or emergency medical services provided, as the case may be;

(d) designate one of its employees to act as the safety coordinator (the "Safety Coordinator"), which Safety Coordinator shall be responsible for ensuring safe working conditions on the Site and compliance with all applicable Laws relating thereto, and the Contractor shall submit the name and qualifications of the Safety Coordinator to the Sponsor prior to the commencement of any Work at the Site;

(e) designate one of its employees to act as the security coordinator (the "Security Coordinator"), which Security Coordinator shall be responsible for ensuring the security of the Site, and the Contractor shall submit the name and qualifications of the Security Coordinator to the Sponsor prior to the commencement of the Work at the Site;

(f) designate one of its employees to act as the quality assurance/quality control manager (the "QA/QC Manager"),[89] which QA/QC Manager shall be responsible for implementing the Contractor's quality assurance/quality control program as set forth in [the Functional Specifications] [Exhibit 5.4(f)][90] and the Contractor shall submit the name and qualifications of the QA/QC Manager to the Sponsor prior to the commencement of any Work and the Contractor shall maintain a quality plan compliant with ISO <<insert number>> or another plan, reasonably acceptable to Sponsor and with the quality assurance plan set forth in Exhibit 5.4(f), detailing:

   (i) specific authorities and responsibilities;

   (ii) procedures, methods and work instructions; and

   (iii) inspection and test plans; and

in the event of a conflict between Exhibit 5.4(f) and ISO <<insert number>>, the higher standard of the two shall prevail; and perform appropriate drug testing on all Contractor Persons having access to the Site upon reasonable written request of Sponsor.

5.5 Cleaning of <Route> <Site>

The Contractor shall at all times keep the <Site> <Route> free from accumulations of waste, including sanitary waste, demolition debris, construction debris, office waste and wastes related to the preparation, testing and commissioning of Facility Systems and Equipment[; provided that the Contractor shall not be required to dispose of Fuel not fit for consumption by the Facility[, ash, sand or limestone]].[91] If, despite <<insert number>> Days' prior written notice from the Sponsor, the Contractor allows waste to accumulate on the <Site> <Route> or hazardous conditions to develop on the <Site> <Route> as a result of the Contractor's inadequate cleaning procedures, notwithstanding any other remedies that the Sponsor may have hereunder, the Sponsor may remove such waste or correct such hazardous conditions at the Contractor's sole expense.

5.6 Scheduling and Progress Reporting[92]

5.6.1   The Contractor shall be responsible for all scheduling and progress reporting to the Sponsor concerning the Work. The engineering and construction of the Facility shall follow the sequence outlined in the Baseline Schedule and Baseline Level 3 Schedule.

5.6.2   Within <<*insert number*>> Days of the <Work Commencement Date> <Contract Date> the Contractor shall prepare, using <<*insert name*>> software (version <<*insert number*>> or higher) a time scaled, [cost loaded][93] precedence diagramming network schedule employing the critical path method. The Baseline Level 3 Schedule shall identify:

(a) all significant construction activities, important shop drawing submittals, procurement activities and receipts of materials and equipment and the schedule shall be divided into separate activities such that no activity, except activities showing only submittal, fabrication or delivery of material or equipment, shall have duration greater than <<*insert number*>> Days or an assigned value greater than <<*insert amount and currency*>>;

(b) completion time and all dates (if any) identified in the Milestone Payment Schedule;

(c) the dependencies between activities so that it may be established what effect the progress of any one activity has on the Baseline Level 3 Schedule;

(d) a unique identification number for each activity;

(e) a unique activity code for each activity that will allow activity sorting in reports and schedules by both specification division number and general process area; and

(f) all activities making up the critical path.

5.6.3   The Baseline Level 3 Schedule will provide a basis for determining the progress of the Work relative to the completion time and specific dates and for determining the acceptability of the Contractor's requests for payment.

5.6.4   Within the period set forth above, the Contractor shall submit for the Sponsor's review and comment a draft Baseline Level 3 Schedule. [{The Sponsor will review the draft Baseline Level 3 Schedule within <<*insert number*>> Days of its receipt.}] If the Sponsor finds that the submitted draft Baseline Level 3 Schedule does not comply with the aforementioned requirements, or is otherwise not in accordance with the terms of the Contract[{, the deficiencies will be identified in writing to the Contractor, and within <<*insert number*>> Days of receipt of the Sponsor's objections}], the Contractor shall correct and resubmit a revised draft Baseline Level 3 Schedule.

5.6.5   The Contractor shall prepare and submit Progress Reports to the Sponsor on a monthly basis. In conjunction with the monthly Progress Report, or otherwise as required upon the occurrence of any disruption or alteration of the Contractor's work plan or the Baseline Level 3 Schedule, the Contractor shall submit to the Sponsor a revised Current Schedule incorporating such changes and showing the current state of progress of each activity and its projected completion date.

5.6.6   All schedule submissions, both initially and with each Progress Report, shall consist of an electronic version of the Current Schedule, transmitted by <<couriered

copy on computer disks>> <<electronic shared drive>> and by electronic mail as appropriate, and the following information on hard copy:

(a) a tabular listing of activities sorted by early start and showing activity description, scheduled duration in working days, early and late start and finish dates, total float, predecessors and/or successors to each activity and the cost assigned to each activity;
(b) a time scaled logic diagram for all scheduled activities;
(c) a critical path report;
(d) separate activities for each approved change order;
(e) in a revision made at least <<*insert number*>> Days prior to the commencement of any Provisional Performance Tests or Performance Tests, all testing activities referred to in Article 8.1; and
(f) a narrative outlining any significant changes in or adjustments made to the schedule.

5.6.7   Should the actual sequence of Work performed by the Contractor deviate [{substantially}{materially}] from the planned sequence indicated in the Baseline Schedule or the Baseline Level 3 Schedule, the Contractor shall prepare and continue to revise a Current Schedule to reflect changes in the actual sequence and/or the future sequence of Work and bring such changes immediately to the attention of the Sponsor.

5.6.8   The Contractor shall be responsible to ensure that all Work, including that of its Subcontractors, as well as work performed by others which may affect the Work, is included on the Baseline Level 3 Schedule. The Contractor's failure to include all such work shall not excuse the Contractor from completing all the Work by the dates required herein.

5.6.9   The Sponsor's review or acceptance of the Baseline Level 3 Schedule or any Current Schedule or Progress Report shall not impose upon the Sponsor any responsibility for the progress, monitoring or scheduling of the Work, and the Contractor shall be fully responsible to provide proper progress of the Work.

5.7 Schedule Delay

5.7.1 If the Contractor shall fail to accomplish (or has reason to believe that it is likely to fail to accomplish) any of the activities on or before the time required for the completion of such activities as set forth in the Baseline Level 3 Schedule for such activities, Contractor shall promptly and in any event within <<*insert number*>> Days deliver to Sponsor a notice of schedule delay (the "Notice of Delay") and [{to the extent {commercially reasonable}{feasible},] accelerate the performance of the Work by providing whatever [{reasonable}{feasible}] means are necessary to recover to the Baseline Level 3 Schedule, including the provision of additional labor (including overtime) and, to the extent necessary, Equipment and Materials, at no additional cost to Sponsor [{(provided that the delay is not caused by Sponsor)}].

5.7.2   If the Facility Provisional Acceptance Date does not occur on or before the Guaranteed Facility Provisional Acceptance Date, then the Contractor shall, within <<*insert number*>> Days of the Guaranteed Facility Provisional Acceptance Date, submit to the Sponsor [{for the Sponsor's express written approval}][94] a detailed technical and narrative analysis of the root causes that prevented the achievement of the Facility Provisional Acceptance Date and a detailed plan (the "<u>Remedial Plan</u>") designating the new expected date for Facility Provisional Acceptance and setting out in detail all the steps necessary for such and the Remedial Plan shall include a feasible schedule demonstrating that Facility Provisional Acceptance can be achieved and by what date such can be expected.[95]

5.7.3   [{Within <<*insert number*>> Days after receiving the Remedial Plan, the Sponsor shall provide its comments to such Remedial Plan Contractor who shall proceed with the Work based on the Sponsor's comments to the Remedial Plan.}]

5.7.4   [{Upon approval of the Remedial Plan,}] Contractor shall diligently perform the Work in accordance with such Remedial Plan and shall deliver written reports to the Sponsor on a weekly basis detailing Contractor's progress under the Remedial Plan. None of the foregoing shall limit Contractor's obligation as set forth in Article 5.10.1(b) to accelerate the performance of the Work [{to the extent commercially reasonable}] by providing whatever [{reasonable}{feasible}] means are necessary to recover to the Baseline Level 3 Schedule [{to the extent reasonable means are available,}] at no additional cost to Sponsor [{provided that the delay is not caused by Sponsor}].

5.7.5   The Contractor shall not adjust the schedule, sequence or method of its performance of the Work in any manner that will result in additional costs to the Sponsor.

5.8 Additional Reports

Contractor shall make available, and upon Sponsor's [{reasonable}] request shall furnish, to the Sponsor [(and, prior to Financial Closing, if [{reasonably}] requested by the Financing Entity for purposes of the financing of the Project, to the Financing Entity and the Financing Entity's Engineer)] such documents as are necessary to review the Work and technical information regarding the design of the Facilities[{; provided, that Contractor shall not be obligated to provide proprietary technical data regarding equipment manufactured by or for Contractor and not provided by Contractor to other Persons}]. Except for the Sponsor's responsibilities set forth in Article 7, Contractor shall be solely responsible for all construction means, methods, techniques, sequences and procedures and for coordinating all portions of the Work under this Contract.

5.9 Compliance with Laws, Etc.

5.9.1   The Contractor shall:

(a) <ensure> <{provide}> that the Facility's engineering, design and construction, and all Equipment furnished and Work performed hereunder, shall fully comply with the requirements of this Contract and all Laws [and the Bid Documents];

(b) provide all necessary code stamps, nameplates and certifications required by the aforementioned; and

(c) <ensure> <{provide}> that the Facility as completed will fully comply with and be capable of operation in accordance with the requirements of this Contract and all Laws.

5.9.2   If Contractor discovers any discrepancy or inconsistency between the terms of this Contract and any applicable Law, the Contractor shall as soon as [{reasonably}] <possible> <{practicable}> provide written notice to the Sponsor. The Contractor is responsible for complying, and shall conform the Facility, the Work and the Contractor's performance hereunder to comply, with any Change in Law occurring prior to the Facility Provisional Acceptance Date. The Contractor may submit an Adjustment Claim based on a Change in Law (provided it does so within <<*insert number*>> Days [<{after it learns}> <should have learned>] of the occurrence thereof) if the Change in Law:

(a) results in additional costs to the Contractor of at least <<*insert currency and amount*>>; or

(b) requires a substantial modification to, or delay in a Critical Path Item.[96]

[Any Adjustment shall only be granted to the extent the Sponsor receives relief therefor under the <<Offtake Agreement>>.]

5.10 Key Contractor Personnel

The Key Contractor Personnel are set forth in Exhibit 5.10 and the names thereof will be furnished in writing to the Sponsor within <<*insert number*>> Days of the Work Commencement Date. [{Without the prior written consent of the Sponsor (which shall not be unreasonably withheld)},][97] no member of the Key Contractor Personnel shall be replaced or reassigned from the Work and if such occurs [{except as a result of termination of employment}], the Contractor shall rebate << *insert percentage* >> percent of the Contract Price to the Sponsor in each case because employment of such individuals was material to the Sponsor's decision to enter into this Contract. [By the Work Commencement Date, the Contractor shall by written notice to the Sponsor nominate a Person to be its project director (the "Contractor's Project Director") [{which approval shall not be unreasonable, delayed, withheld or conditioned}] and such Person's appointment shall be subject to the Sponsor's prior written approval. The Contractor's Project Director shall be the Contractor's single point of contact with the Sponsor in regard to this Contract, the Work and the Facility, and shall:

(a) have the authority to:
   (i) direct the execution of all of the Contractor's obligations hereunder and also enforce all of the Contractor's rights hereunder; and
   (ii) make decisions that shall be binding upon the Contractor, including the execution of Change Orders; and
(b) assign a construction manager to:

(i) provide full time, on Site management and day-to-day supervision of craft personnel, engineering, quality control, Subcontractors, suppliers and Site operations;
(ii) give his continuous personal attention to the execution of the Work and be present during all periods in which Work is in progress at the Site, including overtime and second and third shifts;
(iii) maintain an office on or adjacent to the Site; and
(iv) appoint an alternate construction manager who shall act on behalf of the construction manager in his absence.

5.11 Good Order

5.11.1   The Contractor shall enforce discipline and good order of all Persons on the Site. The Contractor shall remove any Person from the Work:

(a) whose presence is detrimental to the performance of the Work or any Person; or
(b) if such Person is not qualified to perform the portion of the Work assigned to him/her.

5.11.2   The Contractor shall take all <necessary> <{feasible}> precautions in accordance with <Good> <{Prudent}> <Utility> <{Industry}> Practices to <ensure> <{provide}> that no neighboring land, property or property right is injured or damaged by the Contractor's and its Subcontractors' activities.

5.12 [Contemplated Sourcing[98]

The Contractor will <ensure> <{provide}> that the Equipment and Services listed in Exhibit 5.12 to be incorporated into the Facility will be manufactured or rendered as applicable in the countries set forth on Exhibit 5.12 in the values specified therein. The Contractor shall not, and shall prevent its Subcontractors from, importing or using in relation to the Work, any goods or services from any country with which the government of <<*insert country*>> does not maintain diplomatic relations or goods produced, or services provided by any Person organized under the laws of any country with which the government of <<*insert country*>> does not maintain diplomatic relations or by any Person who is a national or resident of any country with which the government of <<*insert country*>> does not maintain diplomatic relations. The Sponsor shall have the right in its sole and absolute discretion to reject any such goods or services that are imported or used in violation of such prohibition and the Contractor shall replace such rejected goods and services at its own risk and expense and shall not be entitled to claim any Adjustment.]

5.12.1   The Contractor shall use construction equipment and other equipment, civil works, materials and products produced and manufactured in <<*insert country*>> such that, in accordance with the provisions set forth in the Bid Documents (the "National Content"); it being understood and agreed by Contractor that (i) the percentage of National Content shall be determined in accordance with the provisions of the Bid Documents and (ii) failure to comply with the National Content requirements shall give rise to the National Content Actual Damages set forth in Exhibit 5.12.1 (it being understood and agreed that the amount of National Content Actual Damages set forth in Exhibit 5.12.1 reflects the amount of liquidated damages payable by Sponsor to <<*insert name*>> under the <<*insert title of agreement*>> for failure to satisfy the National Content requirement).

5.12.2   The Contractor shall deliver to Sponsor a report and supporting documents not later than <<*insert number*>> Days prior to the Guaranteed Facility Provisional Acceptance Date specifying the percentage of total cost of the engineering, procurement and construction of the Facilities qualifying as of <<*insert country*>> origin.

5.12.3   Promptly upon receipt of an invoice from Sponsor for the amount of National Content Actual Damages due by Sponsor to <<*the Offtaker*>> for failure to meet the National Content requirement, Sponsor shall submit to the Contractor an invoice for the amount of such National Content Actual Damages. The Contractor shall pay the full amount of the National Content Actual Damages so invoiced within <<*insert number*>> Business Days after receipt of Sponsor's invoice. Failure by the Contractor to make such payment within such time period shall constitute a Contractor Default under Article 17.

5.13 [Selection of Equipment

The Contractor shall, when selecting machinery and equipment for incorporation as part of the Work, have due regard to the exemptions granted from certain taxes and import duties and to the criteria for exemption from, or reduction of, taxes and such import duties granted by Governmental Authorities in selecting such machinery and equipment. The Contractor shall comply with all amendments and updates to such procedures and requirements in so far as these relate to the exemption from, or reduction of, such taxes and duties.][99]

5.14 Shipping Schedules, Preliminary Shipping Documents and Shipping Documents[100, 101]

At least <<*insert number*>> Days prior to the first shipment involving the Equipment, the Contractor shall submit to the Sponsor a shipping schedule giving the proposed month of shipment of each of the major items of Equipment. The Contractor may from time to time revise such shipping schedule to take account of any changes and supply to the Sponsor each such revised schedule within <<*insert number*>> Days of its revision. Not later than <<*insert number*>> Days before departure of any shipment of a major item, the Contractor shall provide the Sponsor with a written notice (a "Shipment Notice") forwarded by facsimile specifying:

(a) the item to be shipped;
(b) the value of the item to be shipped;
(c) the date of shipment;
(d) the port of loading of each shipment;
(e) the anticipated date of arrival in <<*insert country*>> of each shipment; and

(f) the name of the carrier and the name of the ship (or other means of transport) of each shipment.

The Contractor shall confirm the details referred to in paragraphs (a) through (f) above by facsimile immediately after departure of each shipment. Each item of equipment shall be tagged or marked and complete packing lists and bills of materials shall be included with each shipment. Each piece of every item need not be marked separately provided that all the pieces of each item are packed or bundled together and the packages or bundles are properly tagged or marked.

The Contractor shall:

(a) identify containers, crates, etc., containing spare parts, tools, or instruction books;
(b) affix adequate labels on equipment shipped with bags of desiccant inside, and to indicate existence of dog bolts or other restraints against movement of internal parts; and
(c) include packing lists in duplicate in each crate, package or bundle shipped.
     The Contractor is responsible for all customs clearances and their cost and no delay in clearance [{unless beyond <<*insert number*>> Days from complete application therefor}] shall be considered a Force Majeure Excused Event.
(d) The Contractor shall notify the Sponsor within <<*insert number*>> days of the movement of any Equipment or construction equipment to or from the Site.

5.15 Inspection

5.15.1   No Adjustments shall be made for the Contractor for any failure on the Contractor's part to inspect adequately the <Site> <Route> or the surrounding areas.

5.15.2   The Contractor shall check all quantities and dimensions provided by the Sponsor, and shall be responsible for any errors [{which can be discovered by examination or checking of this Contract}], and the Contractor shall be responsible for the joining and fitting of all parts of the Work with the Existing Plant, and any checking or inspection of such joining or fitting by the Sponsor shall not relieve the Contractor of any responsibility as to the correctness of the Work.

5.16 Interface with Existing Plant

5.16.1   <By the <<*insert number*>> Day after the <Contract Date> <Work Commencement Date>, the Contractor shall notify the Sponsor in writing of the dates and estimated durations of any Existing Plant Outage and the number of days of advance notice of the Existing Plant Outage requested {required} by the Contractor in order for the Contractor to mobilize for the Existing Plant Outage and thereafter the Sponsor shall, acting reasonably, designate expressly in writing each Existing Plant Outage duration (each, an "Existing Outage Duration") and the Existing Plant Outage commencement date (the "Outage Start Date"). [The Contractor shall use its best efforts to complete the Tie-Ins during a single Existing Plant Outage and minimize the duration of such Existing Plant Outage. If any modification or disassembly of the Existing Plant is required for the Work and not expressly described herein, Contractor shall give the Sponsor <<*insert number*>> Days' prior written notice thereof explaining the nature of such. Contractor shall carry out any such work only upon the written permission of the Sponsor provided that any such written permission shall not be unreasonably delayed or withheld.] If the Work may affect the design, configuration, construction or operation of an Existing Plant or system, said Work shall be capable of functioning in a technically and commercially feasible manner and not create any adverse impact on the Facility or Existing Plant operation.

5.17 Protection of Property

5.17.1   The Contractor will take <all necessary> <{provide adequate}> precautions to <ensure> <{protect}> against damage or injury to any roads, buildings, machinery, materials, equipment, bridges, water/fluid/chemical lines, storage tanks, sewage lines, electrical wires and telephone wires, whether above ground, in ground or submerged, by any unloading, loading, transportation or any other actions relating to performance of the Work by any Contractor Person, and in the event of any injury or damage to any of the foregoing then the Contractor shall take immediate corrective action at its own cost to rectify such injury or damage to the extent required to restore such to its original condition immediately prior to such injury or damage.

5.17.2   Contractor shall preserve and protect, in operating conditions, all property and items including, but not limited to, active utilities traversing or within and about the site, piping, conduits, drains, manholes, mains, laterals, catch basins, valve boxes, meters, and other appurtenances and structures. Contractor shall promptly repair any damage to such utilities due to Contractor's work under this Contract, to the satisfaction of the owner's utilities or municipal agencies having jurisdiction.

5.17.3   Contractor shall erect and maintain temporary barricades, warning signs and guards necessary to protect streets and roadway, walkway, personnel, and adjoining properties from damage in accordance with applicable rules and regulations which are required by Contractor to perform the Work.

5.17.4   Contractor shall provide and maintain bracing and shoring as required by applicable regulations for safety specific to the Contractor's work. Contractor shall assume all responsibility for the strength and adequacy of such shoring and bracing.

5.17.5   Contractor shall <ensure><{provide}> all structures are located outside of identified watercourses, all construction equipment shall avoid crossing, or working within, watercourses and storm water runoff is to be diverted away from fill slopes, and other exposed surfaces, to the greatest extent possible. Contractor shall also properly install erosion and sedimentation control measures at the project site.

5.17.6   The Contractor is permitted to be present on the premises of the Sponsor solely for the purpose of performing the activities specified in the Contract, and no Contractor Person shall engage in any activity on the premises of the Sponsor which is outside the purview of this Contract, including but not limited to the solicitation for political, religious or commercial purposes or the sale, display or distribution of any political, religious or commercial materials. No signs shall be used or placed on the premises by the Contractor or its agents or employees without the consent of the Sponsor. The Sponsor will not consent to any signs that are of a political, religious or commercial nature or are otherwise unrelated to the activities specified in the Contract.

5.18 Engagement of Staff and Labor

The Contractor shall make his own arrangements for the engagement of all staff and labor, local or otherwise, and for their payment, housing, feeding, security, and transport including obtaining all visas, work permits, employment permits, dependents' permits, licenses and other permits required for all Contractors' personnel under applicable Law and no delay or denial in the obtention thereof shall be considered a Force Majeure Excused Event.

5.19 Foreign Staff and Labor

The Contractor shall import such staff, artisans, and laborers as are required to execute the Work. The Contractor is solely responsible for <ensuring> <{providing}> that all staff and labor have the required residence visas and work permits and no delay or failure to obtain any of the foregoing shall be deemed to be a Force Majeure Excused Event, and shall be responsible for their safety and security while under their employ. The Contractor shall be responsible for the return to the place where they were recruited or to the domicile of all persons whom the Contractor recruited and employed for the purposes of or in connection with the Work and the Contract. The Contractor shall be responsible for such persons as are to be returned until they have left the <Site> <Route> or, in the case of foreign nationals who have been recruited from outside the Country, shall have left it.

5.20 Rates of Wages and Conditions of Labor

The Contractor shall pay wages and observe conditions of labor not less favorable than those established for the trade or industry where the Works are being carried out. If no such established rate or conditions are applicable, the Contractor shall pay wages and observe conditions not less favorable than the general level of wages and conditions observed by others whose trade or industry is similar to that of the Contractor.

5.21 Persons in the Service of Others

No Contractor Person shall recruit, nor attempt to recruit, employees from amongst persons in the service of any Sponsor Person.

5.22 Labor Laws

The Contractor shall comply, and shall cause all of its Subcontractors to comply, with all the relevant labor Laws applying to his and their employees, and shall duly pay and afford to them all their legal rights under applicable Laws. The Contractor shall require all such employees and Subcontractors to obey all Laws. The Contractor shall submit detailed reports showing the supervisory staff and the numbers of the several classes of labor from time to time employed by the Contractor and Subcontractors. The reports shall be submitted in such form and at such intervals as the Sponsor may prescribe. The Contractor shall submit details in the prescribed forms to the appropriate manpower agencies having jurisdiction over the Work, showing supervisory staff and the numbers of the several classes of labor from time to time employed by the Contractor and the Subcontractors. The forms shall be submitted in such manner and at such intervals as prescribed under the labor laws of the *<<insert country>>* and other applicable regulations, with copies to the Sponsor.

5.23 Working Hours

The Contractor will perform all Work only within the periods specified as permissible by Law. The Contractor shall indemnify and hold harmless all Sponsor Persons against any and all losses, damages, costs, expenses or claims arising out of, related to or in connection with the failure to comply with the foregoing agreement. [The Contractor shall be permitted to work <at> <on> the <Site> <Route> at <any time> <between <<8>> a.m. and <<4 p.m.>> weekday hours>, if permitted by Law.] The Contractor's Representative or a qualified supervisor shall be present during all periods, including overtime and second and third shifts, when work is in progress at the Site. The Contractor shall in dealings with his staff and labor have due regard to all recognized festivals, days of rest and religious or other customs observed by his staff and labor and in *<<insert country>>*. Should the Sponsor consent in writing (which it will be under absolutely no obligation to consider in good faith such consent) to any Work outside of such periods, the Contractor will pay for any Sponsor Person personnel that the Sponsor acting in its solely discretion decides are required on-Site as a result of the Contractor's decision to Work outside of such period, except for payroll costs and expenses of such additional Sponsor Persons.

5.24 Facilities for Staff and Labor

The Contractor shall provide and maintain all necessary accommodation and welfare facilities for his (and his Subcontractor's) staff and labor, as may be required under Applicable Laws or in the Functional Specifications. The Contractor shall also provide the facilities specified in the Functional Specifications for the Sponsor. The Contractor shall not permit any of its employees or any of the employees of Subcontractors to maintain any temporary or permanent living quarters or sleep within the structures forming part of the Works or at the Site [and on the Route].

5.25 [Union Labor]

The Contractor will use only union labor in connection with the Work conducted at the Site and execute all agreements required by *<<insert names of applicable union>>* in connection therewith.[102]

5.25.1   Whenever threatened or actual picketing, slowdowns, work stoppages or other labor unrest may delay or otherwise affect the Work, the Contractor shall immediately notify the Sponsor in writing. Such notice shall include all relevant information regarding the labor issue at hand, its background, and the steps that the Contractor proposes to take to resolve or prevent its occurrence.

5.25.2   In the event of a labor unrest [{directed specifically at the Contractor and not the industry or contractors in general or which are not regional or national}], the Contractor shall promptly take and prosecute all such necessary action, whether to initiate proceedings in such administrative, judicial, or arbitral forum having jurisdiction, or to otherwise resolve, or minimize the impact of the labor dispute. No labor unrest shall constitute a Force Majeure Excused Event or entitle the Contractor to an Adjustment or any other relief [{unless specifically directed at the Contractor}].

5.25.3   In the event of labor unrest [{directed at the Contractor}], the Contractor shall be liable to the Sponsor for all loss or damage incurred by the Sponsor.

5.25.4   In the event of labor [{unrest directed at the Contractor}], the Sponsor shall not be liable for any payment, reimbursement, or other compensation, and any increase in pay benefits, or other terms and conditions of employment shall not be charged to the Sponsor.

5.25.5   [In the event the Contractor is a subscriber to a multi-employer bargaining association or group or party to any union agreement, the Contractor shall, if the Sponsor so directs, participate diligently in the collective bargaining of that group with any of those labor organizations claiming jurisdiction over any portion of the Work under this Contract].

5.26 [Equal Employment Opportunity

5.26.1   The Sponsor is committed to increasing the amount of business it conducts with minority and women owned firms. The Contractor will utilize qualified minority and women Subcontractors wherever possible and if it does so, the Contractor shall identify in writing to the Sponsor the Subcontractor along with the estimated <<*insert currency and amount in words and numerals*>> value of its Subcontract. All services required under the Contract shall be provided on an equal basis without regard to race, religion, creed, color, sex, national origin, age, disability, marital status, sexual orientation, military status, or predisposing genetic characteristics, without any acts of coercion, harassment, intimidation or retaliation.]

5.27 "Whistleblower" Protection

The Contractor shall implement a program and develop procedures to advise its employees and Subcontractors that they are entitled and encouraged to raise safety concerns to the Contractor's management or to the Sponsor without fear of discharge or other discrimination and in the event any allegation is made to the Contractor, the

Contractor shall notify the Sponsor by written notice within <<*insert number*>> Days thereof.

5.28 Coordination with Other Contractors[103]

5.28.1   The Contractor recognizes that other Persons may be working concurrently at the <Site> <Route>. The Contractor shall cooperate with the Sponsor, and other contractors, if any, to ensure that the Work and their work can be performed as scheduled on the Baseline Level 3 Schedule or their own schedule if the Contractor has failed to include their work in the Baseline Level 3 Schedule. The Contractor shall collaborate with any other contractors and coordinate the Work with the work of such other contractor(s), if any, which could affect the Work, and the Contractor shall proceed in such manner as not to interfere or delay their progress.

5.28.2   If any part of the Work depends for proper execution or results upon the work of any other contractor, the Contractor shall inspect and promptly report in writing to the Sponsor any defects in such work that render it unsuitable for such proper execution or results. Failure of the Contractor to do so shall constitute its acceptance of such other work as fit and proper for the reception of the Work.

5.28.3   In cases of disagreement or disputes between the Contractor and another contractor or Subcontractor which could delay or interfere with the Work as a result of the Contractor's failure to collaborate and/or cooperate with another contractor or which cannot be resolved between the Contractor and the other involved contractor, the Sponsor shall be given prompt written notice thereof by the Contractor specifying in detail the disagreement or dispute. In such cases, the Sponsor shall have the right to determine the method of coordinating the Work, and the Sponsor's decisions in this regard shall be final, binding, and conclusive and not ever constitute a Force Majeure Excused Event.

5.28.4   The Contractor acknowledges that correct and complete performance of the Work in accordance with the [Tie-Ins] and coordination with the work of other contractors for the <<Facility and Project>> in particular <<the blast furnace, desalinization plant, continuous casting plant and port and infrastructure facilities as well as the coking and power plant>> is of paramount importance to the Sponsor.

5.28.5   The Contractor undertakes to liaise and cooperate with the Sponsor and other contractors for the various infrastructure and facilities of the Project and to inform them as to all relevant aspects of its Work which may have an impact on the performance of other contractors.

5.28.6   The Contractor further undertakes to inform itself of, and to investigate all details of the scope and progress of work for the other plants and infrastructure relevant for performance of its Work and shall request any necessary information from and carefully review all information received from other contractors and/or the Sponsor.

5.28.7   The Contractor shall:

(a) identify possible points of interface with other contractors and provide interface drawings in sufficient detail to define the method and location of all interfaces of the Work and work to be performed by other contractors;
(b) attend engineering and procurement-related planning, scheduling and coordination meetings with the Sponsor and other contractors who may work at the <Site> <Route>;
(c) provide in timely manner any and all information regarding the Work as may be requested by the Sponsor or any other contractor at no charge and provide personnel in all the relevant appropriate disciplines to attend and participate in interface meetings as may be requested by the Sponsor or other contractors;
(d) continuously review other contractors' schedules, progress reports and drawings to determine interfaces and scheduling;
(e) produce a monthly interdependency schedule that includes, but is not limited to, look ahead, shutdown, tie-in, heavy lift and recovery schedules and interdependency dates and communicate promptly in detail in writing to such other contractors with a copy to the Sponsor, and the Contractor shall coordinate with other contractors on a monthly basis in order to produce an updated Project interdependency schedule; and
(f) plan and coordinate activities so as to minimize interference and delays throughout the performance of the Work and development of the Project.

5.28.8   The Current Schedule to be delivered under Article 5.6.7 and all revisions thereto shall clearly indicate areas of interference or delay that have not been resolved. The Contractor shall include an analysis of alternatives and cost estimates for each interference issue or delay and any anticipated impact on the Baseline Level 3 Schedule or Work.

5.28.9   The Contractor shall provide <<*insert number*>> Days' notice to the Sponsor of any meeting with any contractor. The Sponsor may, at its option, attend any such meeting.

5.28.10   The Sponsor reserves the right to direct the Contractor to schedule the order of performance of its Work in such a manner as not to interfere with the performance of the work of Persons.

5.28.11   [The Contractor is advised that the Work will be performed adjacent to [the Existing Facility] and/or in an operating facility with other contractors present. The Contractor shall cooperate with the Existing Plant personnel and other contractors and allow for interruptions in the Work, required either by other contractors or Existing Plant personnel, so as to benefit the overall progress of the Project and to ensure [provide for] the safe and continuing operations of the Existing Plant.]

5.28.12   Considerable congestion of labor, access routes, loading docks, rail lines, port access and lay down areas may exist in many areas of the <Site> <Route>, and must be allowed for in planning of Work activities. Delays due to congestion shall not be a Force Majeure Excused Event nor entitle the Contractor to <any> <{a Price}> Adjustment <{but solely and exclusively a Time Adjustment}>.

5.28.13   The Contractor shall maintain a flexible plan of equipment, delivery and installation because some equipment of other contractors may arrive later than their targeted delivery dates. {The Sponsor will maintain and issue a current delivery list and will notify Contractor as soon as possible of the expected delivery dates of the equipment of other contractors.}

5.28.14   The Contractor shall not be entitled to, and hereby expressly waives recovery of, any damages suffered by reason of delays of any nature described in this Article 5.28 {and extension of time by means of Time Adjustment shall constitute the sole liability of Sponsor and Contractor's sole remedy for delays}.

5.29 Parts and Spare Parts

Contractor shall only procure Equipment for which the furnishing Subcontractor undertakes that Parts therefor shall be available for not less than <<*insert number*>> Days after the Facility Provisional Acceptance Date. When submitting its recommendations to Sponsor, the Contractor shall use its <best> <{reasonable}> efforts to obtain and submit to Sponsor an undertaking from the Subcontractor in question that the price for the Parts shall be fixed for a period of <<*insert number*>> Days, starting from the Facility Provisional Acceptance Date with such price being subject only to ordinary price escalation not to exceed <<*insert number*>> percent per annum. All spare Parts purchased by the Sponsor hereunder as part of the Work shall be placed by Contractor into storage suitably packed for long-term preservation consistent with <Good> <{Prudent}> <Utility> <{Industry}> Practice no later than Facility Provisional Acceptance.[104] Contractor shall warehouse or otherwise provide appropriate and secure storage for all materials required for permanent and temporary construction and all operating and maintenance spare Parts. All such materials and spare Parts which are stored at a location other than on the Site shall be segregated from any other items.

5.30 Contractor's Responsibilities Regarding Importation

5.30.1   The Contractor shall, with respect to paragraphs (a), (b) and (c) below for the purposes of importation of any of the Work, act <in the capacity of agent of the Sponsor> <for itself> and therefore shall be responsible for obtaining all import Permits in a timely manner and no delay or denial in the obtention thereof shall be considered a Force Majeure Excused Event or entitle the Contractor to an Adjustment for any reason whatsoever, including without limitation:

(a) the preparation, processing, and submission of all documentation to the Governmental Authorities;
(b) the provision of all necessary explanations relating to the nature of the machinery and equipment and the manner in which such will be used; and

(c) any other matter specified in such procedures required by the Governmental Authorities as a condition to the granting of or maintaining of any exemption from, or reduction of, any taxes and import duties (including value added taxes).

5.30.2   The Contractor shall, if the Sponsor so requests at any time and from time to time, consult with the Sponsor before submission of any documentation or the provision of any explanation to any Governmental Authority [{and the Sponsor shall give reasonable assistance to the Contractor in connection therewith, provided, that such assistance shall not relieve the Contractor of its responsibilities referred to above}].

5.31 [Defects and Deficiencies Technician

To provide operations and maintenance support and training for Owner's staff, the Contractor shall provide, without additional cost to Owner, the services of one (1) defects or deficiencies technician who shall remain in the Country for a period of <<insert number>> Days starting on the Taking-Over Date. The defects or deficiencies technician shall work a normal work-week of <<forty (40)>> hours and shall be entitled to reasonable vacations and holidays in accordance with the Contractor's normal employment policies.]

5.32 [Maintenance of Existence[105]

Until the Facility Provisional Acceptance Date, the Contractor shall not:

(a) sell, assign, alienate, encumber or otherwise transfer in any manner, directly or indirectly, all or any [{material}] portion of its assets; or
(b) incur or guaranty any obligation or indebtedness, or merge or consolidate with any Person, if the effect of the foregoing would cause the Contractor's unconsolidated net worth to decrease to less than <<insert number>> percent of the Contractor's unconsolidated net worth on the Contract Date.]

5.33 [Contractor Change of Control

5.33.1   Before any Change of Control shall take effect:

(i) the proposed Change of Control (including the identity of the proposed new controlling Person(s) and such other information concerning the proposed Change of Control as the Sponsor may {reasonably} request) shall be delivered to the Sponsor; and
(ii) the Sponsor's written approval [{such approval, not to be unreasonably withheld or delayed}] must be sought to the proposed Change of Control, and such Change of Control must not proceed with such approval.

5.33.2   {The Sponsor may only withhold its consent if such Change of Control:

(i) will result in a competitor[106] of the Sponsor controlling the Contractor; or
(ii) would[, in the opinion of the Sponsor,] adversely impact upon the ability of the Contractor to complete the Work in accordance with the terms and conditions of this Contract.}

5.33.3   {The above provisions do not require the prior approval of, or (subject to paragraph (i) below) prior to notification to, the Sponsor if a Change of Control results, or would result, from a Person or Persons acquiring control of a Person, the shares of which are listed or traded on an international stock exchange. In such circumstances, the Contractor shall notify the Sponsor in writing of the Change of Control:

(i) as soon as reasonably possible after the public announcement of an offer or other circumstances which, with or without the satisfaction of conditions or the passage of time, may lead to a Change of Control of that type; and
(ii) on the Change of Control taking effect.}]

ARTICLE 6. REVIEW, INSPECTIONS AND AUDITING

6.1   Review, Inspections and Corrections[107]

6.1.1   [{To the extent {reasonably} necessary to determine that the Facility is being designed in accordance with this Contract,}] <T> <t>he Sponsor and its designees shall have the right to review and provide comments on any design document associated with the Work or the Facility [{in accordance with the procedures set forth in the Functional Specifications}].[108] The Contractor shall be responsible for addressing all comments arising from such review. Any such review or comment by the Sponsor and its designees shall not relieve the Contractor of any of its obligations hereunder including compliance with all applicable Law, provision of proper design of the Facility, achievement of the performance requirements hereof, and the schedule requirements hereof. Notwithstanding any such review or comment by the Sponsor, the Contractor shall remain solely responsible for the full and complete performance of this Contract, and any review or comment shall not relieve the Contractor of its responsibility for any errors or omissions in any design and the Contractor shall be liable at its own cost to undertake any and all remedial work necessitated by such error or omission.[109]

6.1.2   The Sponsor, the Sponsor's Project Manager, the Sponsor's Engineer, [each Financing Party, each Financing Party's Engineer,] <<*the Offtaker, the Thermal Customer, the Fuel Supplier, the Fuel Transporter, the Wheeling Company*>>, any of their representatives or agents and any Person authorized in writing by the Sponsor shall at all times have access to the Facility and all Subcontractors' design offices, shops, manufacturing facilities and other facilities. Contractor shall, with respect to each material item of equipment or critical aspect of the Work, submit its own shop inspection plans to Sponsor for comment on or before the date which is <<*insert number*>> days after the Contract Date.

6.1.3   Contractor shall provide Sponsor with a list of the workshops in which such material items of equipment are manufactured and shall furnish all information that may be useful for the checking of the manufacturing, such as material orders, etc.

6.1.4   Contractor shall provide Sponsor with at least <<*insert number*>> Days' prior written notice of the estimated date of performance of each shop test specified as "witness point" or "hold point" in Exhibit 6.1.4. In addition, not later than <<*insert number*>> Days prior to the anticipated date of performance of each shop test specified as "witness point" or "hold point" in Exhibit 6.1.4, Contractor shall provide Sponsor with written confirmation of the date of performance of each such shop test. No Work shall continue on any item which is the subject of a "hold point" until the Sponsor expressly so states in writing. All such Persons shall have the right to be present, witness, and inspect all systems and Facility start-up and testing, whether on Site or off-Site. In order to enable such Persons to be present, the Contractor shall give the Sponsor at least <<*insert number*>> Business Days' advance notice of any system or Equipment check out or testing. If the Sponsor desires access to any places where Work is being performed or from which Equipment is being obtained for the Facility, the Contractor shall provide or arrange access thereto and shall provide the Sponsor at least <<*insert number*>> Business Days' advance notice of any off Site tests that the Sponsor advises the Contractor the Sponsor wishes to observe. The Contractor shall permit the Persons designated above to inspect and review all field work, including Equipment installation, start up and testing. All such Persons shall have the right to review and comment on any and all drawings, plans, specifications, manuals, tests, bids and bid evaluation materials, programs, methods, procedures and other Work. The Contractor shall be responsible for addressing all comments arising from such review. The Sponsor shall be given reasonable opportunity to inspect, prior to shipment at the place of manufacture or fabrication (as the case may be), any items being shipped and referred to in the preliminary shipping documents to evaluate whether they are in accordance with this Contract. [{The approval of the Sponsor following such inspection shall, in the absence of a [{written}] notification of disapproval, be deemed to be provided <<*insert number*>> Days after said inspection.}] No such approval or disapproval by the Sponsor as mentioned above shall relieve the Contractor of any liability under this Contract.

6.1.5   The Contractor shall cooperate with the Sponsor if the Sponsor determines that an inspection of the Facility of any Equipment or Facility System is necessary or appropriate. If the Contractor failed to give the notice required by Article 6.1.4, and in the Sponsor's opinion it is necessary to dismantle any Work for an inspection, the Contractor shall dismantle and reassemble such Work as and when requested by the Sponsor, at the Contractor's expense. If the Contractor has given the notice required by Article 6.1.4, and in the Sponsor's opinion it is necessary to dismantle any Work for an inspection, the Contractor shall dismantle and reassemble such Work as and when requested by the Sponsor:

(a) at the Contractor's expense, if the Sponsor finds the Work to be defective; or
(b) at the Sponsor's expense, by submitting an Adjustment Claim therefor requesting an Adjustment if the Sponsor does not find the Work to be defective provided

that the Contractor must submit such Adjustment Claim within <<*insert number*>> Days after such inspection by the Sponsor.

6.1.6   Before the Warranty Notification Period commences, the Contractor shall promptly correct at its own expense any part of the Work that is defective, not to the Sponsor's satisfaction or not in accordance with this Contract, regardless of the state of its completion or the time or place of discovery and regardless of whether the Sponsor has previously accepted any such part of the Work. For the avoidance of doubt, Article 11 and not this Article 6.1.6 shall apply once the Warranty Notification Period has commenced.

6.1.7   No inspection or review by the Sponsor's personnel or representatives, or any waiver of such inspection or review shall:

(a) constitute an approval, endorsement or confirmation of any drawing, plan, manual, specification, test, bidder, Equipment, material, program, method, procedure or other part of the Work;
(b) constitute an acknowledgment by any such Person that any drawing, plan, manual, specification, test, bidder, Equipment, material, program, method, procedure or other part of the Work satisfies the requirements of this Contract;
(c) relieve the Contractor of any of its obligations to perform the Work or furnish the Equipment so that the Facility, when complete, satisfies all the requirements of this Contract; or
(d) relieve the Contractor from any liability or responsibility hereunder, including for injuries to Persons or damage to property.

6.1.8   The Contractor shall cooperate with the Sponsor and other authorized Persons in making available charge to them the inspection instruments and tools [{available at the Site,}] along with appropriate operating personnel, to the extent [{reasonably}] requested by the Sponsor, to enable the Sponsor and other authorized Persons to carry out their inspection of the Work.

6.2 Project Accounting and Audit[110]

6.2.1   During the performance of the Work, the Contractor shall make available to the Sponsor [and any Financing Party] information relating to the status of the Work, including information relating to the design, engineering, construction and testing of the Facility, the status of any required Permits and any other [{non-financial}] matters as the Sponsor [or any Financing Party] may request. The Contractor shall maintain, in accordance with the Accounting Standards, such financial and other records and books of account with respect to the Work, including:

(a) all payments received by the Contractor hereunder;
(b) all documentation required to support Adjustment Claims;
(c) those records and books that are typically maintained by contractors performing work such as the Work; and

(d) those records and books that will allow the Sponsor and its agents and representatives to monitor and accurately determine the achievement of milestones for which Milestone Payments are due, and any costs to be paid by the Sponsor [{on a cost recoverable or time and materials basis}].

6.2.2 The Sponsor [, each Financing Party] and [its] [their] authorized representatives and agents may inspect and audit such records and books of account pertaining to this Contract or the Work [{Change Orders only}] during normal business hours. In addition, the Contractor shall ensure that such Persons have access for inspection and audit to all records and books of account relating to the Work of each Subcontractor during normal business hours. All such records and books of account of the Contractor and each Subcontractor shall be maintained, and made available for inspection and audit, for at least <<*insert number*>> years after their creation.

6.3 Samples

The Contractor shall provide samples if, as and when required in the Functional Specification and if instructed by the Sponsor. Each sample shall be labeled as to origin and intended use in the Work.

ARTICLE 7. SPONSOR'S RESPONSIBILITIES AND RIGHTS[111]

7.1 General

The Sponsor shall:

(a) provide the Sponsor Permits [{on a timely basis so as not to delay the Contractor's execution of the Work}];
(b) [{supply all necessary information in the Sponsor's possession required by the Contractor for the Contractor to procure the Contractor Permits};]
(c) provide the Contractor with possession of the <Site> <Route> on the [Unit One] Work Commencement Date [and provide [{unrestricted}] access to the <Site> <Route> for the Contractor after the Contract Date to the extent Contractor needs access to the <Site> <Route> to perform the <Site> <Route> Assessment];
(d) [{not, by itself or through any Sponsor Person, interfere with the conduct of the Work in any material regard};][112]
(e) provide, or cause to be provided [{qualified}] operating and maintenance personnel to be trained by Contractor, which personnel shall operate the Facility during the [start up, commissioning, and] performance testing of the Facility and while conducting Performance Tests of the Facility under the Contractor's direction and supervision;
(f) [make arrangements with the Grid for the exportation of electricity from the Facility not later than [*insert calendar date*] for pre-commissioning, commissioning, start up and the Performance Tests;]
(g) provide all [raw water for demineralization] in the Facility;

(h) purchase and supply [reimburse Contractor for the cost of] [backfeed electricity to start the Units] from the Grid by [insert calendar date] for [construction] [start up and commissioning] [and] [performance testing];

(i) [commence the Existing Plant Outage on the Outage Start Date and provide the Contractor with [{unrestricted}] access to the Existing Plant during such period;]

(j) [<<*insert number*>> date] Day after the Work Commencement Date, provide [Start Up Fuel, limestone, sand, Fuel] and make arrangements with the [Wheeling Company and Offtaker] for the exportation of electricity from the Facility for pre-commissioning, commissioning, start up and the Performance Tests, provided that, in the case of Fuel:

    (i) the Contractor gives the Sponsor at least <<insert number>> Days' prior written notice of the first date on which Fuel will be required and the quantities that will be required each month thereafter;[113] [and

    (ii) the Sponsor will dispose of any ash related thereto;][114] [and

    (iii) commence the Existing Plant Outage on the Outage Start Date and provide the Contractor with access to the Existing Plant during such period; and]

(k) [cause the <<*Offtaker, Dispatcher and the Wheeling Company*>> to] comply with the [reasonable] requests of the Contractor so that the Contractor can perform all its obligations hereunder;

(l) [provide any electricity and water for the Facility not required to be supplied by the Contractor under Article 6.1;] and

(m) provided that the Contractor has complied with its obligations under Article 5.1, make the Sponsor's personnel available to be trained by the Contractor in accordance with the schedule provided by the Contractor under Article 5.1 and make arrangements such that a sufficient number of the operating personnel who have completed the Contractor's training program conducted in accordance with the Functional Specifications are available to operate the Facility during the Performance Tests.

7.2 Adjustment[115]

To the extent the Sponsor's failure to perform any of its obligations specified in <{Article 7.1}> <Contract> [materially and adversely] affects the Contractor's ability to perform the Work, the Contractor shall be entitled to Adjustment; provided that:

(a) the Contractor shall notify the Sponsor promptly [(but in any event within <<*insert number*>> Days) after any such failure by the Sponsor]; and

(b) any Adjustment Claim shall be submitted within <<*insert number*>> Days of the Sponsor's remedy of any such failure and shall be accompanied by detailed documentation of such failure, the effect of such failure on the Contractor's performance of the Work and the Contractor's plan to reschedule its performance of the Work in order to overcome the effects of such failure.

The Contractor's right to an Adjustment pursuant to this Article 7.2 shall be the Contractor's sole and exclusive remedy and the Sponsor's sole and exclusive liability for any failure by the Sponsor to perform any of its obligations under this

[<Contract>] [<{Article 7.1}>] except as provided in Article 17.3 and Article 17.4, and no such failure shall constitute a breach of, or default under, this Contract.

7.3 [{Sponsor's Project Manager}]

The Sponsor <may> <{must}> appoint a project manager (the "Sponsor's Project Manager") who shall act as the Sponsor's [{single point}] of contact with the Contractor with respect to this Contract and the Work[{, and shall be responsible for administering all aspects of this Contract}]].[116]

7.4 Sponsor's Right to do Work

7.4.1   If the Contractor fails to commence to furnish sufficient workers of the required skill, or materials of the required quality or quantity, or perform this Work in accordance with the requirements hereof or the Baseline Schedule and falls behind the Baseline Schedule[{, for <<insert number>> days after written notice specifying such failure,}] the Sponsor shall have the option to supply workers and/or materials and perform the Work.[117] The Sponsor shall deduct expenses incurred in engaging other contractors, and supplying workers and material from payments due or which may become due to the Contractor or Retainage (including, without limitation, drawing under the Retainage LC). If such expenses exceed the balance due or which becomes due to the Contractor, the Contractor shall pay the excess to Sponsor immediately upon written demand therefor.

7.4.2   The Sponsor shall have the right to perform work with its own employees or by engaging other contractors and to permit other entities to do work during the progress and within the limits of, or adjacent to, the <Site> <Route> and all Contractor Persons shall conduct their Work and cooperate with all other Persons as necessary so as to mitigate any possible interference. The Contractor shall allow other Persons access to their work within the <Site> <Route>. The Contractor shall make no claims against the Sponsor for additional payment due to delays or other conditions created by the [{reasonable}] operations of such other Persons.[118]

7.4.3   Any Work provided by the Sponsor as a result of the Contractor's failure to perform shall be charged to the Contractor at the Sponsor's cost, plus a fee of <<insert number>> percent to cover overhead and burdens.

7.5 Work Provided by Others

The Sponsor has the right to contract with other Persons. The Contractor shall coordinate its Work with that of any other Persons so that all work can be as is completed to the Sponsor's satisfaction.[119]

7.6 The Sponsor's Right to Suspend Work[120]

The Sponsor, in its sole discretion, may suspend or slow the Work, in whole or in part, for any reason, at any time and from time to time, upon express written notice to the

Contractor of such suspension or deceleration (any such suspension, or deceleration an "Optional Suspension"). Promptly after receipt of such a notice (and in any event within <<*insert number*>> Days thereof), the Contractor shall suspend or slow the Work to the extent specified by the Sponsor. During any Optional Suspension, the Contractor shall (unless directed otherwise in writing by the Sponsor):

(a) place no further commitments, orders or Subcontracts relating to the suspended Work;
(b) protect and care for all Work, Equipment, components and materials relating to the suspended work at the <Site> <Route> or at storage areas for which it is responsible; and
(c) give the Sponsor copies of all outstanding orders and contracts for Equipment, materials and services and take any action with respect to such orders and contracts as the Sponsor may direct.

Immediately upon cessation of an Optional Suspension, and after being directed expressly in writing to do so by the Sponsor, the Contractor shall resume performance of the Work that is no longer subject to suspension as promptly as possible (but in any event within <<*insert number*>> Days of receiving such notice). The Contractor shall be entitled to an Adjustment for all the actual costs and damages and the schedule impact of any Optional Suspension within <<*insert number*>> Days of the cessation thereof, but the Contractor waives any and all claims for damages of any nature whatsoever against the Sponsor as a result of such Optional Suspension, including consequential damages and loss of anticipated profit for Work not yet performed.

7.7 Optional Termination by the Sponsor[12]

7.7.1   The Sponsor, in its sole discretion, may terminate this Contract, in whole or in part, for any reason at any time by written notice thereof to the Contractor (any such termination, an "Optional Termination"). Upon receipt of any such notice, the Contractor shall, unless the notice directs otherwise:

(a) immediately discontinue the Work on the date and to the extent specified in such notice;
(b) undertake no further commitments and place no further orders or Subcontracts for Equipment or services except as may be necessary for completion of such portion of the Work as is not discontinued;
(c) deliver to the Site, or such other location as the Sponsor may direct, all Equipment (whether or not completed), components and materials for which Milestone Payments have been made;
(d) promptly make every effort to procure cancellation upon terms satisfactory to the Sponsor of all commitments, orders, Subcontracts and other agreements to the extent they relate to the performance of the Work that is discontinued; and
(e) thereafter:
   (i) execute only those portions of the Work that are not discontinued or may be necessary to preserve and protect Work already in progress; and
   (ii) protect all Equipment at the Site or in transit thereto.

7.7.2   The Contractor waives any claims for damages, including loss of anticipated profits for Work not yet performed, on account of an Optional Termination. As the Contractor's sole remedy for such Optional Termination, the Sponsor shall pay the Contractor [in accordance with Exhibit 7.7.2] [:

(a) all Milestone Payments due and unpaid relating to Work performed by the Contractor before the Optional Termination (without deducting Retainage therefrom);
(b) an amount equal to all accumulated and unpaid Retainage;
(c) all of the Contractor's reasonable costs of termination, including Subcontractor cancellation charges, that are substantiated in writing by the Contractor; and
(d) if, at the date of such termination:
  (i) the Contractor has properly fabricated any Equipment off Site for subsequent incorporation at the Site; and
  (ii) the Contractor delivers such Equipment or that portion of Equipment prepared or fabricated to the Site or to such other place as the Sponsor shall direct in writing, an amount equal to the value of such Equipment, less the amounts of Milestone Payments theretofore made in respect of such Equipment.]

7.8 Adequate Assurances of Further Performance[122]

The Sponsor shall have the right to require the Contractor to provide the Sponsor adequate assurance that the Contractor will perform its obligations in a timely fashion in accordance with this Contract. Should the Sponsor request the Contractor to provide adequate assurance of future performance, the Contractor shall, within <<*insert number*>> Days of said request, provide the Sponsor such assurance in writing. The Contractor's failure to provide either said adequate assurance or a statement by the Contractor that the Contractor cannot perform in a timely fashion in accordance with the Contract, or any act or omission of the Contractor which makes it apparent, at the time, that the Contractor will not perform in accordance with this Contract, shall entitle the Sponsor to take over the Work[123] and be a Contractor Event of Default.

7.9 Sponsor Directed Changes[124]

The Sponsor may, at any time, by written notice to the Contractor, propose a Change Order requesting that changes be made in the Work. On receipt of such notice from the Sponsor, the Contractor shall, within <<*insert number*>> Days thereafter, submit to the Sponsor a proposal (in the format of Exhibit 1A) including a detailed description of changes in the Work that would be necessitated by the proposed Change Order and any resulting changes that would be required in the Project Variables. Following receipt of the Contractor's proposal, the Parties shall in good faith attempt, within <<*insert number*>> Days thereafter, to agree upon the cost of the Sponsor's proposed changes and the necessary changes, if any, to the Project Variables. If the Parties agree upon all of the foregoing and any other terms and conditions which would be required by such Change Order, they shall execute a Change Order reflecting a revised Contract Price (based upon the agreed cost), the agreed upon changes (if any) to the Project Variables and the other agreed upon terms and conditions, if any. At any time before the Parties reach an agreement upon the cost, the changes (if any) to the Project Variables and any other terms or conditions of the Sponsor's proposed Change Order, the Sponsor may direct, in writing, the Contractor to proceed with the proposed Change Order [{if such can reasonably be expected not to exceed <<*insert currency and amount* >> in value}],[125] in which event, notwithstanding any other provision of this Contract, the Contractor:

(a) shall implement the Sponsor's directed change as set forth in such proposed Change Order; and
(b) may, within <<*insert number*>> Days of such written direction from the Sponsor submit an Adjustment Claim in respect of the consequences of clause (a) above.

Any directed change which reduces the quantity of Work to be provided shall not entitle the Contractor to an Adjustment or a claim for damages or for anticipated or lost profits on the Work involved in such reduction, and the Sponsor shall be entitled to a credit on the Contract Price for the value of the omitted Work, but subject to an allowance to the Contractor for any actual loss incurred by it in connection with the purchase, rental, delivery, and subsequent disposal of materials planned, but which could not be used in any of the Work actually performed as a result of such changes.

7.10 [Additional Facilities Option

The Contractor agrees that the Sponsor may, by written notice thereof to the Contractor, at any time within <<*insert number*>> Days of the <Contract> <Work Commencement> Date, elect to include in the scope of the Work, the <<*describe work or facilities*>> described in [the Functional Specifications] [and] [in Exhibit 7.10] (the "Additional Scope Option") at a cost of <<*insert currency and amount*>> (the "Additional Facilities Price") on the terms specified in Exhibit 7.10.]

7.11 Sequence of Work.

Contractor shall commence Work when requested in accordance with the approved schedule. If in the Sponsor's opinion, it would be advantageous to perform any part of the Work earlier than planned, the Contractor shall, upon written notice, make every effort to meet such requirement at no additional cost to the Sponsor.

7.12 [Change in Ownership of Sponsor

If the ownership of more than <<*fifty (50)*>> percent of the common voting equity the Sponsor is transferred to a Person not affiliated with the Sponsor and such transfer occurs after the Contract Date; and a notice to proceed with the Work has been issued to a contractor other than the Contractor to engineer, procure and construct the Facility; then the Sponsor will pay to the Contractor the sum of <<*insert currency and amount*>>.][126]

ARTICLE 8. TESTING AND COMPLETION[127]

8.1 Performance Tests

The [Sponsor Information and the] Functional Specifications [and Exhibit 8.1] set forth the Performance Tests and the program for testing [each Unit and] the Facility [{in order to evaluate the Work's compliance with this Contract}]. [The Performance Tests must be carried out on two separate occasions; on one such occasion, the ambient air temperature at the Site must be below <<*insert number*>> degrees Fahrenheit (the "Cold Weather Test") and on the other such occasion, the ambient air temperature at the Site must be above <<*insert number*>> degrees Fahrenheit (the "Hot Weather Test"). At any time before the required ambient conditions are extant for the required testing period, the Contractor may choose to conduct provisional performance tests (the "Provisional Performance Tests") at any time at which the results therefor may be extrapolated to Reference Conditions using the Correction Curves to determine (on a provisional basis) if the Performance Guarantees have been achieved but the conduct of such Provisional Tests Performance shall not release the Contractor from its obligations to pass successfully the Cold Weather Test and the Hot Weather Test.] The Contractor shall determine and submit in writing the raw data and completed results of all Performance Tests.

8.2 Contractor's [{Undue}] Delay in Carrying Out Performance Tests.

If the Performance Tests are being [{unduly}] delayed by the Contractor, the Sponsor may, by written notice, require the Contractor to carry out any Performance Test within <<*insert number*>> Days after the receipt of such notice. The Contractor shall carry out the Performance Tests on such Days within that period as the Contractor may fix and of which the Contractor shall give written notice to Sponsor. If the Contractor fails to carry out the Performance Tests within <<*insert number*>> Days of such notice then the Sponsor may proceed with the Performance Tests. All Performance Tests so made by the Sponsor shall be at risk and cost of the Contractor and the cost thereof shall be deducted from the Contract Price. The Performance Tests shall be deemed to have been made in the presence of the Contractor if it fails to attend such and the results of the Performance Tests shall be accepted as accurate.

8.3 [Correction of Test Results

All Performance Test and Provisional Performance Test results shall be extrapolated to Reference Conditions by using the Correction Curves.]

8.4 Safety

If during any Performance Test or Provisional Performance Test it is discovered that the Facility [or a Unit] cannot be operated in a safe manner in accordance with <Good> <{Prudent}> <Utility> <{Industry}> Practice, such Performance Test or

Provisional Performance Test shall be terminated and the defective system or component replaced or repaired by the Contractor at its sole expense, whereupon such Performance Test or Provisional Performance Test shall be reinitiated and reperformed in its entirety and all previous acquired data disregarded.

8.5 Advance Notice of [Facility] [Unit] Provisional Acceptance.

The Contractor shall provide, at least <<*insert number*>> Days prior to the date upon which the Contractor believes that [the Facility] [and each of Unit One and Unit Two], will achieve [Facility] [Unit] Provisional Acceptance, written notice thereof to the Sponsor.

8.5.1   "[Facility] [Unit] Commissioning Completion" with respect to [the] [a] [Facility] [Unit] shall occur when the Sponsor issues to the Contractor a Certificate of [Facility] [Unit] Commissioning Completion acknowledging that all of the following conditions have been satisfied:[128]

(a)   the Work has been [{substantially}] completed in accordance with this Contract[, including all Work that is necessary to allow the [Facility] [Unit] to safely and reliably [handle fuel and generate and export electrical power [and thermal energy [steam, etc.]]];[129]

(b)   the Work has been inspected for completeness using the Specifications and Drawings as the basis of such inspection and completed in accordance therewith;

(c)   [the Contractor has conducted hydrostatic, pneumatic, electrical and all other necessary field tests, and has restored systems and components thereof to operating condition;]

(d)   all safety and fire protection requirements have been met, and all safety and fire protection systems have been installed and are operable in accordance with all applicable Laws;

(e)   all Vendors' and other manufacturers' instructions and drawings relating to Equipment have been transmitted to the Sponsor in a format consistent with the Functional Specifications;

(f)   special tools, spare parts and other items provided by the Contractor that are necessary for the operation of the [Facility] [Unit] are available [at <the Site> <*name other appropriate location*>];

(g)   all test and inspection certificates and reports applicable to the Work, including those relating to relay settings and instrument calibration, have been submitted to the Sponsor;

(h)   all necessary insulation and fireproofing of Equipment, including piping and instruments has been completed;

(i)   [painting of the [Facility] [Unit] has been nearly completed, at least to the extent that no excessive scaffolding remains standing in areas in which operators must work to start up and operate the [Facility] [Unit], and any portions of the Facility that would be inaccessible during [Facility] [Unit] operation have been painted;]

(j)   all chemical cleaning and/or flushing materials, rust preventatives and oils used to protect the Equipment during construction have been removed to prevent interference during operation, and a record thereof shall be provided to the Sponsor;

(k) [the initial charge of all manufacturer supplied lubricants (including any required lube oil system flushing and the installation and cleaning of temporary screens) has been installed;]

(l) [all mechanical seals, permanent packing and accessories have been installed as required, except as restricted by commissioning activities;]

(m) [all temporary supports, bracing or other foreign objects that were installed in vessels, transformers, rotating machinery or other Equipment to prevent damage during shipment, storage or erection have been removed;]

(n) [alignments of all rotating Equipment have been made to manufacturers' tolerances, and verification reports relating thereto have been submitted to the Sponsor;]

(o) [rotating machinery has been checked for correct direction, rotation and freedom of moving parts before connecting drivers, and verification reports relating thereto have been submitted to the Sponsor;]

(p) [the Tie-ins and the Site Electrical Interconnection Facilities have been completed by the Contractor and approved by the Sponsor in writing for compliance with the Functional Specifications [and the Sponsor Information];]

(q) [all mechanical and electrical safety devices have been tested, adjusted and sealed where necessary and a list of proper settings and certification records for such devices has been provided to the Sponsor;]

(r) [all Facility Systems pertaining to the [Facility] [applicable Unit] have been cleared and are free of construction debris and have been satisfactorily chemically cleaned and flushed where required, and the Contractor shall have provided all chemicals, lubricants, Consumables and any temporary piping and modifications required for steam blow, hydrotesting and chemical cleaning and flushing activities;]

(s) [all steam blows have been completed;]

(t) [the Contractor has replaced with fresh reserves thereof all Consumables used during commissioning, including lube oil, gaskets, and filter cartridges;]

(u) [the Contractor has installed any required temporary equipment to permit start up, such as tanks, heaters, piping and instruments;]

(v) [all pipe supports and spring hangers have been properly installed and set per Vendors' specifications;]

(w) [initial vibration data has been taken to ensure compliance with Vendor recommended tolerances on all applicable rotating Equipment;]

(x) [all temporary filters, strainers, blinds and screens that were required during commissioning have been removed and equipment restored;]

(y) [all waste materials, including used cleaning fluids, have been disposed of in compliance with applicable Laws;]

(z) [all inner packing materials, such as sand, gravel, balls and saddles, have been properly installed in vessels;]

(aa) [all materials such as chemicals, resins, desiccants, catalysts and other similar materials have been installed in vessels;]

(bb) [all mixed beds containing chemicals, resins, desiccants, catalysts or other operating materials has been installed;]

(cc) cleanup of the Site has been completed to the extent that it does not interfere with the operations of the [Unit] [Facility];

(dd) [all emission control systems have been installed and certified as required by all applicable Laws;]

(ee) [<[originals]> <[{copies}]> of all the Contractor Permits have been submitted to the Sponsor;]

(ff) all Specifications and Drawings (current as of such time) have been submitted to the Sponsor;

(gg) final construction turnover packages and functional tests have been completed for all Facility systems pertaining to the [Facility] [applicable Unit], as more particularly described in the Functional Specifications;

(hh) all activities in the construction and construction testing phases for all Facility systems pertaining to the [applicable Unit] [Facility], as more particularly described in the Functional Specifications, have been successfully completed;

(ii) [equipment and Facility Systems pertaining to the [Facility] [applicable Unit] have been cleaned, leak checked, lubricated and point to point checked to verify that they have been correctly installed so as to respond to simulated test signals which are equivalent to actual signals which would be received during operation;]

(jj) [{the Parties have agreed in writing to a list and schedule for all Work and testing which was to be carried out by [Facility] [Unit] Commissioning Completion of the [Unit in question] but has not yet been carried out};][130]

(kk) all Facility System operations manuals have been submitted to the Sponsor;

(ll) [functional tests on each Facility System pertaining to the [Facility] [applicable Unit] have been satisfactorily completed;]

(mm) Facility operation and maintenance procedures and manuals have been prepared by the Contractor and submitted to and approved by the Sponsor;

(nn) [Facility] [Unit] testing procedures have been approved in writing by the Sponsor in its reasonable judgment;

(oo) [the Contractor's training program for the Sponsor's operating personnel has been completed;]

(pp) the Contractor has removed all of its construction equipment, material and support personnel (except to the extent they are needed to perform warranty obligations hereunder) from the Site[; and][.]

(qq) [add any other project specific conditions.].

8.5.2   The following items of Work are not included in [Facility] [Unit] Commissioning Completion[:

(a) erection of the permanent perimeter fence, provided that a temporary security fence shall remain in place until the permanent perimeter fence is completed;

(b) landscaping;

(c) painting, except as required by Article 8.5.1(i);

(d) paving; and

(e) final cleanup].

8.5.3   When, at any time, the Contractor believes that [Facility] [Unit] Commissioning Completion has occurred [with respect to a Unit], the Contractor shall provide written

notice thereof to the Sponsor, including such documentation and certifications necessary to verify the conditions for [Facility] [Unit] Commissioning Completion [for
such Unit]. Thereafter, the Sponsor shall within <<*insert number*>> Business Days:

(a) notify the Contractor of any reason why [Facility] [Unit] Commissioning
    Completion has not occurred [for such Unit]; or
(b) issue to the Contractor a Certificate of [Facility] [Unit] Commissioning
    Completion acknowledging that [Facility] [Unit] Commissioning Completion
    has occurred [for such Unit] on the first date on which all the conditions there-
    for were satisfied, provided that such date shall be no earlier than the date of the
    Contractor's last written notice to the Sponsor under the previous sentence.

8.5.4   Upon being advised by the Sponsor that [Facility] [Unit] Commissioning
Completion has not occurred [for such Unit], the Contractor shall perform all
necessary Work therefor, and the foregoing notice procedure shall be repeated
until [Facility] [Unit] Commissioning Completion occurs [for such Unit] and the
Sponsor has issued a Certificate of [Facility] [Unit] Commissioning Completion
therefor.

8.6 *[Unit Provisional Acceptance]*

8.6.1   "Unit Provisional Acceptance" with respect to a Unit shall occur when the
Sponsor issues to the Contractor a Certificate of Unit Provisional Acceptance for
such Unit acknowledging that all of the following conditions have been satisfied:

(a) the Sponsor has issued a Certificate of Unit Commissioning Completion for
    such Unit;
(b) all activities in the commissioning phase have been successfully completed,
    as more particularly described in the Functional Specifications [and Sponsor
    Information];[131]
(c) all Facility System tests of such Unit have been successfully completed and
    accepted by the Sponsor, as more particularly described in the Functional
    Specifications [and Sponsor Information];
(d) all of the Performance Tests have been successfully completed for such Unit
    [and the Offtaker Facility Acceptance with respect to such Unit has occurred];
(e) each of the Unit Minimum Performance Levels for the Unit in question has been
    achieved;
(f) such Unit has passed all Performance Tests relating to it, demonstrating full com-
    pliance with all Unit Performance Guarantees, or the Contractor has paid all
    Unit Buy Down Amounts relating to such Unit; and
(g) the Parties have agreed in writing to the Punch List.

8.6.2   When, at any time, the Contractor believes that Unit Provisional Acceptance
has occurred for a Unit, the Contractor shall provide notice thereof to the Sponsor,
including such operating data, documentation and certifications necessary to verify
the conditions for Unit Provisional Acceptance. Thereafter, the Sponsor shall within
<<*insert number*>> Days:

(a) notify the Contractor of any reason why Unit Provisional Acceptance has not occurred with respect to such Unit; or

(b) issue to the Contractor a Certificate of Unit Provisional Acceptance acknowledging that Unit Provisional Acceptance has occurred for such Unit on the first date on which all the conditions therefor were satisfied provided that such date shall be no earlier than the date of the Contractor's last written notice to the Sponsor under the previous sentence.

8.6.3   Upon being advised by the Sponsor that Unit Provisional Acceptance has not occurred for the Unit in question, the Contractor shall perform all necessary Work therefor and, in addition to paying any Delay Damages due, bear all the Sponsor's costs incurred as a result of additional testing, and the foregoing notice procedure shall be repeated until Unit Provisional Acceptance occurs for such Unit and the Sponsor has issued a Certificate of Unit Provisional Acceptance for such Unit.]

8.7 Facility Provisional Acceptance

8.7.1   "Facility Provisional Acceptance" with respect to the Facility shall occur when the Sponsor issues to the Contractor a Certificate of Facility Provisional Acceptance acknowledging that all of the following conditions have been satisfied:

(a) the Sponsor has issued a Certificate of [[Unit] Commissioning] [Unit Substantial] Completion [for both Units];

(b) all activities in the commissioning phase have been successfully completed, as more particularly described in the Functional Specifications [and Sponsor Information];

(c) all Facility System tests have been successfully completed and accepted by the Sponsor, as more particularly described in the Functional Specifications [and Sponsor Information];

(d) no Adjustments Claims or Change Orders are pending {in excess of <<*insert currency and amount*>> in the aggregate};[132]

(e) each of the Facility Minimum Performance Levels has been achieved;

(f) the Facility shall have passed all Performance Tests, demonstrating full compliance with all Performance Guarantees or the Contractor has paid all Facility Buy Down Amounts [provided that the Contractor need not pay such amounts until the end of the cure period expressly provided herein];

(g) the Contractor shall have paid to the Sponsor any Delay Damages due {(provided, however, that any payment of Delay Damages shall not be construed as a waiver of the Contractor's right to dispute whether or not such Delay Damages had actually accrued)};

(h) all insurance requirements and requirements of Law {which are expressly included hereunder as part of the Work} for placing the Facility into commercial operation have been satisfied;

(i) [the Offtaker Completion has occurred;][133]

(j) [all of the Contractor's representations and warranties contained in Article 12 are true and correct;][134]

(k) the Project has received releases and waivers of all Liens against the Facility and the Site and the Sponsor and its property, employees and agents from the Contractor and each Subcontractor, and such other documentation as the Sponsor [or any Financing Party] may request establishing proof thereof;

(l) the Contractor has removed all of its construction, equipment, material and support personnel (except to the extent they are needed to perform warranty obligations hereunder) from the Site;

(m) [no Contractor Event of Default exists;][135]

(n) the Contractor has completed all Work except as set forth on the Punch List;

(o) the Parties have agreed in writing to the Punch List; and

(p) [<<*insert any other Project specific conditions*>>].

8.7.2   [If all of the conditions in clauses << *insert applicable clause number* >> of this Article 8.7.1 have been satisfied but it is not possible to determine whether the conditions set forth in clauses << *insert applicable clause number* >> of this Article 8.7.1 have been satisfied because there has been no period since the satisfaction of the conditions set forth in Article 8.7.1(c) during which the ambient air temperature at the Site permits the Cold Weather Test and/or the Hot Weather Test as applicable, then the conditions for Facility Provisional Acceptance set forth in clauses << *insert applicable clause number* >> of this Article 8.7.1 shall be deemed satisfied if the Facility passed the Provisional Performance Tests, but only until the ambient air temperature at the Site permits the Cold Weather Test and/or the Hot Weather Test as applicable to be conducted. In the case that the Facility has passed the Provisional Performance Tests, Milestone Payment No. <<__>> shall be <<*insert currency and amount*>> instead of <<*insert currency and amount*>>, and a new Milestone Payment No. <__A> shall be added to the Milestone Payment Schedule in the amount of <<*insert currency and amount*>> and the corresponding milestone event therefor shall be the satisfaction by the Contractor of all the conditions set forth in clauses << *insert applicable clause number* >> of this Article 8.7.1 during the Provisional Performance Tests. If the Provisional Performance Tests have been successfully conducted and, other than as the result of any action or inaction by any Contractor Person, the Cold Weather Test and/or the Hot Weather Test as applicable is delayed more than:

 (i) <<*insert number*>> Days, the Sponsor shall pay to the Contractor one half of Milestone Payment No. __A, provided that the Performance Tests must still be conducted; and

(ii) more than <<*insert number*>> Days, the Sponsor shall pay the Contractor all of Milestone Payment No. __A but {and} Facility Provisional Acceptance must still {will} be {deemed} achieved [{without conducting the Performance Tests}]].

8.7.3   When, at any time, the Contractor believes that Facility Provisional Acceptance has occurred, the Contractor shall provide notice thereof to the Sponsor, including such operating data, documentation and certifications as necessary to verify the conditions for Facility Provisional Acceptance. Thereafter, the Sponsor shall within <<*insert number*>> Days:

(a) notify the Contractor of any reason why Facility Provisional Acceptance has not occurred; or

(b) issue to the Contractor a Certificate of Facility Provisional Acceptance acknowledging Facility Provisional Acceptance on the first date on which all the conditions therefor were satisfied provided that such date shall be no earlier than the date of the Contractor's last written notice to the Sponsor in the previous sentence.

8.7.4   Upon being advised by the Sponsor that Facility Provisional Acceptance has not occurred, the Contractor shall perform all necessary Work therefor and in addition to continuing to pay any Delay Damages due as a result of the Sponsor's costs of additional testing, and the foregoing notice procedure shall be repeated until Facility Provisional Acceptance occurs and the Sponsor has issued a Certificate of Facility Provisional Acceptance therefor.

8.8 [{Sponsor Delay of Testing For [Facility] [and/or Unit] Provisional Acceptance}]

[{In addition to any relief to which the Contractor may be entitled under Articles 19.1 and 19.2 for a Sponsor-caused delay of the Performance Tests,}] [If] [if] the Contractor has satisfied all the requirements of [Article 8.6 and/or] Article 8.7.1 <<*insert applicable clause number*>> and the Facility is [or a Unit], as adjudged by <Good> <{Prudent}> <Utility> <{Industry}>Practices, ready for the Performance Tests but the Sponsor has delayed such Performance Tests for more than <<*insert number*>> Days in the aggregate}, [and such delays have not previously been addressed through an Adjustment] then:

(a) the Contractor's obligations under Article 9 shall be suspended except with respect to Delay Damages accruing prior to the date that the Facility [and/or Unit] was ready, if so determined above, for the Performance Tests;

(b) the Sponsor shall pay to the Contractor the Milestone Payment due upon Facility Provisional Acceptance [and/or Unit Provisional Acceptance]; and

(c) the Contractor shall, unless expressly directed otherwise in writing by the Sponsor, take all necessary steps (at the Sponsor's expense by means of an Adjustment) to maintain the Facility in a state such that it will be ready for the Performance Tests upon <<*insert number*>> Days' prior written notice from the Sponsor (the date that is <<*insert number*>> Days after such notice, the "New Test Date") and [{,so long as such notice is provided to the Contractor within <<*insert number*>> Days of the date that the Contractor was prepared to conduct the Performance Tests}], the Contractor shall perform the Performance Tests, at which point all of the Contractor's obligations under Article 9 shall be reinstated and the Guaranteed [Unit] Completion Date shall be extended under Articles 18.1 and 18.2 by at least the number of Days that the Sponsor delayed the Performance Testing {plus <<*insert number*>> Days}.]

8.9 Facility Final Acceptance[136]

8.9.1   "Facility Final Acceptance" shall occur when the Sponsor has issued to the Contractor a Certificate of Facility Final Acceptance acknowledging that all of the following conditions have been satisfied:

(a) the Sponsor has issued a Certificate of Facility Provisional Acceptance;
(b) all items on the Punch List have been completed [to the Sponsor's [entire] satisfaction];
(c) the Sponsor has received all As Built Drawings and operation and maintenance manuals for the Facility including all thereof to be provided by any Subcontractors under the original terms of their Subcontracts as executed;
(d) the Sponsor has received an Affidavit of Payment and Final Release, in the form set forth in Exhibit 5.10.1(d)B from the Contractor;
(e) the Sponsor has received releases and waivers of all Liens against the Facility and the Site and the Sponsor and its property, employees and agents from the Contractor and each Subcontractor and such other documentation as the Sponsor [or any Financing Party] may request to establish proof thereof; and
(f) the Contractor has completed all Work (and for the avoidance of doubt Work includes correction of any defect or deficiency detected by a Certificate of Facility Final Acceptance has been signed by the Sponsor).

8.9.2   When the Contractor believes that Facility Final Acceptance has occurred, the Contractor shall provide written notice thereof to the Sponsor, including such operating data, documentation and certifications as necessary to verify the conditions for Facility Final Acceptance. Thereafter, the Sponsor shall within <<*insert number*>> Days:

(a) notify the Contractor of any defects or deficiencies in the Work or of any other reason why Facility Final Acceptance has not occurred; or
(b) issue to the Contractor a Certificate of Facility Final Acceptance acknowledging that Facility Final Acceptance has occurred on the first date on which all conditions therefor were satisfied; provided that such date shall be no earlier than the date of the Contractor's last written notice to the Sponsor under the previous sentence.

8.9.3   Upon being advised by the Sponsor that Facility Final Acceptance has not occurred, the Contractor shall perform all necessary Work therefor, and the foregoing notice procedure shall be repeated until Facility Final Acceptance occurs and the Sponsor has issued a Certificate of Facility Final Acceptance therefor.

8.10 Punch Lists

At least <<*insert number*>> Business Days before the start of any Performance Tests and each of the Contractor's notices pursuant to [each of Article 8.7.2] and Article 8.9.2, the Contractor shall submit to the Sponsor the Punch List with respect to [the applicable Unit or] the Facility. The Sponsor shall notify the Contractor in writing that it accepts or rejects the Punch List or any part thereof,[{ and, if the Sponsor rejects the Punch List or any part thereof, its reasons for disagreement in reasonable detail}]; provided that the Sponsor's acceptance or rejection of any Punch List shall not relieve the Contractor of its liability to complete all Work. The Sponsor shall be entitled to add items to the draft or agreed upon Punch List as the Sponsor evaluates the degree of completion of the Work. The Sponsor shall have the right, in its sole discretion, without additional cost to the Sponsor, to suspend or delay Work on any items on the Punch List at any time that the Sponsor determines such Work may [{adversely}] affect its operation of the Facility, in which case the Sponsor and the Contractor shall coordinate with each other to schedule the timely completion of the items on the Punch List so as to minimize disruption of the Facility's operations [{and the necessity and cost of remobilization by the Contractor}]. The Sponsor may complete at the Contractor's expense any item on the Punch List that has not been completed within <<*insert number*>> Days after the Sponsor determined it should be placed it on the Punch List.[137]

8.11 Refusal to Accept the Facility [or a Unit]

If the requirements of this Contract have not been fulfilled or documentation required hereunder has not been completed, the Sponsor may refuse to accept care, custody and control of the Facility [or a Unit] [provided further that, if there is a defect or deficiency which deprives the Sponsor of substantially the entire benefit of the Works, the Sponsor may recover all sums paid to the Contractor for them, plus all the Sponsor's financing and hedge breakage costs and the cost of dismantling the Work, clearing and restoring the <Site><Route> and returning all plant and materials to the Contractor if the Contractor fails to do the foregoing].[138]

8.12 Operating Revenues[139]

Any and all revenues generated by the operation of the Facility or a Unit at any time shall be solely for the account of the Sponsor.[140]

8.13 Early Operation[141]

The Sponsor may, at its option, place the Facility or a Unit or any Facility System into service prior to the Facility Provisional Acceptance Date. Such placement into service shall not constitute acceptance thereof. The Contractor shall cooperate with the Sponsor and allow the Sponsor sufficient access to the Facility or a Unit so that the Sponsor may operate the Facility or a Unit or any Facility System that is actually in service. The Contractor shall remain liable for the Facility or a Unit or any Facility System that the Sponsor places into service under this Article 8.13 [{and the Sponsor shall follow proper safety procedures and conform to the Contractor's directions in the receipt, handling, storage, protection, installation, maintenance, inspection and operation of the Facility or any Facility System in service}]. If the Contractor shall fail or refuse to give the Sponsor directions as aforesaid, the Sponsor shall follow accepted operating <{Good}> <Prudent> <{Utility}> <Industry> Practices with respect to its operation of the Facility or a Unit or Facility System actually placed in service. The number of hours that the Sponsor <uses> operates <a <Unit> <the Facility> <a Facility System> to produce <<*electricity or hot water* >>pursuant to this Article 8.13> shall be subtracted from the end of the Warranty Notification Period with respect to the <Unit> <Facility>. [To the extent that the Contractor needs to perform Work on any common facilities to achieve Facility Provisional Acceptance and such is most easily accomplished with Unit One not operating, the Sponsor agrees not to operate Unit One for any period that the Contractor reasonably so requests in writing]. To the extent that the Sponsor's exercise of any of its rights hereunder causes:

(a) the Contractor to suffer delays or increased costs, the Contractor shall be entitled to submit an Adjustment Claim therefor so long as the Contractor submits
such within <*insert number*> Days after such circumstances arise[; or

(b) results in damages to the Facility for which the Contractor is responsible hereunder, the Sponsor shall pay any deductibles required to be paid in connection with
any insurance proceeds received therefor].[142]

ARTICLE 9. GUARANTEES AND LIQUIDATED DAMAGES[143]

9.1 Completion Guarantees and Delay Damages

9.1.1 The Contractor covenants that:

(a) [Unit] [Facility] Commissioning Completion [for Unit One] shall occur (the
"Guaranteed [Unit One] [Facility] Commissioning Completion Date"):
  (i) on or before <<insert date>> {if the Work Commencement Date occurs by
  <<insert date>>}[; or
  (ii) on or before <<insert date>>, if the Work Commencement Date occurs by
  <<insert date>>; or
  (iii) on or before the date mutually agreed to by the Parties in a Change Order,
  if the Work Commencement Date occurs after <<insert date>> (the last
  date in (ii) above)];

(b) [Unit Commissioning Completion for Unit Two shall occur on or before the date
which is <<*insert number*>> Days after the [Unit Two][144] Work Commencement
Date (the "Guaranteed Unit Two Commissioning Completion Date");]

(c) [Unit Provisional Acceptance for Unit One shall occur on or before the date
which is <<*insert number*>> Days after the Work Commencement Date (the
"Guaranteed Unit One Provisional Acceptance Date");]

(d) [Unit Provisional Acceptance for Unit Two shall occur on or before the date
which is <<*insert number*>> Days after the Work Commencement Date[145] (the
"Guaranteed Unit Two Provisional Acceptance Date");]

(e) Facility Provisional Acceptance shall occur on or before the date which is <<*insert
number*>> Days after the Work Commencement Date (the "Guaranteed Facility
Provisional Acceptance Date"); and

(f) Facility Final Acceptance shall occur on or before the date which is <<*insert number*> Days after the Work Commencement Date (the "Guaranteed Facility Final
Acceptance Date").

9.1.2 If Unit Commissioning Completion for Unit One does not occur on or before
the Guaranteed Unit One Commissioning Completion Date, then the Contractor shall
pay the Sponsor liquidated damages in the amount of <<*insert currency and amount*>> for
each Day following the Guaranteed Unit One Commissioning Completion Date until
the Unit Commissioning Completion Date of Unit One[{, provided that the Contractor
shall be excused from paying such liquidated damages on any particular Day with
respect to which the Offtaker does not assess liquidated damages against the Sponsor
for the Sponsor's failure to meet its construction completion obligations for such Unit,
or the Sponsor would not have operated the Facility for any {market} reason}].[146]

9.1.3   If Unit Commissioning Completion for Unit Two does not occur on or before the Guaranteed Unit Two Commissioning Completion Date, then the Contractor shall pay the Sponsor liquidated damages in the amount of <<*insert amount and currency*>> for each Day following the Guaranteed Unit Two Commissioning Completion Date until the Unit Commissioning Completion Date of Unit Two[;{; provided that the Contractor shall be excused from paying such liquidated damages on any particular Day with respect to which the Power Purchaser does not assess liquidated damages against the Sponsor for the Sponsor's failure to meet its construction completion obligations for such Unit, or the Sponsor would not have operated the Facility for any {market} reason}].

9.1.4   If the Unit Provisional Acceptance Date for Unit One does not occur on or before the Guaranteed Unit One Provisional Acceptance Date, then the Contractor shall pay the Sponsor liquidated damages in the amount of <<*insert amount and currency* >> for each Day following the Guaranteed Unit One Provisional Acceptance Date until the Unit Provisional Acceptance Date of Unit One[{; provided that the Contractor shall be excused from paying such liquidated damages on any particular Day with respect to which the Offtaker does not assess liquidated damages against the Sponsor for the Sponsor's failure to meet its construction completion obligations for such Unit, or the Sponsor would not have operated the Facility for any {market} reason}].

9.1.5   If the Unit Provisional Acceptance Date for Unit Two does not occur on or before the Guaranteed Unit Two Provisional Acceptance Date, then the Contractor shall pay the Sponsor liquidated damages in the amount of <<*insert amount and currency*>> for each Day following the Guaranteed Unit Two Provisional Acceptance Date until the Unit Provisional Acceptance Date of Unit Two [{, provided that the Contractor shall be excused from paying such liquidated damages on any particular Day with respect to which the Offtaker does not assess liquidated damages against the Sponsor for the Sponsor's failure to meet its construction completion obligations for such Unit, or the Sponsor would not have operated the Facility for any {market} reason}].

9.1.6   If the Facility Provisional Acceptance Date does not occur on or before the Guaranteed Facility Provisional Acceptance Date, then the Contractor shall pay the Sponsor liquidated damages in the amount of <<*insert amount and currency*>> for each Day following the Guaranteed Facility Provisional Acceptance Date until and including the Facility Provisional Acceptance Date[; {provided that the Contractor shall be excused from paying such liquidated damages on any particular Day with respect to which the Offtaker does not assess liquidated damages against the Sponsor for the Sponsor's failure to meet its construction completion obligations for such Unit, or the Sponsor would not have operated the Facility for any {market} reason}].

9.1.7   If the Facility Final Acceptance Date does not occur on or before the Guaranteed Facility Final Acceptance Date, then the Sponsor shall be entitled (but shall not be required) to complete any or all Punch List items remaining uncompleted at such time (either itself or through other contractors) and to apply to the cost thereof any or all of the unpaid Contract Price or Retainage [(including drawing upon any Retainage LC)] held by the Sponsor, and, if such Retainage is insufficient to cover the cost of completing all Punch List items, the Contractor shall pay the excess cost thereof to the Sponsor promptly following demand therefor.[147]

9.1.8   The liquidated damages described in Articles <<9.1.2, 9.1.3, 9.1.4 and 9.1.5>> shall be referred to as "Delay Damages." The Sponsor may issue an invoice to the Contractor for any Delay Damages accrued hereunder <at any time and from time to time.>

{(i) during the first <<*insert number*>> Days of any month; and
  (a) after the Contractor issues a notice to the Sponsor requesting a Certificate of [Facility] [Unit] Commissioning Completion, Certificate of Unit Provisional Acceptance, or a Certificate of Facility Provisional Acceptance}.

9.1.9   [If the Contractor required more than <<*insert number*>> Existing Plant Shutdown Hours, the Contractor shall pay liquidated damages ("Shutdown Liquidated Damages") to the Sponsor of an amount equal to <<*insert number*>> percent of the Contract Price for each Existing Plant Shutdown Hour in excess of <<*insert number*>> Existing Plant Shutdown Hours.]

9.1.10   Any Delay Damages [and/or Shutdown Liquidated Damages] shall be due to the Sponsor within <<*insert number*>> Days after the Sponsor submits to the Contractor an invoice therefor which the Sponsor may submit from time to time as often as it so desires. [{If Contractor recovers from a Vendor, by settlement, judgment or otherwise, performance-liquidated damages exceeding the amount of the Delay Damages Contractor has paid Sponsor, Contractor shall promptly pay that excess amount to Sponsor}; provided that Contractor may first use some or all of the excess to make up any delay in the schedule caused by the Vendor. The amount that Contractor recovers from a Vendor shall include all value that Contractor obtains the right to receive, directly or indirectly, including cash, property, discounts on future purchases, credits, equipment and parts.]

9.2 Performance Guarantees and Facility Buy Down Amounts.

9.2.1   The Performance Guarantees and Impaired Performance Liquidated Damages for impaired performance are set forth in Exhibit 9.2.1.

9.2.2   [The Performance Guarantees relating to availability are set forth in Exhibit 9.2.2.]

9.3 [Escrow Arrangement]

Should the Contractor so request at any time, an escrow agreement in the form of Exhibit 9.3 shall be signed by the Parties and an escrow agent agreed upon that will execute such escrow agreement as well, and thereupon any payments which come due under this Article 9 after the time the escrow agent executes such escrow agreement must be deposited into such escrow established thereunder until resolution of any Dispute thereover].[148]

ARTICLE 10. CONTRACT PRICE, PAYMENT AND CREDIT SUPPORT[149]

10.1 Contract Price.[150] [151]

10.1.1 As full consideration to the Contractor for the full and complete performance of the Work, including the Preliminary Services and all costs incurred in connection with all of the foregoing, the Sponsor shall pay, and the Contractor shall accept, the sum of <<*insert amount and currency*>> (the "Base Price"), [unless the Work Commencement Date occurs after <<*insert calendar date*>>, in which case such amount will be multiplied by the Escalator on the Work Commencement Date to yield the new sum,] as such sum may be adjusted [in this Article 10.1.1] [and for the Cost Plus Work Amount] [or pursuant the election of the Additional Facilities Option for the Additional Facilities Price by any Price Adjustment incorporated into a Change Order] (as so [escalated or] adjusted, the "Contract Price").

10.1.2   [The Sponsor shall make an advance payment, as an interest-free loan for mobilization and design, when the Contractor posts an Advance Payment LC. The advance payment shall be repaid by the Contractor through <<*equal*>> deductions in the Milestone Payments.]

10.1.3   [At any time and from time to time, the Sponsor may upon written notice to the Contractor elect to assign the full responsibility for payment of all or any portion of the unpaid price of any or all of the Assigned Agreements to the Contractor, at which time:

(a) Contractor agrees to assume full responsibility therefor and indemnify the Sponsor from any claims by Vendors related thereto except with respect to the payment of any portion of the price of the Assigned Agreements not assigned to the Contractor for which the Sponsor will indemnify the Contractor; and
(b) the Contract Price will be increased by {<<*insert factor*>>}[152] multiplied by {the price of the Assigned Agreements assigned to the Contractor to be paid by the Sponsor to the Contractor in a manner such that the Contractor can meet its payment obligations to Vendors in respect of the price of the Assigned Agreements on a timely basis but not billed by the Contractor before such come due thereunder.}]
(c) [The Contractor acknowledges that the payment by the Sponsor of the sum of <<*insert currency and amount*>> under the Assigned Agreements constitutes part of the payment of the Contract Price and all of such amounts are hereby deemed paid against the Contract Price and the amount payable under the Contract Price is hereby paid in the amount of such sum. Any amounts paid by the Sponsor under the Assigned Agreements during the period between the Contact Date and the Work Commencement Date shall also be deemed payment of the Contract Price and shall be credited against the first Milestone Payment to be paid hereunder to serve to reduce the amount due in connection

therewith.] [The Contractor will, upon the request of the Sponsor, furnish a breakdown of the Equipment and the other components of the Work, and the prices thereof, to be used for the Sponsor's taxation and accounting purposes.] [In addition to the Contract Price, the Contractor may invoice on a current basis, and Sponsor will pay, any non-Contractor Taxes which the Contractor must collect from the Sponsor, because such are not covered by any tax exemption certificate delivered to the Contractor.] [A list of the Equipment and other components of the Work and the price for each used in the calculation of the Contract Price is shown in Exhibit 10.1.3(c).][153,154] [If the Sponsor has elected the Additional Scope Option, the sum of <<*insert currency and amount*>> shall be added to the Base Price.]

10.1.4   If the Sponsor has directed a change pursuant to Article 7, the portion of the cost therefor which must be paid to non-affiliates of the Contractor shall not be added to the Contract Price but shall be <paid> <denominated in <<*insert currency*>> (and paid to the Contractor in up to <<*insert number*>> percent <<*insert currency*>> converted at the Exchange Rate on the date of payment if so requested by the Contractor)> to the Contractor on a time and material basis therefor until any Adjustment related thereto is resolved, and the Sponsor shall pay any such amounts due pursuant to invoices from the Contractor which shall be rendered no more often than monthly.[155] Time and materials Work shall be performed on a cost plus basis where the Contractor's compensation for performing the Work shall not exceed the Contractor's reasonable costs plus markup, all as set forth below.

(a) **Direct Labor Costs**—For Work self-performed by the Contractor, actual costs based on the build-up labor rate sheets shown in Exhibit 1A for all craft directly involved in the Work, including foremen and below. The markup for direct labor costs shall not exceed <<*insert number*>> of the straight-time base wage rate shown in Exhibit 1A.

(b) **Indirect Labor Costs**—Actual costs based on the built-up labor rate sheets shown in Exhibit 1A for all craft labor indirectly involved in the Work and Exhibit 1A for all non-manual supervision or professional services indirectly involved in the Work. The Contractor shall only be compensated for indirect labor costs associated with an extension of time to complete the Work or an increase in either manual or non-manual staffing to complete the Work.

(c) **Subcontractor Costs**—Actual and reasonable costs for Subcontracted labor or material. The markup for Subcontractor costs shall not exceed <<*insert number*>> percent. The Subcontractor's markup on their direct labor costs shall not exceed <<*insert number*>> percent of the straight-time base wage rate shown in Exhibit 1A.

(d) **Small Tools and Consumables**—Small tools and consumables are tools and consumable items with a purchase value of <<*insert number*>> or less. The Contractor's total compensation for these shall not exceed <<*insert currency and amount*>> per hour for each direct craft man-hour worked and for each indirect general foreman man-hour worked for each estimated direct craft man-hour and indirect general foreman man-hour.

(e) **Equipment for Maintenance and Protection of Traffic** ("MPT")—The Contractor shall be compensated for its actual cost of equipment for MPT. The markup for MPT shall not exceed <<*insert number*>> percent.

(f) Equipment Rentals—Actual costs based on the equipment real rates shown in Exhibit 1A. For equipment not covered by Exhibit 1A, the Contractor's actual costs shall apply. The market for equipment rentals shall not exceed <<*insert number*>> percent.

The markups shown in paragraphs (a) through (f) above include fees, overhead and profit. Cost shall mean the lowest price that a buyer, comparable to the Sponsor, can obtain after using best efforts to obtain the lowest price for such, provided that labor costs shall never exceed those set forth in Exhibit 1A. Said costs shall be tracked independently and reported to the Sponsor on a monthly basis and when performing this Work, the Contractor shall submit daily timesheets to the Sponsor for review and approval.

10.1.5 [Irrespective of what Consortium Member furnished the Work in question, the Sponsor's liability for payment of the portion of the Contract Price with respect to such Work shall be extinguished when such payment is made to the Consortium Member submitting the Payment Application for such Work, and the other Consortium Member[s] shall not be entitled to submit a Payment Application with respect to Work.][156]

10.1.6 The Contract Price shall not be subject to change for any reason except as expressly provided in this Article 10.1.6. Payment of the Contract Price by the Sponsor shall constitute full and complete compensation to the Contractor for all of the Work to be performed hereunder. Except as provided herein[157] [for the <Site> <Route> <<Assessment>> <<*insert appropriate work to be paid for by Sponsor irrespective of whether Notice to Proceed is given*>>] the Sponsor shall have no obligation to compensate or otherwise reimburse, and shall incur no liability of any type to the Contractor or any Subcontractor under this Contract or otherwise, unless and until the [Unit One] Work Commencement Date shall have occurred and the Sponsor has received a Performance LC meeting the requirements of this Article 10 and a duly executed original of <each> <the> Parent Guarantee. The Contractor shall not be entitled to receive aggregate payments in excess of the Contract Price as compensation for Work performed.[158] The costs of the Preliminary Services are included in the Contract Price, and no Adjustment shall be granted as a result of the Contractor's performance of the Preliminary Services. [Any and all payments made to the Contractor in respect of the Preliminary Services shall be considered payments on account of the Contract Price, and shall be credited against, and reduce, the first payment or payments required to be made by the Sponsor hereunder.][159] {If the Work Commencement Date has not occurred within <<*insert number*>> Days after the Contract Date, and the Contractor has properly performed and delivered the <Site> <Route> Assessment to the Sponsor, Sponsor shall pay the Contractor for its actual cost of the <Site> <Route> Assessment to the Sponsor up to <<*insert currency and amount*>> upon the Contractor's invoice therefor (which payment by the Sponsor shall be credited against the Contract Price).}

10.1.7 The Contract Price includes all Contractor Taxes. [{<< *insert country* >> VAT [for construction contracts on a lump sum, turnkey basis, which is currently <<*insert percentage*>>percent], shall be added to any << *insert currency* >> amount invoiced by <<*insert name of Governmental Authority*>>}.] Without regard to the passage of title under

Article 14, the Contractor shall be liable for, and shall hold the Sponsor harmless against, and indemnify the Sponsor for, any Contractor Tax. [The Sponsor shall pay (and indemnify and hold Contractor harmless against) or obtain an exemption from sales taxes due to <<*insert name of Governmental Authority*>>.] The Sponsor shall provide to the Contractor an exemption certificate with regard to all personal property, materials and taxable services not less than <<*insert number* >> Days before the Contractor purchases any personal property, materials or taxable services exempt by applicable Law. In all circumstances in which the Contractor has relied upon Sponsor's exemption certificate, the Contractor shall be reimbursed by the Sponsor for all documented sales or use taxes, interest and penalties that are collected from the Contractor with respect to such personal property. The Contractor shall pay all payroll and other related employment and compensation taxes for the Contractor's employees, and all federal, state and other taxes which may be assessed on the Contractor's income from the Work, as well as any and all engineering and business license costs, excise, and other similar taxes which may be assessed on the Work. The Contract Price includes sales and use tax on all equipment, material, building supplies, and tool and equipment rentals, but specifically excludes <<*insert Governmental Authority and/or country*>> sales and use tax on machinery, apparatus, or equipment, including related materials to be installed under this Contract. The Contract Price shall include all sales and use taxes [other than sales and use tax for purchases covered by the Sponsor's tax exemptions),] excises, customs and import duties, charges and levies, assessments or other charges of any kind levied by any Governmental Authority on or because of the Work, or any item thereof, or on or because of Contractor's income in the performance of this Contract, and Contractor shall be responsible for all foreign, <<*insert country*>> federal, state, city and other local income, license, net worth, privilege, personal, gross receipts or any other taxes arising out of or related to this Contract or the performance of the Work. Contractor is not responsible for any <<*insert name of Governmental Authority*>> sales and use taxes for permanent plant production machinery and equipment, except in the event and to the extent Sponsor is required to pay any such taxes as a result of Contractor's failure to comply with Sponsor's correct written instructions regarding tax exemptions and credits, which, if followed, would be effective to relieve Sponsor of liability for such taxes.

10.2 Spare Parts[160]

All spare Parts which are, in the opinion of the Sponsor, necessary for operation and maintenance of the Facility from the Facility Provisional Acceptance Date until the <<*insert number*>> anniversary thereof (a partial list thereof is furnished under the heading "Operational Spares" in Exhibit 5.29.1 and which list shall be updated promptly by the Contractor from time to time as the Contractor's selection of Vendors continues) are included in the Work and the Contract Price, and must be [on Site] [at the <<*insert name*>> depot] by the Facility Provisional Acceptance Date[161] [except that any spare Parts used by the Sponsor pursuant to operation under Article 8.13 need not be replaced by the Contractor]. The Contractor agrees to make available for purchase by the Sponsor at the [price of each Part set forth in Exhibit 5.29.1{plus escalation of <<*insert number*>> percent}] and for delivery no later than the Facility Provisional Acceptance Date each spare Part listed under the heading

"Strategic Spares" in Exhibit 4.29.1, until <<*insert number*>> Days, after the Work Commencement Date. If the Contractor uses any of the Sponsor's spare Parts, the Contractor, at its expense, shall replenish such spare Part within <<*insert number*>> Days, or as soon as practicable if such spare Part is not readily available within such <<*insert number*>> Day period.

10.3 Milestone Payments[162]

10.3.1   Not more than once per month during the performance of the Work and after the [Unit One] Work Commencement Date, if such event occurs, [each Consortium Member] the Contractor may [if the other Consortium Member is simultaneously submitting][163] submit to the Sponsor a properly completed and detailed Payment Application for the Milestone Payment set forth in the Milestone Payment Schedule, upon reaching each milestone listed in the Milestone Payment Schedule (other than for Work for which the Contractor has already been paid), together with a duly executed Affidavit of Payment and Partial Release of Claims for Payment in the form of Exhibit 5.10.1(d)A, with respect to the previous Milestone Payment made by the Sponsor, and such documentation as the Sponsor may require to substantiate the Contractor's progress; provided, however, that the Sponsor shall not be obligated to make payment in respect of any milestone on a date which is earlier than the calendar date, if any, set forth next to such milestone on the Milestone Payment Schedule.

10.3.2   For the avoidance of doubt, no Milestone Payment shall be due until all the items required to be achieved therefor have been achieved.

10.3.3   Within <<*insert number*>> Days after the Sponsor's receipt of a Payment Application, the Sponsor shall determine:

(a) whether the Work has been done as described by the Contractor;
(b) whether the Work performed conforms with the requirements of this Contract;
(c) whether the Payment Application has been properly submitted with a duly executed Affidavit of Payment and Partial Release of Claims for Payment, in the form of Exhibit 5.10.1(d)A, with respect to the previous Milestone Payment made by the Sponsor; and
(d) the amount due to the Contractor therefor.

10.3.4   Subject to such favorable determination or adjustments by the Sponsor, and except for disputed portions of any Payment Application, within <<*insert number*>> Days after the Sponsor's receipt of Payment Application, the Sponsor shall pay the Contractor the difference between:

(a) the total amount as determined pursuant to this Article 10.3.3; minus
(b) the Retainage[164] relating to any amount of the Contract Price; minus
(c) any amounts withheld under Article 10.5.

[Contractor acknowledges and agrees that the Financing Parties' Engineer has to approve payment of Contractor's invoices by Sponsor. Any payment which the

Financing Parties' Engineer refuses to authorize as a result of defective, deficient or incomplete Work shall be deferred until it is agreed by the Sponsor[, the Financing Parties' Engineer] and the Contractor that the Work for which such payment is due has been completed and/or corrected properly. If it is subsequently determined that any such deferred payment should have been made earlier, Contractor shall be entitled to receive interest at the Delayed Payment Rate on all such deferred amounts from the date payment should have been paid to and including the date of payment. Sponsor shall promptly inform Contractor of the reason which the Financing Parties' Engineer refuses to authorize payment of Contractor's invoice].

10.3.5   If the Contractor delivers to the Sponsor a letter of credit in the form of Exhibit 10.3.5 (a "Retainage LC"), issued by an Acceptable Issuer, and which expires no earlier than <<*thirty (30)*>> Days after the Facility Final Acceptance Date, the Sponsor must refrain from withholding Retainage in the form of cash, as permitted in Article 10.3.3, in an amount equal to the available and undrawn amount of the Retainage LC.

10.3.6   The Sponsor shall notify the Contractor of the reason that the Sponsor is withholding any portion of a Milestone Payment pursuant to Article 10.5. Upon receipt of such notice, the Contractor shall, at its sole expense, promptly take any and all steps necessary to remedy any condition identified by the Sponsor as the basis for its withholding payment, including the bonding of Liens. The Sponsor shall pay the disputed portion of the Payment Application within <<*insert number*>>[165] Business Days following any agreed upon written resolution of the Sponsor's claims.

10.4 Payment of the Retainage.[166]

10.4.1 The Sponsor shall pay the Retainage to the Contractor within <<*insert number*>> Days after the last to occur of:

(a) the Facility Final Acceptance Date;
(b) the resolution of all items for which the Sponsor withheld payments under Article 10.5;
(c) [the delivery to the Sponsor of an effective Warranty LC;] and
(d) execution of the Affidavit of Payment and Final Release by the Contractor in the form of Exhibit 5.10.1(d)B.

10.4.2   The Retainage LC may be drawn upon by the Sponsor to protect or reimburse the Sponsor, and such funds may be held or applied by the Sponsor:

(a) for the Contractor's noncompliance with any provisions of this Contract; or
(b) if the Retainage LC is set to expire within the next <<*thirty (30)*>> Days, and the Facility Final Acceptance Date has not occurred.

10.5 Payments Withheld.[167]

10.5.1   The Sponsor may withhold payment, in addition to the Retainage, in respect of any Milestone Payment or any other amount due to the Contractor hereunder, in an amount and to such extent as may be necessary to protect the Sponsor from loss as a result of:

(a) defective Work not remedied in accordance with this Contract, including any Punch List items that are incomplete;
(b) the Contractor's failure to comply with any General Warranty;
(c) the Contractor's failure to perform Work in accordance with this Contract;
(d) Liens against the Facility or the Site;
(e) claims, suits, stop notices, attachments, levies or Liens against the Facility, the Work or either Party;
(f) damage to the Sponsor or any Subcontractor that results from the Contractor's failure to obtain or maintain insurance on the terms required to be maintained by it hereunder;
(g) costs incurred by the Sponsor as a result of the Contractor's failure to coordinate adequately with the Sponsor or other contractor with regard to the Work, or Contractor's delay in carrying out work previously scheduled in coordination with the Sponsor or another Contractor;
(h) the Contractor's failure to provide, on a timely basis, the documentation required under Article 5;
(i) the Contractor's failure to pay, when due, any Liquidated Damages;
(j) claims filed or anticipated to be filed by any Subcontractor that has not paid {in accordance with its Subcontract furnished to the Sponsor};
(k) the Contractor's failure to make proper payment for any materials or labor or other obligations incurred as a result of activities carried out under this Contract;
(l) [the cost to complete the Work as determined by <<*the Sponsor*>> exceeding the unpaid balance of the Contract Price];[168]
(m) reasonable evidence that any prior Milestone Payment (together with the previously requested amounts) exceeds the amount payable with respect to Work actually performed, as determined in accordance with the Milestone Payment Schedule; or
(n) the Sponsor's estimate of the cost to complete any work on the Punch List using contractors other than the Contractor exceeding the sum of twice the cost of such Work listed therefor on the Punch Lists.

10.5.2 The Sponsor may apply any funds withheld (including the Retainage) or moneys to become due to the Contractor to satisfy, discharge or secure the release of any Claims [{that the Contractor is not faithfully defending}]. Any such application shall be deemed payment to the Contractor. Any additional expense incurred by the Sponsor as a result of the Contractor's default hereunder shall be deducted from the Contract Price. No action by either Party during the above activities shall serve as a basis to make an Adjustment.

10.5.3 The Sponsor may withhold from the Milestone Payment for Facility Provisional Acceptance the amount of any Impaired Performance Damages for which the Contractor is liable hereunder.

10.6 Payment of Subcontractors

The Contractor shall promptly pay, or cause to be paid, each Subcontractor, materialman and employee the amount to which such Person is entitled. The Contractor shall promptly notify the Sponsor of any dispute with, or claim by (relating to payment or otherwise), any Subcontractor or any other Person furnishing labor, materials, services or equipment included in the Work {if such exceeds <<*insert currency and amount*>>}. The Contractor shall promptly pay the undisputed portions of such claims.

10.7 Right to Verify Progress[169]

In order to verify the progress made by the Contractor in connection with making Milestone Payments, the Sponsor [and representatives of the Financing Parties] may inspect the Work [{in accordance with the provisions of this Contract}].

10.8 Disputed Payments[170]

If there is any dispute about any amount due to the Contractor, the amount not in dispute shall be promptly paid to the Contractor, as described in Article 10.3.3, and any disputed amount that is ultimately determined to have been payable shall be paid with interest, computed at the Delayed Payment Rate from the date such was due until the date of payment.

10.9 Payments Not Acceptance of Work

No payment made hereunder shall be considered an approval or acceptance of any Work or constitute a waiver of any claim or right that the Sponsor may have at that time or thereafter, including claims and rights relating to warranty and indemnification obligations of the Contractor. The Contractor's acceptance, by endorsement or otherwise, of final payments shall constitute a waiver of any and all claims, including, but not limited to any and all lien rights, claims, or notices of any kind, against the Sponsor or its property.

10.10 Interest

Amounts not paid by either Party to the other when due under this Contract shall bear interest from the date payment was due until and including the date of payment at the Delayed Payment Rate.

10.11 Overpayment

If either Party makes any overpayment to the other Party, the amount of such overpayment, together with interest calculated at the Delayed Payment Rate from the date of such overpayment until the date it is repaid or credited, shall be promptly repaid by such Party.

10.12 Payment Currency[171]

Unless otherwise specified herein, all payments hereunder shall be in <<*insert currency*>>. Any payments to be made hereunder by the Contractor:

(a) in <<*insert currency*>> or any currency other than <<*insert currency*>> shall be paid by <<*fill in proper Consortium Member*>>; and
(b) in <<*insert currency*>>, shall be paid by <<*fill in proper Consortium Member*>>.[172]

10.13 [Suspension of Milestone Payments[173]

The Sponsor shall not be obligated to make payment on the Contract Price so long as a Contractor Event of Default exists.]

10.14 [ECA Payments[174]

Should any ECA so require, Contractor agrees that the Sponsor may pay any portion of the Contract Price, on Contractor's behalf, directly to any Subcontractor for such Subcontractor's services or products, and the Contractor will promptly prepare and certify the nature and costs of such services or products to any ECA within <<*insert number*>> Days of the Sponsor's request therefor. Any such payment shall be deemed to be a payment on account of the Contract Price.]

10.15 [Deferred Payment[175]

10.15.1   The Contractor recognizes that breach of its agreement in Article 10.14 may cause the Work to be no longer eligible to be financed by the ECAs, and there may be a cancellation of or shortfall in the amount of financing available to the Sponsor to fund the payments due to the Contractor under the Contract, in respect of the ineligible Work (such canceled amount or shortfall shall be referred to as an "Ineligible Amount"). If and to the extent that an Ineligible Amount arises as a result of the actions or omissions of any Contractor Person, then the Contractor agrees to effect or procure any of the following remedies at the <Sponsor's> <{Contractor's}> election:

(a) the Contractor shall make available to the Sponsor, on terms and conditions acceptable to the Sponsor, loan facilities from a third party, of the same value as the Ineligible Amount, on terms to the Sponsor no less favorable than those which it would have received if the Ineligible Amount had been financed by the ECA-supported loan facilities, as originally envisaged; or
(b) the Contractor shall defer payment of an amount of the Contract Price equal to the value of the Ineligible Amount, which shall become a debt of the Sponsor to the Contractor, so that the Sponsor receives terms no less favorable than those which it would have received had such Ineligible Amount been financed by ECA loan facilities as originally documented.

10.15.2   The conversion of any deferred payments into debt, as provided in Article 10.15.1(b) above, shall be unsecured by any collateral whatsoever, and the Contractor hereby expressly waives any lien arising by the operation of law, or which it may otherwise have by reason of such subordination or deferral. To the extent and in the manner set forth herein, the payment of the principal of, and interest on, the subordinated deferral of the Contract Price (including, for all purposes of these subordinated terms, all other amounts payable on or in respect thereof) is expressly made subordinated and subject in right of payment to the prior payment in full of all obligations owed by the Sponsor to any Financing Parties (the "Senior Debt"). The Contractor agrees that it will not ask, demand, sue for, take or receive, from the Sponsor, by set-off or in any other manner, or retain payment (in whole or in part) of such subordinated deferment of the Contract Price, other than payments made at the times, in the amounts and to the extent provided, herein.

10.15.3   In the event of:

(a) any insolvency or bankruptcy case or proceeding, or any receivership, liquidation, reorganization or other similar case or proceeding, in connection therewith relative to the Sponsor or to its assets,
(b) any liquidation, dissolution or other winding up of the Sponsor, whether partial or complete and whether voluntary or involuntary and whether or not involving insolvency or bankruptcy, or
(c) any assignment for the benefit of creditors or any other marshaling of assets and liabilities of the Sponsor,

then, and in any such event, any Financing Parties shall be entitled to receive payment in full of all amounts due or to become due on or in respect of all Senior Debt before the Contractor shall be entitled to receive any payment on account of the subordinated deferral of the Contract Price (whether in respect of principal, interest, or otherwise), and to that end, any payment or distribution of any kind or character, whether in cash, property or securities, which may be payable or deliverable in respect of the subordinated deferral of the Contract Price in any such case, proceeding, dissolution, liquidation or other winding up or event shall, instead be paid or delivered directly to the Financing Parties for application first to Senior Debt.

10.15.4   The Contractor shall not, without the prior written consent of the agent for the Financing Parties:

(a) commence any proceeding against the Sponsor in bankruptcy, insolvency, receivership or similar Law; or
(b) commence enforcement proceedings with respect to, or against, or otherwise take, any collateral as security for the subordinated deferral of the Contract Price.

10.15.5   In the event that the Contractor receives, on account or in respect of the subordinated deferral of the Contract Price, any distribution of assets by the Sponsor, or payment by or on behalf of the Sponsor, of any kind or character, whether in cash, securities or other property, other than in accordance with the terms hereof, the Contractor shall hold in trust (as property of the Financing Parties) for the benefit of, and shall, immediately upon receipt thereof, pay over, or deliver, to the agent for the Financing Parties such distribution or payment in precisely the form received (except for the endorsement or assignment by the Contractor where necessary). The Contractor irrevocably waives all rights of subrogation or similar rights under applicable Law that may arise by virtue, or in connection with any payment by the Contractor, pursuant to, these subordination provisions. Interest shall accrue on the amount of the deferred Contract Price at the interest rate which would have otherwise been payable by the Sponsor on the Ineligible Amount had the Contractor not violated its agreement in Article 10.12. The Contractor shall enter into such subordination agreements or other agreements requested by the Sponsor or any Financing Party to confirm and elaborate upon the provisions of this Article 10.15.]

10.16 <Performance Bond><Performance LC>

No later than <<*insert number*>> Business Days after the <Contract> <Work Commencement> Date, the Contractor shall deliver to the Sponsor a Letter of Credit in substantially the form of Exhibit 10.16 (the "Performance Bond") posted by an Acceptable Bondsman, and which shall expire no earlier than <<*thirty (30)*>> Days after the Facility Final Acceptance Date. The cost of obtaining and maintaining the Performance Bond shall be borne by the Contractor. If any bondsman that has issued a Performance Bond fails to be an Acceptable Bondsman, the Contractor shall, within <<*insert number*>> Days of such failure, cause another Performance Bond to be posted by an Acceptable Bondsman, or the Sponsor may call such Performance Bond, and a Contractor Event of Default will be deemed to exist hereunder.>[176] [177] No later than <<*insert number*>> Business Days after receiving written notice to do so from the Sponsor {but in no event sooner than <<*insert number*>> Days after the <Contract> <Work Commencement>Date}, the Contractor shall deliver to the Sponsor an irrevocable standby letter of credit in substantially the form of Exhibit 10.16 (the "Performance LC") executed by an Acceptable Issuer, in a stated amount equal to <<*insert currency and amount*>>) which shall expire no earlier than <<*thirty (30)*>> Days after the end of the Warranty Notification Period, provided that, if, after the Facility Final Acceptance Date, the Contractor delivers to the Sponsor an effective Warranty LC, which expires no earlier than <<*thirty (30)*>> Days after the end of the Warranty Notification Period, issued by an Acceptable Issuer in an amount not less than <<*insert currency and amount*>>) the Sponsor shall return the Performance LC to the Contractor. The cost of obtaining and maintaining such Performance LC and the Warranty LC, if any, shall be borne by the Contractor. The Performance LC and/or the Warranty LC may be drawn upon by the Sponsor at any time and from time to time in the case the Contractor has not fulfilled any of its obligations hereunder and proceeds thereof may be applied by Sponsor as it sees fit.

10.17 Parent Guarantee[178]

Concurrently with the execution and delivery of this Contract, the Contractor shall deliver to the Sponsor a guarantee executed by [each of] the Parent Guarantor[s][179] in substantially the form of Exhibit 10.17 (the "Parent Guarantee").

10.18 [Sponsor Parent Guarantee

On the Work Commencement Date, the Sponsor shall deliver to the Contractor a guarantee executed by the Sponsor's parent guarantor in substantially a form to be agreed (the "Sponsor Parent Guarantee"). At its option, the Sponsor may substitute the Sponsor Parent Guarantee with an irrevocable standby letter of credit in a stated amount equal to <<*insert currency and amount*>> to support payment of the Contract Price until such time as Milestone Payment number <<*insert number*>> has been paid and thereafter such may be automatically reduced to a stated amount equal to <<*insert currency and amount*>>.[180]]

ARTICLE 11. FORCE MAJEURE, HAZARDOUS MATERIALS AND SITE CONDITIONS

11.1 Force Majeure Excused Events[181]

11.1.1   If a Party (the "Asserting Party") is rendered wholly or partially unable to perform its obligations under this Contract as a result of a Force Majeure Excused Event, the Asserting Party shall be excused from whatever performance is affected by such Force Majeure Excused Event to the extent so affected, provided that:

(a) the Asserting Party shall <immediately> <{within a reasonable period}> notify the other party of its inability to perform as a result of such Force Majeure Excused Event, which notice shall be confirmed in writing in detail as soon as possible but in any event within <<*insert number*>> Days of the occurrence of such Force Majeure Excused Event and which notice shall contain reasonable proof of the nature, anticipated duration and expected impact on compliance with this Contract of such Force Majeure Excused Event;
(b) the excusal from performance shall be of no greater scope and of no longer duration than that caused by such Force Majeure Excused Event; and
(c) no obligation of the Asserting Party that arose prior to such Force Majeure Event shall be excused as a result of such Force Majeure Excused Event.

11.1.2   The Asserting Party shall, at its own expense, use <its best> <{commercially reasonable}> efforts to mitigate and minimize the effects of the impact and consequence of any Force Majeure Excused Event and to resume its performance under this Contract upon the cessation of such Force Majeure Excused Event.[182] Upon the Asserting Party's failure to take all necessary and appropriate steps to remove the cause(s) and/or mitigate the effects of a Force Majeure Excused Event, the other Party, in its sole discretion and upon written notice to the Asserting Party (if such notice is practical), at the Asserting Party's expense, may initiate any measures in order to alleviate the causes(s) or mitigate the effects of any Force Majeure Excused Event. Such measures may include the hiring of Persons to take the actions that the other Party in its sole [opinion] deems appropriate or necessary. Upon the termination of the circumstances preventing performance created by such Force Majeure Excused Event, the Asserting Party shall resume performance of its excused obligations hereunder.

11.1.3   The Asserting Party shall provide to the other Party written reports:

(a) by the <<*insert number*>> Day of each month during a Force Majeure Excused Event describing the effect thereof on the performance of its obligations here-under and its plan to overcome and "work around" the consequences thereof (including preparation of a contingency plan therefor); and
(b) notifying it of the cessation of such Force Majeure Excused Event.

11.1.4   If the parties do not agree upon the occurrence of a Force Majeure Excused Event, the burden of proof of such occurrence of a Force Majeure Excused Event shall rest with the Asserting Party.

11.1.5   If the Contractor is the Asserting Party, the Contractor shall be entitled to submit Adjustment Claims for Time Adjustments in respect of Force Majeure Excused Events, provided that all requests for Time Adjustments are submitted within <<*insert number*>> Days after the cessation of the Force Majeure Excused Event and are accompanied by detailed documentation of the Force Majeure Excused Event in question and the Contractor's plan to reschedule its performance of the Work that was delayed by such Force Majeure Excused Event. Any Adjustment Claim for a Time Adjustment based upon a Force Majeure Excused Event which:

(a) occurred outside of <<*insert country or region*>>; or
(b) did not result in a loss for which compensation is available under the insurance maintained under Article 15.7 or Article 15.8;

shall <only be granted to the extent the Sponsor receives time relief therefor under the Offtake Agreement> <not be granted>.

11.1.6   If the Contractor is the Asserting Party, the Contractor may [submit an Adjustment Claim for a Price Adjustment only for its additional costs arising out of its excusal from performance or stoppage of work as a result of a Force Majeure Excused Event. The Contractor may not submit an Adjustment Claim for a Price Adjustment in connection with any costs it incurs to satisfy its obligations under Article 14.2 as a result of a Force Majeure Excused Event.[183] Any Price Adjustment to which Contractor is entitled as a result of a Force Majeure Excused Event shall only be paid by the Sponsor to the extent the Sponsor receives monetary relief therefor from the Offtaker under the Offtake Agreement] [not submit an Adjustment Claim for a Price Adjustment under any circumstances ever].[184]

11.2 <Site> <Route> Conditions[185]

The Sponsor shall have no responsibility with respect to ascertaining for the Contractor facts concerning physical or other characteristics or constraints at [on] the <Site> <Route>. Borings, test excavations and other sub-surface investigations, if any, made by the Sponsor prior to the start of the Work, the records of which may be available to the Contractor, are made available solely for informational pur-poses. Any such borings, test excavations and other sub-surface investigations are not warranted to show the actual sub-surface conditions and the actual conditions encountered may not conform to those indicated by said borings, test excavations and other sub-surface investigations. The Sponsor shall not be liable for any mis-interpretation or misunderstanding of this Contract on the part of the Contractor, or for any failure by the Contractor to acquaint itself fully with all circumstances relating to the Work. The Sponsor may provide the Contractor, for informational purposes only, certain documents purported to describe potential sub-surface con-ditions <at> <on> the <Site> <Route> [and the Existing Plant]. [Except as provided in Article 11.3,] [{and the next sentence},]<The> <the> Contractor shall not be entitled to any Adjustment as a result of the existence of any <Site> <Route> con-dition unless it makes such Adjustment Claim within <<insert number>> Days of the Contract Date.[186] [{The Contractor shall be entitled to an Adjustment as a result of the existence of any impediment, obstacle (including riverbed debris) or otherwise disruptive site or sub-surface condition (including, but not limited to, geological, seismic, hydrological, topographical or otherwise), on the Site which was in exist-ence as of the Contract Date ("Differing Site Conditions"). With regard to Differing Site Conditions, the Sponsor's maximum liability with respect to an Adjustment for Differing Site Conditions is as set forth below:

On Land—the Sponsor will pay only the first <<insert currency and amount>> of cost related thereto [provided that with respect to piling the Sponsor shall not be obli-gated to pay for more than <<insert number>> piles at <<insert depth>> to be paid based upon the Unit costs set forth in Exhibit 1A].

In Water – the Sponsor will pay only the first <<insert currency and amount>> of costs related thereto; and the Contractor shall absorb any costs thereabove in both cases}.]

Except as provided in this Article 11.2 or in Article 11.3, the Contractor shall not be entitled to any Adjustment as a result of the existence of any <[Site]> <[Route]> condition noted in this Contract nor shall the existence of any [{usual apparent}] <Site> <Route> condition be considered a Force Majeure Excused Event. Except as a result of an earthquake, in excess of the magnitude set forth in the Functional Specifications as such is measured at the nearest seismic station to the Site without being adjusted to take in to account the distance therefrom to the Site (a "Threshold Earthquake"),[187] the Contractor shall make good at its own expense any damage or loss to the Facility occurring as a result of any surface or sub-surface condition which changes or develops before the end of the Warranty Notification Period.

11.3 Hazardous Materials[188]

11.3.1 Immediately after discovering any material on the <Site> <Route> believed or considered to be a Hazardous Material, or learning of the use, existence or stor-age of a Hazardous Material which does not conform to, or is not conducted accord-ing to the standards prescribed under applicable Laws, the Contractor shall stop all Work involving such Hazardous Material and any area affected thereby, and promptly report, in writing, the condition to the Sponsor before further disturbing such Hazardous Material.

11.3.2 The Contractor shall be responsible for recommending to the Sponsor the legal alternatives for removal and disposal of such Hazardous Material, identification of disposal sites in the area, and shall assist the Sponsor in developing and implement-ing a health and safety plan for the removal and disposal of such Hazardous Material, including obtaining Permits (if necessary) in its or the Sponsor's name. The Sponsor shall review such recommendations and direct the Contractor which, if any, to implement, and all of such implementation shall be at the Sponsor's expense. The Contractor shall provide to the Sponsor copies of all documents relating to such Hazardous Material, including hazardous waste manifests, approvals and/or authorizations from landfill or other treatment, storage or disposal facilities and all laboratory data.[189]

11.3.3   The Contractor shall, upon Sponsor's request, <{consider negotiation of}> <enter into> a separate agreement with the Sponsor to perform the necessary clean up, removal and disposition services with respect to such Hazardous Materials.

11.3.4   The Work in the affected area shall resume only by written direction of the Sponsor.

11.3.5   If any Hazardous Material was brought onto the <Site> <Route> by any Contractor Person, the Contractor shall:

(a) be responsible for the removal of such Hazardous Materials at its expense in compliance with all applicable Laws and without further written direction from the Sponsor; and
(b) not be entitled to an Adjustment, and such shall not be considered a Force Majeure Excused Event.

11.3.6   The Contractor may submit an Adjustment Claim in connection with its discovery and disposal of such Hazardous Materials if:

(a) such Hazardous Materials were not disclosed in, or were not or could not reasonably have been discovered by the Contractor or the relevant Subcontractor during its investigation of the <Site> <Route> for the <Site> <Route> Assessment;
(b) such Hazardous Materials were not brought onto the <Site> <Route> by any Contractor Person;
(c) the presence of such Hazardous Materials has resulted in an increase in the Contractor's cost of any Work, or the time required for any Critical Path Item; and
(d) the Contractor has performed all its obligations under this paragraph.

11.3.7   If any condition described in paragraph (a), (b), (c) or (d) above is not met, the Contractor shall:

(a) be responsible for the removal of such Hazardous Materials at its expense; and
(b) not be entitled to an Adjustment, and no Force Majeure Excused Event shall be deemed to have arisen.

11.4 Man-Made <Site> <Route> Conditions

In the event that, while carrying out the Work at the <Site> <Route>, the Contractor discovers a man-made structure that could not [{reasonably}] have been expected to become known to the Contractor at the Contract Date through a diligent review of available resources and not described in the Sponsor Information or contained in documents delivered to the Contractor, it shall immediately notify the Sponsor and provide written notice to the Sponsor within <<*insert number*>> Days of such discovery. The Contractor shall use its <best> <{commercially reasonable}> efforts to continue the performance of the Work notwithstanding the effects of such discovery. If such discovery adversely affects the Contractor's ability to perform the Work, such shall be considered a Force Majeure Excused Event, and the Contractor, subject to Article 19, shall be entitled to an Adjustment, provided that the Contractor has provided to the Sponsor the notices required by the first sentence of this paragraph, within the time periods required thereby and the provisions of Article 11.4.

ARTICLE 12. REPRESENTATIONS AND WARRANTIES[190]

12.1 Mutual Representations and Warranties[191]

<The Contractor> <Each Consortium Member> hereby represents and warrants to the Sponsor, and the Sponsor hereby represents and warrants to the Contractor that:

(a) it is duly organized and validly existing under the laws of its jurisdiction of incorporation and is qualified to do business in each jurisdiction in which the nature of the business conducted by it makes such qualification necessary, and has all requisite legal power and authority to carry on its business and to execute this Contract and perform the terms, conditions and provisions hereof;

(b) the execution, delivery and performance by it of this Contract have been duly authorized by all requisite action of its governing bodies;

(c) this Contract constitutes the legal, valid and binding obligation of it, enforceable in accordance with the terms hereof;

(d) neither the execution nor delivery nor performance by it of this Contract nor the consummation of the transactions contemplated hereby will result in:
   (i) the violation of, or a conflict with, any provision of the organizational documents of it;
   (ii) the contravention or breach of, or a default under, any term or provision of any indenture, contract, agreement or instrument to which it is a party or by which it or any of its property may be bound; or
   (iii) the violation by it of any applicable Law;

(e) it is not in violation of any applicable Laws; and

(f) there is no action, suit or proceeding now pending or (to the best of its knowledge) threatened against it (or any Subcontractor, in the case of the Contractor) before any court or administrative body or arbitral tribunal that could reasonably be expected to adversely affect the ability of such Party (or any Subcontractor, in the case of the Contractor) to perform its obligations hereunder (or under any Subcontract).

12.2 Contractor's Representations and Warranties.

The Contractor represents, warrants and covenants that:

(a) all of the representations and warranties made by the Parent Guarantor[s] made in [each] [the] Parent Guarantee are true and correct;

(b) [the statements and information contained in its responses to the Bidding Documents are true;]

(c) no officer or employee of any Government Authority is or shall become interested directly or indirectly as a contracting party, partner, stockholder, surety, or otherwise, in this bid or in the performance of the Contract to which it relates, or in any portion of the profits thereof;

(d) no principal, general partner, employee, agent or affiliate of the Sponsor has received (or is expecting to receive) any compensation, directly or indirectly, or has (or will have) any financial interest as a contracting party, partner, stockholder, surety or otherwise in this Contract or in any portion of the profit hereof;

(e) it has or will be the holder of all Permits required to allow it to operate or conduct its business as contemplated hereby and, to the best of its knowledge after due inquiry, it has no reason to believe that any Permit not obtained by it by the Contract Date will not be readily obtained upon due application therefor;

(f) it has full experience and proper qualifications to perform the Work and to construct the Facility, and its designers and design Subcontractors, including all those involved in the preparation of the preliminary and design, all have the engineering and design experience and capability necessary for the design of the Work;

(g) it and each Contractor Person has ascertained the nature and location of the Work, the character and accessibility of the <Site> <Route> and its surrounding areas, availability of lay down areas for Equipment and tools, the existence of all [{visually apparent}][192] obstacles to construction (including underground obstacles, if any, referred to in this Contract), the availability of facilities and utilities, the location and character of existing or adjacent work or structures, the conditions of roads, waterways and railroads in the vicinity of the <Site> <Route> and in <<*insert country*>>, including the conditions affecting shipping and transportation (such as the limitations of bridges, tunnels and turning radii along the transportation route of equipment (all of which have been revealed in a transportation and traffic study by an experienced and expert Subcontractor which is attached as Exhibit 12.2(g) and which does not suggest removal or relocation of any facilities or obstacles whatsoever)), access, disposal, handling and storage of materials, the surface conditions and other general and local conditions, including those involving labor, safety, weather, the environment, geology, water supply, water quality, and all other matters that might affect its performance of the Work or its costs or the construction of the Facility;[193]

(h) it and each Contractor Person has investigated the <Site> <Route> and surrounding locations necessary for completing the Work and is familiar with[{, to the extent possible and as set forth in this Contract,}] the general and local conditions with respect to environment, transportation, access, waste disposal, handling and storage of materials, availability and quality of electric power, availability and condition of roads, climatic conditions and seasons, physical conditions at the <Site> <Route> and the surrounding area as a whole, topography and ground surface conditions, location of underground utilities, and equipment and facilities needed prior to and during performance of all Contractor's obligations under this Contract (the foregoing, collectively, the "<u>Site</u> <u>Route</u> Conditions", and, for the avoidance of doubt, <Site> <Route> Conditions include wastewater, storm water and sewage disposal capacity of existing municipal systems, municipal and well water supply capacity, tunnels, caves, cavities or other structures or geological conditions of any nature);

(i)   it and each Contractor Person is, and its Subcontractors are, and shall be, at all times, fully qualified and capable of performing every phase of the Work in order to complete the Facility in accordance with the terms of this Contract;

(j)   all services provided and procedures followed by the Contractor hereunder to engineer, design, procure, construct, commission, start up and test the Facility shall be done in a workmanlike manner and in accordance with:
  (i) <Good > <{Prudent}> <Utility> <{Industry}> Practices;
  (ii) all applicable requirements of all applicable Laws in effect [or proposed] before the Facility Provisional Acceptance Date;
  (iii) all requirements of this Contract; and
  (iv) all instructions of Vendors and manufacturers of Equipment, including instructions relating to storage, erection and testing;

(k) it and each Contractor Person is familiar with all necessary facilities for delivering, handling and storing all Equipment and other parts of the Work;

(l)   it and each Contractor Person is familiar with all labor conditions and agreements relating to the performance of the Work;

(m) it and each Contractor Person will design the Facility so that the useful life thereof may [{reasonably}] be expected to exceed <<*insert number*>> years;

(n) it is not aware of any facts, conditions or events which would affect its ability to complete the Work in accordance with the Baseline Schedule or Baseline Level 3 Schedule;

(o) [neither] it [nor the Parent Guarantor[s]] [are][is] immune in any country from any judicial claim, pre-judgment attachment, attachment in aid of execution, execution of judgment or other procedural acts arising out of or in relation to this Contract;

(p) it and each Contractor Person has satisfied itself as to the means of communication with an access to and through the <Site> <Route> and accommodations it may require, the possibility of interference by Persons with access to or use of the <Site> <Route> after the Contractor shall have been given possession thereof, and the precautions, times and methods of working necessary to prevent any Contractor Person from creating any nuisance or interference, whether public or private, which might give rise to any legal action;

(q) it and each Contractor Person has thoroughly examined this Contract and all applicable Laws and has become familiar with their terms and the Laws in effect [and proposed] on the Contract Date and all Laws in existence on the Contract Date that, by their terms, became or will become effective and applicable to either Party, the Facility or the Work after the Contract Date, have been taken into account in its calculation of the Contract Price and shall be reflected in the design of the Facility and the Work;

(r) the Contractor does not discriminate against its employees or applicants for employment and is in compliance with all of the Laws against discrimination;

(s) all raw data delivered to the sponsor from the Performance Tests and of the conversion of such raw data into test results will be accurate and correct;

(t) it and each Contractor Person has ascertained all the facts concerning conditions to be found at the location of the Work, including all physical characteristics above, on and below the surface of the ground and all administrative, organizational, procedural, regulatory and other obstacles and constraints, and considered fully these and all other matters which could in any way affect the Work and made the necessary investigations relating thereto;

(u) [<it> <the Parent Guarantor[s]> [are] [is] owned by only natural Persons;>]

(v) prior to submitting any bid or proposal for the Work, it received and carefully and completely read, examined and compared all parts of this Contract including all plans, drawings, specifications and all exhibits thereto and other documents referenced herein, and found them to be complete, accurate, consistent and appropriate for the Work and that there are no defects, errors, inconsistencies or omissions, whether subtle or obvious, in this Contract, of which the Contractor is or should be aware;

(w) <it> <the Parent Guarantor[s]> [has] [have] the Required Rating; and

(x) it has inspected the <Site> <Route> and surrounding premises where the Work is to be performed, before entering into this Contract, and has satisfied itself as to the conditions under which it will be obligated to perform the Work, or any matter that could affect the Work under this Contract.

12.3 No Representation by any Sponsor Person

The Contractor acknowledges that no Sponsor Person has made or makes any representation and each of them disclaims any warranty whatsoever, whether express or implied, including with regard to the Work, the Facility, Sponsor or this Contract.

ARTICLE 13. CONTRACTOR'S WARRANTIES[194]

13.1 General Warranty

The following warranties (the "General Warranties"), shall continue during the Warranty Notification Period [unless expressly specified otherwise herein]. The Contractor hereby warrants that:

(a) for the life of the Facility, any drawings, specifications, engineering, Computer Programs and other data and documents prepared or furnished by the Contractor to the Sponsor under this Contract shall be capable of being used in connection with the Facility by the Sponsor for the life of the Facility at no charge;

(b) all Computer Programs, in whatever form or format furnished hereunder, including firmware,

(i) will execute and perform within a reasonable time all functions described for it in the Functional Specifications and in applicable user manuals or other written description of performance or function for it provided to the Sponsor by or on behalf of the Contractor or the creator or supplier thereof; and

(ii) will not contain any instructions or code which, without the intent of the Sponsor, will damage, impair or destroy, the functioning of any computer equipment, computer software, database or data storage device such as, but not limited to, code commonly known as a virus, worm, Trojan horse, time bomb, or lock out and will not allow entry or access to the Sponsor's software without the Sponsor's knowledge or consent by such devices (sometimes known as a trap door);

(c) the Equipment incorporated in the Facility and the Work shall:

   (i) be new when incorporated into the Facility;

  (ii) be fit for its purpose [{of <<*insert function*>>}] when operated in accordance with the Contractor's written instructions or in the absence thereof in accordance with <Good > <{Prudent}> <Utility> <{Industry}> Practices]; and

 (iii) be free of defects and deficiencies [[{in design, materials, construction and workmanship}] [{normal wear and tear (that can be expected of properly designed, machined, assembled and installed items made of proper material and alloys under the circumstances give expected duty) excepted}]];

(d) the Facility and each component thereof (including Computer Programs delivered to the Sponsor) shall be:

   (i) built strictly in accordance with the Specifications and Drawings, the Functional Specifications and the other requirements of this Contract and applicable Law;

  (ii) capable of operating and achieving the levels contained in the Performance Guarantees (or such levels which formed the basis for the Facility Buy Down Amounts if the Contractor paid any Facility Buy Down Amounts) [{so long as the Sponsor operates and maintains the Facility in accordance with the written and reasonable operating and maintenance manuals furnished to the Sponsor by the Contractor}].

(e) prior to the transfer of title thereof to the Sponsor, it has and will have good and marketable title to, and ownership of, all Equipment and other portions of the Work incorporated into the Facility, and title to all Equipment and other portions of the Work incorporated into the Facility shall pass to and vest in the Sponsor free and clear of any Lien or right of any Person; [any system designed or provided as part of the Work shall operate properly and dependably and be compatible with other existing or connection system(s) in the Existing Plant without negatively impacting [the] existing system and any materials provided as part of such system shall be compatible with the system and components of the Existing Plant;] and

(f) all civil structures and foundations will withstand [(i)] Threshold Earthquakes (and any soil liquefaction resulting therefrom) for a period until the <<*insert number*>> anniversary of the Warranty Period end [; and (ii) <<*insert any other weather conditions or anything else applicable*>>].

13.2 Remedies for Breach of General Warranty[195]

If, during the Warranty Notification Period or such longer period as is expressly specified herein, the Sponsor shall notify the Contractor in writing of any breach of, or non-conformity with, any General Warranty[{, then, unless in such case the Sponsor's failure to comply generally with the Warranty Stipulations has prejudiced the Contractor's right}], the Contractor shall, within <<*insert number*>> Days of such notice, correct any defect or deficiency and perform or re-perform, repair or replace, re-work or re-test, as appropriate, any Equipment and/or Work not in conformance with any of the General Warranties, and construct any changes, modifications or additions to the Facility that are dictated by <Good> < {Prudent}> <Utility> <{Industry}> Practices as a result of the failure to perform any Work or furnish the Equipment which shall be covered by the General Warranties. All costs incurred in connection with the foregoing, including the removal, replacement and reinstallation of Equipment necessary to gain access to any portion of the Facility and all other costs incurred by the Contractor as a result thereof, and all of the Sponsor's costs and expenses incurred in connection therewith [{as a result of the Sponsor's efforts to mitigate the effects of the situation or restore safety}], shall be borne solely by the Contractor and reimbursed to the Sponsor promptly upon written request therefor. The Contractor shall at its own expense search for the cause of any defect, under the direction of the Sponsor. [{The Contractor shall be excused from its warranty obligations hereunder with respect to any spare part that the Sponsor has not stored in accordance with the Contractor's reasonable written instructions therefor (provided that the Contractor supplied such written instructions to the Sponsor at the time the spare part in question arrived at the Site).}] [{In cases of breach of a General Warranty set forth in this Article 13, the Sponsor's sole remedies therefor shall be those listed in this Article 13.2 and Article 13.3 or damages for Contractor's breach of its obligations herein.}] The Sponsor shall have the right, without waiving its right to require the Contractor's compliance with the Warranties, to operate any and all Equipment and use and occupy all the Work as soon and as long as it is in operating condition whether or not such Equipment or Work has yet been accepted as complete and satisfactory by the Sponsor and/or the Contractor. [{This shall not be construed, however, to permit continued operation of Equipment which may be materially damaged by such operation before a required repair has been made.}] If the operation or use of the Equipment or Work, after installation and/or repair, is unsatisfactory to Sponsor, Sponsor shall have the right to operate and use such Equipment for such time as the Sponsor deems necessary until it can be taken out of service for repairs or replacement in whole or part by the Contractor.

13.3 Latent Defects

The Contractor warrants that the Work is free from latent defects and latent deficiencies for a period of <<*insert number*>> Days from the Provisional Acceptance Date[; provided that the Contractor shall not have any liability until an individual breach is for an item or service whose cost to repair, replace, or re-perform exceeds <<*insert currency and amount*>> and until the aggregate amount for all breaches exceeds <<*insert currency and amount*>> for all such claims for such latent defects and deficiencies and thereupon the Contractor shall have liability for all such breaches including any breaches for which it previously did not have, liability under this Article 13.3 [, {provided that the warranty for latent defects and deficiencies in any Work supplied pursuant to the <<*Vendor Agreement*>> shall not exceed the warranty for latent defects and deficiencies provided pursuant to the <<*Vendor Agreement*>>}]]. Upon discovery by the Sponsor of latent defect or latent deficiency for which a claim is to be made under this Article 13.3, the Sponsor shall promptly provide the details of the latent defect or latent deficiency to the Contractor by written notice. Any such latent defects or latent deficiencies shall be corrected and made good by the Contractor at its sole cost as prescribed in Article 13.2 and Article 13.4. By way of illustration and not exclusion, latent defects and latent deficiencies include defects or deficiencies which were in existence during the Warranty Notification Period but were not visually apparent to the naked eye during the Warranty Notification Period. In all cases, a defect or deficiency is latent if it is of a nature such that it is either (i) patent; (ii) not readily apparent upon a visual inspection by the naked eye, (iii) not discernible without disassembly of Work; or (iv) it would not reasonably be expected to manifest itself until after the Warranty Notification Period ends.

13.4 Failure to Correct Defects and Deficiencies

If the Contractor fails after <<*insert number*>> Days' written notice to correct any defect or deficiency <{in a timely manner}> within <<*insert number*>> Days, the Sponsor shall be entitled, but shall not be required, to correct (itself or through any Person) any such defect or deficiency and to apply to the costs and expenses that it incurs on account thereof, any or all Retainage then held by the Sponsor [and any amounts available under the Warranty LC], and if the Retainage then held [and the amounts under the Warranty LC] are insufficient to cover such costs and expenses, the Contractor shall pay the excess of such costs and expenses to the Sponsor within <<*insert number*>> Days following written demand therefor. [Thereafter, any Work so corrected by any Person shall be covered by the Contractor's General Warranties as if the Contractor had actually performed such remedial Work.][196] If any failure of the Facility or the Work or any portion thereof is caused by breach of or non-conformance with any General Warranty, and the <Sponsor> <{Contractor}> elects in its sole discretion not to have the non-conforming item repaired or replaced, then the Contractor shall pay to the Sponsor an amount equal to the excess of the cost of a comparable item conforming to all General Warranties minus the actual cost of the non-conforming item.[197] In the event of a Dispute regarding whether replacement or repair is required under Article 13.2 and/or Article 13.3, the onus shall be on the Contractor to prove that no defect or deficiency exists.

13.5 Subcontractor Warranties[198]

Without limiting the General Warranties, the Contractor shall, in procuring any part of Work and entering into any Subcontract, obtain from each Subcontractor a warranty that:

(a) shall be consistent with the General Warranties; and
(b) (to the extent such is available at no additional cost to Contractor) is for a longer duration than that of the General Warranties.[199]

If this Contract is terminated pursuant to Article 13, then all Subcontractor warranties concerning the Work are hereby assigned to the Sponsor to the full extent of the terms thereof, effective:

(a) as of such termination; and
(b) as of the expiration of the Warranty Notification Period in any case they are still in effect.

13.6 Contractor Principally Responsible

The Contractor shall be liable with respect to all General Warranties, whether or not any deficiency or defect is also covered by a Subcontractor warranty, and the Sponsor need only seek recourse from the Contractor for corrective action as required hereby.

13.7 Scheduling of Warranty Work[200]

The Contractor shall perform the repairs, rectification, replacements and corrective Work required hereunder in conjunction with the Sponsor's operations schedule, which may require that such Work be performed during scheduled maintenance [and outages of the Unit in question or the Facility] and/or nights, off-peak hours, weekends and holidays and Contractor must absorb all overtime charges for its work. The Contractor agrees that the Contractor will perform such work required hereunder even if the Warranty Notification Period has expired but such work is scheduled to be performed after the expiration of the Warranty Notification Period.

13.8 Acceptance or Approval.

Acceptance or approval by the Sponsor of the Contractor's quality assurance/quality control program set forth in [the Functional Specifications] [Exhibit 5.4(f)] shall not be construed to diminish any of the Contractor's obligations under this Article 13.

13.9 Shutdown

If, during the Warranty Notification Period [any or all Units of] the Facility [are] [is] not capable of producing <[electricity] [or thermal energy]> <*insert name of product*> in accordance with <Good> <{Prudent}> <Utility> <{Industry}> Practices as a result of the breach of, or non-conformance of the Work or the Facility with, any General Warranty, then the Contractor, if the Contractor is not using <{reasonable}> < its best> efforts [{(but not more than those which are commercially reasonable taking into account the severity of the situation)}] to rectify such situation, shall pay the Sponsor Shutdown Liquidated Damages in the amount of <<*insert currency and amount*>> [per Unit] for each Day [such Unit] [the Facility] is not capable of producing <<*electricity and/or thermal energy*>> in accordance with <Good> <{Prudent}> < Utility > <{Industry}> Practices as a result of such breach or non-conformance.

13.10 [{Limitation of Warranties}[201]

Except as expressly provided in this Contract, there are no warranties relating to the Work and the Contractor disclaims any [{implied warranties or}] warranties imposed by law (including warranties of merchantability or fitness for a particular purpose)}.]

13.11 [{Warranty Exclusions}[202]

The duties, liabilities and obligations of the Contractor under this Article 13 do not extend to Consumables beyond their normal useful life or to any repairs, adjustments, alterations, replacements or maintenance which may be required as a:

(a) result of normal wear and tear in the operation of the Facility which can be expected from properly designed, manufactured and installed items made from proper materials; or

(b) result of Sponsor's failure to operate or maintain the Facility in accordance with reasonable written manuals[203] provided by the Contractor to the Sponsor or in accordance with <Prudent> <{Good}> < Industry> <{Utility}> Practices in the absence of the Contractor's provision thereof}.]

13.12 Warranty for Civil Works.

Notwithstanding the provisions above, Contractor shall remain liable for a period of <<*insert number*>> Days commencing on the Facility Provisional Acceptance Date, if a breach of Contractor's General Warranties or obligations under this Contract results in partial or total collapse of all or any portion of the Facility or affects the integrity or stability of the physical structure, civil works and/or safety in the occupation of the Facility.

13.13 Patent and Title Warranties

Notwithstanding any conflicting provision hereof, there is no expiration of the Warranty Notification Period for breach of the warranties relating to use of intellectual property or title to any portion of the Work.

13.14 The Warranty Notification Period Extension

The Warranty Notification Period shall be extended by a period equal to the number of days [{in excess of <<*insert number*>> [{consecutive}] Days}], in the aggregate, during which the Facility, while under General Warranties, cannot operate [{and produce <<*electricity*>> <<*product*>> for sales}] by reason of a defect, deficiency, damage or warranty repairs [counting from the Day of Notification thereof to the Contractor] [{; provided, however, that Contractor shall not be responsible for any such extension of the Warranty Notification Period to the extent that such defect or damage results from any default under this Contract as a result of the gross negligence or the willful misconduct of any Sponsor Person; and provided further that such extended Warranty Notification Period shall not exceed the <<*insert number*>> Days}]. [The Sponsor may [{before <<*insert number*>> Days after the Contract Date}] purchase an extension of the Warranty Notification Period [{relating to <<*insert items of Work*>> only}] for up to <<*insert number*>> Days at a cost of <<*insert currency and amount*>> per Day.][204]

ARTICLE 14. TITLE AND RISK OF LOSS[205]

14.1 Care, Custody and Control[206]

The Contractor shall have care, custody and control of all Work until the Facility Provisional Acceptance Date, at which time care, custody and control of the Facility shall pass from the Contractor to the Sponsor.[207]

14.2 Risk of Loss[208]

14.2.1   The Contractor shall assume the risk of loss or of damage to real or personal property comprising the Work (and repair and replace such at its sole cost and expense without being entitled to a Price Adjustment or any other relief other than a Time Adjustment if expressly provided for in another provision of this Contract) until the Facility Provisional Acceptance Date. The Contractor shall bear the risk of loss, and be responsible for, and obligated to replace, repair or reconstruct, all at its expense and as promptly as possible, any portion or all of the Work that is lost, damaged or destroyed (including any damage or loss that has occurred as the result of a Force Majeure Excused Event) prior to the Facility Provisional Acceptance Date, irrespective of how such loss, damage or destruction shall have occurred except to the extent that:

(a) [the Sponsor, other than as a result of the action or omission of any Contractor Person, has vitiated the Contractor's coverage therefor under an insurance policy required to be maintained by the Sponsor under Article 15 or allowed coverage thereunder for such loss to lapse;][209]
(b) [{such is the result of a Sponsor Risk Event;}][210]
(c) [{such is a result of the willful misconduct or gross negligence of any Sponsor Person or any other non-Contractor Person expressly invited onto the Site by the Sponsor and the Sponsor shall bear the cost of any deductible assessed under any insurance policy to be maintained under this Article 14 or any such part of such loss not covered by such insurance}][; or]
(d) [{the Insurance Proceeds, if any, from the insurance policy or policies taken and maintained by the Sponsor in accordance with Article 15 to cover such loss are not paid, within <<*insert number*>> Days after they have been deposited into the Sponsor's or any Financing Party's accounts, to the Contractor if the Contractor has any obligation under this sentence [in the case that such Insurance Proceeds exceed <<*insert currency and amount*>> and within <<*insert number*>> Business Days in all other cases]}]

14.2.2   [{In the case that any loss referred to in the previous sentence is the result of the negligent acts or omissions of the Sponsor or any non-Contractor person invited on the Site by the Sponsor, the Sponsor shall bear the cost of any deductible assessed under any insurance policy to be maintained under this Article 15}.]

14.2.3   The Contractor shall be responsible for any damage to the Facility caused by any Contractor Person during the Warranty Notification Period[, the Optional Facility Performance Guarantee Cure Period, the Optional Unit One Performance Guarantee Cure Period and the Optional Unit Two Performance Guarantee Cure Period, if any] [, {except to the extent that, at the time such loss occurs, the Sponsor (other than as a result of an act or omission of any Contractor Person) has allowed the insurance set forth in this Article 14 to lapse during the period such was required to be maintained or otherwise be vitiated, and as a result, the Contractor is unable to recover under that insurance for the relevant loss}].

14.2.4   All construction tools and equipment used by or on behalf of Contractor or any Subcontractor for its performance hereunder shall be brought to and kept at the Site at the sole cost, risk and expense of Contractor or such Subcontractor, and the Sponsor shall not be liable for loss or damage thereto.

14.2.5   The Contractor shall be responsible for any damage any Contractor Person causes to the Facility after Facility Provisional Acceptance.

14.2.6   The Contractor shall take all {reasonable} precautions necessary to protect all Work, owned, leased, or rented by it, from any loss or damage to same including from fire, theft, accident, failure of parts, improper handling, incompetent operators, vandalism, strikes, depreciation, wear and tear, careless operations, neglect, failure to lubricate properly, or lack of protection from weather (other than a Force Majeure Excused Event), and the Contractor agrees that the Sponsor shall not be liable for any loss, repairs, or replacement made necessary by any or all such causes and such shall not be a Force Majeure Excused Event.

14.3 Title to Work; Licenses; No Liens[211]

14.3.1   Title to, and ownership of, all Equipment and other portions of the Work, whether during engineering, manufacturing or shipment, or when incorporated in the Facility or in storage on the <Route><Site> or elsewhere, shall pass to and vest in the Sponsor upon the earliest of:

(a) delivery thereof to the <Site><Route>;
(b) payment therefor by the Sponsor to the Contractor;
(c) the occurrence of any event by which title passes to the Contractor from the Subcontractor providing such Equipment or portion of the Work;
(d) payment therefor by the Contractor to a Subcontractor; or
(e) incorporation thereof into the Facility.

The transfer of title hereunder shall in no way affect the Contractor's responsibility for, and its obligation to take proper steps and precautions to protect all Equipment and other portions of the Work until the Facility Provisional Acceptance Date.

14.3.2   The Contractor shall:

(a) keep (and shall ensure that each Subcontractor keeps with respect to its portion of the Work) the Facility, the Site and the Equipment free and clear of any Lien (other than those created by the Sponsor), and shall not record or file this Contract or any portion hereof with any Governmental Authority;
(b) indemnify, defend and hold harmless the Sponsor from and against all [{direct}] losses and expenses incurred by the Sponsor as a result of any breach of the foregoing provisions of this Article 14.3;
(c) pay, in advance, all costs and expenses, including bonding costs and legal fees, in connection with any Claim resulting from the Contractor's breach of the provisions of this Article 14.3; and
(d) secure the removal of any Lien within <*insert number*>> Days of its obtaining notice thereof.

14.3.3   If any Lien is not discharged, satisfied or released within the <<*insert number*>> Days referred to in the previous sentence or such earlier time as may be necessary in order for the Sponsor to avoid foreclosure of any such Lien or other financial loss or risk resulting from such Lien, the Sponsor may in addition to any other remedy provided under this Contract, upon notice to the Contractor of its intention to do so, apply any funds withheld or moneys to become due to the Contractor under this Contract to satisfy, discharge or secure the release (including by posting a bond) of such Lien. Any such application by the Sponsor shall be deemed a payment of the Contract Price to the Contractor. Any additional expense incurred by the Sponsor as a result of the Contractor's breach of any provision of this Article 14.3 shall be borne by the Contractor.

14.4 [No Technology Transfer[212]

For the avoidance of doubt, Sponsor understands that the foregoing is not and shall not be deemed to be a technology transfer to Sponsor of any technology relating to the design, and the Contractor shall retain all rights, title and interest in the Contractor's and its affiliates' proprietary technology, patents, trade secrets and shop drawings.]

14.5 Title to Drawings; Intellectual Property

Title to copies and other tangible embodiments of drawings, specifications and like materials (collectively, the "Contractor Document Deliverables") shall be transferred to the Sponsor upon payment for such Work and shall remain the exclusive property of the Sponsor. Any intellectual property rights relating to the Contractor Document Deliverables Work or the Facility shall remain the property of the Contractor Person entitled to such but the Contractor and the Contractor on behalf of each Contractor Person hereby grants to the Sponsor and its successors and assigns an irrevocable, fully paid-up, royalty free, non-exclusive license (including the right to grant sub-licenses and to transfer the license freely to third parties) to modify, use and reproduce all such intellectual property rights [{for the purpose of completing construction of, operating, transferring, financing, designing, maintaining, commissioning, testing, operating, servicing, letting, promoting, advertising, extending, modifying, reinstating, repairing, rebuilding, expanding, selling, decommissioning or demolishing any part of the Work or the Facility (each, a "Specified Purpose"). Without limiting the generality of the foregoing, the Sponsor shall not make any use of any Contractor Document Deliverable, except in connection with a Specified Purpose without the prior written consent of Contractor, which consent Contractor shall not unreasonably withhold, delay or condition}]. The Sponsor shall be entitled to assign the entire benefit of such license to any Person that leases, licenses, uses and/or owns (or has a mortgage or security interest in) the Facility.

14.6 Provision of Sponsor Drawings

The drawings or other documents provided by the Sponsor to the Contractor shall remain the sole property of the Sponsor.

14.7 Return of Drawings and Documents.

At the end of the Warranty Notification Period, the Contractor shall hand over to the Sponsor all drawings, documents, specifications and designs relating to the Work and return to the Sponsor all documents provided under the Contract[{, except that the Contractor may maintain a confidential file copy of such documents until such time as it has discharged all its obligations under the Contract whereupon it must destroy such or return it to the Sponsor}].

ARTICLE 15. INSURANCE

15.1 Insurance Standards[213,214]

All insurance described in this Article 15 shall be written by major insurance companies of good repute and standing and rated at least <<*insert rating*>> by <<[*insert agency*]>> or if not permitted by Law in the country of the <Site> <Route>, reinsured by such an insurer with a written "cut through" clause given by such insurer to the Sponsor [and the Financing Parties]. The Sponsor, the Contractor and all Subcontractors shall not violate or [knowingly] permit any violation of any conditions or terms of the policies of insurance described herein. Terms used in this Article 15 and not otherwise defined in this Contract shall have the meaning generally ascribed to them in the <United States> <British> <international> insurance industry.

15.2 Employer's Liability and Worker's Compensation Insurance

The Contractor shall, at its cost, maintain worker's compensation insurance to the extent required by law and employer's liability insurance with a minimum cover of <<*insert currency and amount*>>. The Contractor shall maintain (or cause its

Subcontractors to maintain) this insurance to cover any Work undertaken by the Contractor or any Contractor Person.

15.3 Third-Party Liability Insurance[215]

The <Sponsor> <Contractor> shall, from the date upon which [the Contractor notifies the Sponsor in writing, that] the Contractor will commence Work at the Site [(which date shall be no earlier than <<*insert number*>> Days after the Work Commencement Date)][216] until the end of the Warranty Notification Period, at its cost, maintain commercial general liability insurance [as further set forth in Exhibit 15] with insuring as co-insureds, the Sponsor, [the Financing Parties,] each Subcontractor, each Contractor Person and any other Persons named by the Sponsor in an amount of not less than <<*insert currency and amount*>>) for each and every occurrence for any liability for death or personal injury to any Person or damage to any property [including the Existing Plant] (other than property forming part of the Facility) occurring prior to the Facility Provisional Acceptance Date (but including a provision for Contractor warranty repair visits after the Facility Provisional Acceptance Date) due to or arising out of the execution of the Work. This insurance policy shall include provisions for "cross liability" (covering all Insured Parties from claims against one another as though a separate policy has been issued to each Insured Party), "contingent motor," "tool of trade" and "pollution" (sudden and accidental) and a policy deductible of no more than (<<*insert currency and amount*>>) for each and every occurrence. Coverage must include liability arising from operations, subcontractors, blanket contractual liability to support indemnification obligations of the Contractor hereunder and tort liability of another assumed by Contractor hereunder, explosion, collapse, subsidence. Coverage may not exclude damage caused by Subcontractors.

15.4 Erection All Risk Insurance.[217]

The <Contractor> <Sponsor> shall from the date upon which <the Contractor notifies the Sponsor in writing that> the Contractor <will commence> <commences> Work <on the Route> <at the Site> [(which date shall be no earlier than <<*insert number*>> Days after the Work Commencement Date)][218] until the Facility Provisional Acceptance Date [(but also to the extent necessary to cover the Contractor's visits thereafter)], at its cost, maintain builder's all risk insurance [as further set forth in Exhibit 15] for full replacement value of the Work and covering all Work. The Sponsor, the Contractor, all Subcontractors and any other interested parties designated by the Sponsor shall be identified as named insureds thereunder. The Sponsor, all Sponsor Persons, the Contractor, and all Subcontractors agree to waive all rights of recovery against one another for damages covered thereunder. The Contractor shall cause its insurance underwriters to waive all rights of recovery against its Subcontractors and Sponsor Persons for any loss or damage covered by said insurance. Any and all losses not covered by reason of deductible or exceedence of policy limit or otherwise in the insurance policies required by this Article 15 shall be borne by, for the account of, and at the sole risk of the Contractor. The cover and/or deductibles stated in this clause may be broadened and/or lowered at the cost of the Contractor upon the express prior written agreement of the Sponsor[, which agreement shall not be unreasonably withheld]. The foregoing insurance must be issued on an "all risks" basis covering direct physical loss or damage to the Facility and associated property including risks associated with startup and testing of the Facility and all materials, supplies and equipment comprising the Facility or intended for installation into the Facility or temporary works and covering all such materials, supplies or equipment during transit to or from the <Site> <Route> during construction. Such insurance maintained shall be written on a full replacement cost basis for new property and an appropriate provision for existing property and shall include "consequences design cover" (at least to <DE3/LEG2/96>), "offsite storage", "local authority", "debris removal", "full terrorism", "extended maintenance extension" (MRe 004), "expediting expenses" (limited to <<*insert number*>> percent of the claim) and a "50/50 clause" for "marine risk." The deductibles shall be:

(a) "design, testing, commission or maintenance": <<*insert currency and amount*>>); and
(b) "other risks": <<*insert currency and amount*>> for all, each and every occurrence.

Erection all risk shall include coverage for removal of debris, coverage for buildings, structures, boilers, turbines, machinery, Equipment and other properties constituting part of the Work, with peril sublimits to insure the full replacement value of any key equipment item, off site coverage with peril sublimits sufficient to insure the full replacement value of any property or Equipment not stored on the Site, and coverage for operational testing in a minimum amount equal to the full replacement cost of the Work, which shall include:

(a) earthquake insurance with a <<*insert currency and amount*>> minimum annual aggregate limit;
(b) flood insurance with a <<*insert currency and amount*>> minimum annual aggregate limit;
(c) inland transit for locally produced materials and offsite insurance in an amount sufficient to cover the full replacement cost of any item in transit or temporarily off site, but in no event less than <<*insert currency and amount*>>; and
(d) Hazardous Material cleanup for matters resulting from the Contractor's action or inaction with a minimum limit of <<*insert currency and amount*>>.

No such policies may have deductibles of greater than <<*insert currency and amount*>> for turbine [boiler] related equipment, <<*insert currency and amount*>> for the period of operational testing, <<*insert currency and amount*>> for the perils of flood and earth movement and <<*insert currency and amount*>> for all other property and perils.

15.5 Moment of Start-Up Insurance

The Contractor shall provide moment of start-up insurance as part of its builders all risk insurance in accordance with <Exhibit 15> <Good> <{Prudent}> <Utility> <{Industry}> Practices. The Contractor, all Sponsor Persons, Subcontractors and any other interested parties designated by the Sponsor in writing shall be identified as named insured thereunder.

15.6 Ocean Marine Cargo Insurance[219]

The <Contractor> <Sponsor> shall, <within <<*insert number*>> Days after the Contractor notifies the Sponsor of> <from> the date upon which the first shipment of Equipment will occur [(but in no event earlier than <<*insert number*>> Days after the Work Commencement Date)] until the Facility Provisional Acceptance Date, at its cost, maintain ocean marine cargo insurance (naming the Sponsor[, the Financing Parties], each Subcontractor, the Contractor and any other parties designated by the Sponsor in writing as insureds) covering any and all materials and Equipment originating outside <<*insert country name*>> while they are in transit to the Site and existing equipment and materials removed from the Site while they are in transit by wet marine conveyances or by air transportation and/or road or rail connecting conveyances, with a policy limit not less than <<*insert currency and amount*>> until the Facility Provisional Acceptance Date. The insurance policy shall include a "50/50" clause (erection all risks). The deductible shall not be more than <<*insert currency and amount*>> for any one occurrence.[220]

15.7 Contractor's Owned or Rented Equipment

The Contractor shall maintain or cause to be maintained insurance to cover loss or damage to any construction tools and equipment used in connection with the Work.

15.8 Comprehensive Automobile Liability Insurance[221]

The Contractor shall obtain automobile liability insurance on vehicles used in connection with the Work (whether owned, hired or otherwise) for third-party property damage with a limit of at least <<*insert currency and amount*>> per occurrence and personal injury without limit. Coverage must include sudden and accidental pollution coverage for hauled materials resulting from accident.

15.9 Delayed Opening Insurance[222]

Contractor shall obtain delayed opening insurance for the Work with a minimum coverage sufficient to cover the Sponsor's fixed expenses and debt service for a minimum period of <<*insert number*>> Days for any delay in the operation of the Work caused by damage or loss to the Facility or any Work, which policy may have a deductible of not greater than <<*insert number*>> Days.

15.10 Pollution/Environmental Liability Insurance[223]

The Contractor shall, from the date upon which the Contractor will commence Work at the Site until the Facility Provisional Acceptance Date, at its cost, maintain pollution/environmental liability insurance for any pollution liability exposure, (i) covering (A) the liability of the Contractor or Subcontractor during the process of construction, removal, storage, encapsulation, transport and disposal of hazardous waste and contaminated soil and/or asbestos abatement, (B) on-Site and off-Site bodily injury and loss of damage to, or loss of use of property, directly or indirectly arising out of the discharge, dispersal, release or escape of smoke, vapors, soot, fumes, acids, alkalis, toxic chemicals, liquids or gas, waste materials or other irritants, contaminants or pollutants into or upon the land, the atmosphere or any water course or body of water, whether it be gradual or sudden and accidental, including no exclusion for mold or asbestos, and (C) defense and cleanup costs and (ii) includes (A) limits of <<*insert amount and currency*>> each occurrence and in the aggregate with a maximum deductible of <<*insert amount and currency*>> and (B) contractual liability coverage aligned with indemnification obligation of Contractor hereunder.

15.11 Medical, Accident and Travel Insurance

The Contractor shall obtain an insurance policy that covers all Contractor personnel who will travel in connection with performance of the Work to the Site or other location, including coverage of any cost associated with comprehensive emergency medical evacuation, treatment, and repatriation, including repatriation of mortal remains and any costs related thereto.

15.12 Marine Hull and Protection and Indemnity Insurance

If Contractor is chartering a vessel in connection with performance of the Work for full loss or damage coverage of not less the value of the vessel in use.

15.13 Subcontractors' Insurances

The Contractor shall <ensure> <require> that all Subcontractors providing Equipment, materials or services in connection with the Work obtain, maintain and keep in force, during the time which they are involved in performance of the Work, workers' compensation, commercial and auto liability and comprehensive general liability insurance and insurance providing cover against loss or damage to any construction tools and equipment owned by the Subcontractor or for which the Subcontractor accepts responsibility. The Contractor must purchase insurance covering the insolvency of all [{Major}] Subcontractors. The Contractor shall require each Subcontractor to deliver to the Sponsor a release and waiver any and all rights of recovery against each Sponsor Person which such Subcontractor may otherwise have or acquire, in or from, or in any way connected with, any loss covered by policies of insurance required to be maintained by the Subcontractors in connection with the Work. Written evidence of each such insurer's awareness of this waiver shall be provided by the Contractor to the Sponsor within <<*insert number*>> Days after the execution of each and every Subcontract.

15.14 Aircraft Liability Insurance[224]

If the performance of the Work requires the use of any aircraft that is owned, leased or chartered by any Contractor Person, the Contractor shall maintain aircraft liability insurance with a <<*insert currency and amount*>> minimum limit per occurrence for property damage and bodily injury, including passengers and crew.

15.15 Professional Indemnity/Errors and Omissions Liability Insurance

An insurance policy that (i) covers liability for financial loss or damage due to an act, error, omission, breach of duty, or negligence resulting from errors or omissions in the delivery of professional services with a minimum limit of <<*insert amount and currency*>> each occurrence and in the aggregate with a maximum deductible of <<*insert amount and currency*>>, (ii) includes coverage for contractual liability, and (iii) remains in effect for the statute of repose.

15.16 Umbrella Liability or Excess Insurance

Contractor shall purchase umbrella liability or excess liability insurance on an "occurrence" basis covering claims in excess of and following the terms of the underlying insurance as set forth in the foregoing Articles 15.2, 15.3, 15.4, 15.5, 15.6, 15.7, 15.8, 15.9, 15.10, 15.11, 15.12, 15.13, 15.14, 15.15 and 15.16, with a <<*insert currency and amount*>> minimum limit per occurrence and a <<*insert amount and currency*>> minimum limit per occurrence and a <<*insert currency and amount*>> aggregate annual limit; provided that aggregate limits of liability, if written under a policy covering more than the Work, shall apply specifically to claims occurring with respect to the Work and the Equipment.

15.17 Endorsements

All policies of insurance hereunder (other than employer's liability insurance and automobile liability insurance) shall provide for waivers of each insurer's subrogation rights given in favor of each Sponsor Person, and the Contractor Person (collectively, the "<u>Insured Parties</u>"). All policies of insurance hereunder (other than employer's liability insurance and automobile insurance) shall also be endorsed as follows:

(a) in cases of liability insurance, to provide a severability of interests or cross-liability clause covering all Insured Parties from claims against one another (as though a separate policy has been issued to each Insured Party) (cross-liability being mandatory in the event of more than one named insured); and

(b) that the insurance shall be primary and not excess to, or contributing with, any insurance or self-insurance maintained by any Insured Party.[225]

15.18 Certain Insurance Policy Provisions

All policies of insurance required to be maintained pursuant to this Article 15 shall provide that:

(a) the insurer is required to provide the Sponsor, Contractor [and the Financing Parties] with at least <<*insert number*>> Days' (or<<*insert number*>> Days', in the case of non-payment of premium), prior notice of reduction in coverage or amount (other than a reduction in coverage or amount resulting from a payment thereunder), cancellation or non-renewal of any policy; and

(b) the proceeds of all policies (collectively, "Insurance Proceeds") required to be maintained under Articles 15.3, 15.4, 15.5, 15.6, 15.9 and 15.10 shall be payable without contribution by any claimant [and to the Financing Parties' or] Sponsor's accounts [provided that no Insurance Proceeds paid under a policy maintained under Articles 15.7 and 15.8 in excess of [<<*insert amount and currency*>>] which result from claims for a single occurrence or series of related occurrences shall be paid to the Contractor unless permitted under the Sponsor's agreements with the Financing Parties or their agents].

15.19 Insurance Certificates

Each Party shall provide evidence reasonably satisfactory to the other Party (prior to the date on which such is to be procured) that the insurance which such Party is obliged to procure under this Article 15 is in force. The Parties shall provide each other with certificates showing that the said insurance is in force, the amount of the insurer's liability thereunder, and further providing that the insurance will not be canceled or changed or not renewed until the expiration of at least <<*insert number*>> Days (or <<*insert number*>> Days in the case of cancellation due to non-payment of premiums) after written notice of such cancellation, change or non-renewal has been received by the each Party. All copies of certificates of insurance submitted under this Article 15.14 shall be in form and content reasonably acceptable to each Party. At the request of a Party, the other Party shall deliver duplicates, certified by such Party's independent insurance broker, of each policy of insurance required to be in effect hereunder (including, without limitation, each renewal policy). Prior to the beginning of any Work at the Site and within <<*insert number*>> Days after the close of each calendar year, each Party shall deliver to the other Party a certificate of insurance prepared by such Party's independent insurance broker:

(a) confirming that all insurance policies required to be maintained by such Party pursuant to this Article 15 are in force on the date thereof;
(b) confirming the names of the insurers issuing such policies;
(c) confirming the amounts and expiration date or dates of such policies; and
(d) confirming that all premiums then due and payable have been paid.

15.20 Expiration

Not less than <<*insert number*>> Days prior to the expiration date of any policy of insurance required to be in effect hereunder, the Party responsible therefor shall deliver to the other Party a certificate of insurance with respect to each renewal policy, certified by such Party's independent insurance broker, bearing a notation that all premium then due and payable have been paid.

15.21 Right to Procure Insurance

If the Contractor fails to procure or maintain the full insurance coverage required by this Article 15, the Sponsor, upon <<*insert number*>> Days' prior notice (unless such insurance coverage would lapse within such period, in which event notice should be given as soon as reasonably possible) to the Contractor of any such failure, may (but shall not be obligated to) take out the required policies of insurance and pay the premium thereon. All amounts so advanced therefor by the Sponsor shall become an obligation of the Contractor to the Sponsor, and the Contractor shall forthwith pay such amounts to the Sponsor, or at the Sponsor's option reduce the next upcoming payment of the Contract Price until paid and, thus, the Sponsor may deduct such amounts from the Milestone Payments.

15.22 Notice of Event of Loss or Change in Insurance Coverage

Each Party shall promptly notify the other Party of:

(a) any actual or, upon obtaining knowledge thereof, potential claim under the insurance maintained under this Article 15; and
(b) each written notice received by it with respect to any potential event of loss or the cancellation of, adverse change in, or default under, any insurance policy required to be maintained in accordance with this Article 15.

15.23 Cost Responsibility[226]

No insurance company shall have recourse against any Insured Party for payment of any premium or assessment under any policy unless such Insured Party is responsible for maintaining such insurance under this Contract.

15.24 Cancellation

Irrespective of the requirements as to insurance to be carried as provided for in this Article 15, the insolvency, bankruptcy or failure of any insurance company to pay any claim accruing shall not excuse any Party from its obligations to carry or arrange to be carried insurance as herein required. In case of cancellation of any policy required to be carried by this Article 15, or the insolvency, bankruptcy or failure of any such insurance company that has issued a policy hereunder, the Party responsible therefor shall within <<*insert number*>> Days obtain new insurance policies in the amounts and coverage required hereby.

15.25 Other Insurance

Any insurance other than that specified in this Article 15, which the Contractor may be required by Law to carry or may desire for its protection, shall be secured and maintained at its own expense. Any such policies of insurance shall contain waivers of subrogation as provided in Articles 15.4 and 15.143

15.26 Cooperation

The Contractor shall promptly comply with the recommendations of the insurance carriers of the Sponsor so that said insurance carriers will continue to provide the coverage maintained by the Sponsor at reasonable premiums.

15.27 Claims[227]

Unless a policy otherwise requires, the Contractor will be responsible for notifying insurers, and filing and prosecuting all insurance claims pursuant to any insurance policy concerning the Facility or the Work that arises before the Facility Provisional Acceptance Date. The Contractor and its Subcontractors shall assist and cooperate in every manner possible in connection with the adjustment of all claims arising out of the operations conducted under, or in connection with, the Work and shall cooperate with the insurance carrier or carriers of the Sponsor and of the Contractor and its Subcontractors in all litigated claims and demands which arise out of the Work and which the said insurance carrier or carriers are called upon to adjust or resist.

15.28 Procedures and Services

The Contractor shall fully comply with, and shall require its Subcontractors to comply fully with, all procedures and services including completion of all necessary applications for insurance coverage, prompt and full compliance with all audit requests and claim reporting procedures, and full participation in and compliance with safety and loss control programs implemented by, or at the request of, the Sponsor. Any claims with respect to any event covered by the insurance listed in Articles 15.3, 15.4, 15.5, 15.6, 15.9 and 15.10 shall be made on the basis that such insurance is the primary insurance therefor (and not on a shared loss basis with any other insurance Contractor may have) and thereby will not jeopardize the availability of coverage under any advance loss of profits or delayed start up or similar type insurance that the Sponsor may have.

15.29 Disclosure

The Sponsor and the Contractor shall disclose all information material to the risks covered by the insurances procured by the Sponsor under this Article 15.[228]

15.30 Manufacturer's Risks Exclusion

None of the insurance policies taken out pursuant to this Article 15 shall contain any provision for "Manufacturer's Risks Exclusions."

15.31 [{Waiver}]

The Sponsor shall obtain from any insurers providing it with "advance loss of profits" or "delayed start up" or similar type insurance waivers of subrogation against the Contractor and its Subcontractors.}][229]

ARTICLE 16. INDEMNIFICATION

16.1 General Indemnity[230]

The Contractor shall defend, indemnify and hold harmless [each Sponsor Person] [{the Sponsor}][231] from and against all Losses including those[232] resulting from:[233]

(a) injury to or death of any Person, resulting from events at the Site or relating to the Work [{caused by the fault, negligence or failure (in whole or in part) of any Contractor Person except to the extent that at the time of the incident giving rise to such loss, the Sponsor has failed to maintain the insurance required under Article 15 (other than as a result of any act or omission of a Contractor Person), and as a result, the Contractor is unable to recover under that insurance for the relevant loss}];

(b) damage to or loss of property caused by [{the fault, negligence or failure (in whole or in part) of}] any Contractor Person [{except to the extent that at the time of the incident giving rise to such loss, the Sponsor has failed to maintain the insurance coverage it is required under Article 15 with respect to the loss in question if the Sponsor is so required by Article 15 to obtain coverage for the loss in question (other than as a result of any act or omission of a Contractor Person), and as a result, is unable to recover under that insurance for the relevant loss}];

(c) any [{material}] misrepresentation by the Contractor or [{material}] breach by the Contractor of any representation or warranty made herein;

(d) any violation of any Law by any Contractor Person;

(e) any failure of any part of the Work;

(f) breach of any covenant, agreement or obligation of the Contractor contained in this Contract;

(g) any Contractor Person's [fault or] negligent acts or omissions; and

(h) any consequential, special, incidental or indirect damages for which any Sponsor Person becomes liable [{but only to the extent caused in whole or in part by or arising out of:
   (i) any Contractor Person's acts or omissions;
   (ii) breach of this Contract by the Contractor; or
   (iii) failure of any of any Contractor Person's Work}].

16.2 Contractor Taxes

The Contractor shall defend, indemnify and hold harmless each Sponsor Person from and against all claims by any Governmental Authority claiming Contractor Taxes.

16.3 Proprietary Rights

16.3.1   The Contractor shall defend, indemnify and hold harmless each Sponsor Person against all Losses arising from any Claim for unauthorized disclosure or use of any trade secret, proprietary right, intellectual property right or of patent, copyright or trademark infringement arising from any Work [{and asserted against any Sponsor Person that:

(a) concerns the Equipment, the Facility or the Work;

(b) is based upon the performance of the Work by the Contractor or any Subcontractor, including the use of any tools, implements or construction by the Contractor or any Subcontractor; or

(c) is based upon the design or construction of the Equipment or the Work}].

16.3.2   If the Sponsor is enjoined from completion of the Facility or any part thereof or from the use, operation or enjoyment of the Facility or any part thereof as a result of any Claim, the Contractor shall promptly arrange to have such injunction removed at no cost to the Sponsor, and the Sponsor may, at its option and without thereby limiting any other right it may have hereunder or at law or in equity, require the Contractor to supply, temporarily or permanently, items that function to meet the requirements of this Contract and that do not subject to such injunction and not infringing any such patent, or to remove all such offending facilities and refund the cost thereof to the Sponsor, or to take such steps as may be necessary to ensure compliance by the Sponsor with such injunction, all to the satisfaction of the Sponsor and without cost or expense to the Sponsor.

16.4 Employee Claims

In any and all Claims against any Sponsor Person by any Contractor Person or by anyone directly or indirectly employed by any Contractor Person, or anyone for whose acts any Contractor Person may be liable, the indemnification obligation stated in Article 16, shall not be limited in any way by any limitation on the amount or type of damages, compensation or benefits payable by or for any Contractor Person under any applicable workers' compensation Law, disability Law or any other Law.

16.5 Notice of Claim.

[The] [Each] Sponsor [Person] shall, <reasonably promptly> <within <<*insert number*>> Days> after the receipt of notice of the commencement of any legal action or of any Claim against it in respect of which indemnification, pursuant to the foregoing provisions of this Article 16, may be sought, notify the Contractor in writing thereof; provided that the failure of [any] [the] Sponsor [Person] to provide any such notice as required above shall not relieve the Contractor of any of its obligations hereunder, but shall only reduce the liability of the Contractor by the amount of any damages directly attributable to the failure of the Sponsor [Person] to give such notice in such manner [but if another Sponsor Person has given such notice, that must be taken into account in the foregoing estimation].

16.6 Conduct of Proceedings.

[The] [Each] Sponsor [Person] shall have the right, but not the obligation, to contest, defend and litigate (and to retain legal advisers of its own choice in connection with) any Claim alleged or asserted against it arising out of any matter in respect of which it is entitled to be indemnified hereunder [{to the extent that the Contractor has not undertaken the reasonable defense thereof}], and the costs and expenses thereof shall be covered by such indemnity.

16.7 No Waiver

The Sponsor's acceptance of the Contractor's engineering designs or proposed or furnished Equipment and Work shall not be construed to relieve the Contractor of any obligation or liability under this Article 16.

16.8 Payment Delay.

If the Contractor must reimburse any Loss suffered by [any] [the] Sponsor [Person] in <the currency of the Contract Price> <the currency such Loss was incurred in which case such Loss shall be increased by the change in [the << *insert appropriate index*>>] from the date the Sponsor [Person] suffered such Loss until the date of reimbursement therefor by the Contractor.[234]

16.9 No Prejudicial Admissions By Contractor

The Contractor shall not make any admission which might be prejudicial to any Sponsor Person or its reputation in connection with any Claim.

ARTICLE 17. DEFAULT,[235] TERMINATION AND SUSPENSION

17.1 Contractor Events of Default[236]

The following shall be events of default with respect to the Contractor (each, a "Contractor Event of Default"):

(a) the Contractor does not pay all Liquidated Damages outstanding within <<*insert number*>> Days after the due date established therefor by this Contract:
  (i) a proceeding is instituted against [<the Contractor>] [<or any Consortium Member>] <or [the] [any] Parent Guarantor> seeking to adjudicate it bankrupt or insolvent and such is not dismissed within <<*ninety (90)*>> Days thereof;
  (ii) [<the Contractor>] [<or any Consortium Member>] <or [the] [any] Parent Guarantor> generally fails to pay its debts as they become due;
  (iii) a receiver is appointed on account of the insolvency of [<the Contractor>] [<any Consortium Member>] <or [the] [any] Parent Guarantor> or any of <their> <its> <respective> properties;
  (iv) a petition for winding up is instituted against [<the Contractor>] [<any Consortium Member>] <or [the] [any] Parent Guarantor> in any jurisdiction and, in the case of any such proceeding instituted against any [the Contractor] [Consortium Member] or [the] [any] Parent Guarantor, such proceeding is not dismissed within <<*ninety (90)*>> Days of such filing; or
  (v) [the Contractor] [any Consortium Member] or [the] [any] Parent Guarantor takes any organic or other action to authorize the filing of, or files, a petition seeking to take advantage of any Law relating to bankruptcy, insolvency, reorganization, winding up or composition or readjustment of debts;
(b) <the Contractor> <or any Consortium Member> assigns all or any part of this Contract;
(c) [the Contractor refuses or fails to supply enough properly skilled workmen or proper materials to perform the Work in accordance with this Contract and/or the Baseline Level 3 Schedule;][237]
(d) the Contractor [{without reasonable justification}] abandons the Work or any portion thereof for more than <<*insert number*>> consecutive Days;

(e) the Contractor or any of its Subcontractors either:
  (i) fails to make prompt payment of any undisputed invoice due to any Subcontractor or otherwise for materials or labor; or
  (ii) repudiates, or is in default with respect to, any of its obligations to any Subcontractor under a [{Major}] Subcontract [{with a value in excess of <<*insert currency and amount*>>}];
(f) the Contractor fails [{in a material respect}] to prosecute the Work diligently or fails [{in a material respect}] to make the anticipated by the Baseline Level 3 Schedule;
(g) the Contractor fails within <<*insert number*>> Days after being notified thereof by the Sponsor to correct [{or submit a written plan acceptable to the Sponsor acting <in its sole discretion> <{reasonably}>} for the prompt correction of}] any defective or deficient Work;
(h) any representation or warranty made by <[any Contractor Person]> <any [Consortium Member]> <or the Parent Guarantor> to the Sponsor shall prove to be false or misleading in any [{material}] respect as of the time made, confirmed or furnished [{and such has a material adverse impact on the Work}];
(i) [either] [the] Parent Guarantee [or the Performance LC, Retainage LC or Warranty LC] shall cease, for any reason, to be in full force and effect, or the Parent Guarantor [or the issuer of the Performance LC, Retainage LC or Warranty LC, as the case may be,] shall so assert in writing [{and, in the case of the Parent Guarantee only, the Contractor fails, within <<*insert number*>> Days thereafter, to furnish additional security for the Contractor's performance hereunder in form and substance acceptable to the Sponsor and each Financing Party in their sole discretion}];
(j) the Contractor or [either] [the] Parent Guarantor fails, after being notified thereof, to comply in any [{material}] respect with any [{material}] provision of this Contract or [the] [its] Parent Guarantee;
(k) the Contractor fails to achieve Facility Provisional Acceptance within <<*insert number*>>[238] Days of the Guaranteed Facility Provisional Acceptance Date [or invokes the protection of the Delay Damage Cap];[239]
(l) the Contractor becomes involved in litigation or labor problems[{, which will delay or adversely affect the Work}] unless the Contractor cures such failure within <<*insert number*>> Days after the Contractor has been given written notice thereof by the Sponsor [{provided that if the Contractor has commenced to cure such failure within <<*insert number*>> Days of such notice but a longer time period is reasonably necessary to effectuate such a cure, then the Contractor shall have a period of <<*insert number*>> Days to effect such a cure;}][240]
(m) the issuer of the [Performance LC] [Warranty LC] or [Retainage LC] fails to be an Eligible Bank [{and such is not replaced within <<*insert number*>> Days thereof with a Performance LC, Warranty LC or Retainage LC (as the case may be) issued by an Eligible Bank}][241];
(n) [the Parent Guarantor fails to maintain the Required Rating;][242]
(o) a Contractor Event of Default arises under Article 5.12.4 or Article 7.8;
(p) [a Change of Control occurs in violation of Article 21.12;] or
(q) [the Contractor fails, after being notified thereof by the Sponsor, to comply with any provision of this Contract].[243]

17.2 Sponsor's Remedies[244]

17.2.1   If any Contractor Event of Default has occurred [{and is continuing}]:

(a) the Sponsor may:
   (i) terminate <any> or <all of> the Sponsor's obligations under [Articles <<*insert desired Articles*>> of][245] this Contract effective <<*insert number*>> Days after giving written notice of such termination to the Contractor;
   (ii) suspend payment under this Contract in whole or in part;
   (iii) take the Facility, or any part thereof, wholly or partially out of the control of the Contractor [{in which case the risk of loss with respect to the part of the Facility taken out of the Contractor's control shall pass to the Sponsor}], in which event the Sponsor or any Replacement Contractor may take over such of the Contractor's tools, equipment, materials and supplies as the Sponsor deems necessary to complete such Facility, and may award any uncompleted or improperly completed Work to a Replacement Contractor to finish such by whatever method that it or the Sponsor may deem expedient, all at the Contractor's expense; and/or
   (iv) exercise any other remedy it may have hereunder, under Law or in equity, including seeking the recovery of damages;[246]
(b) the Contractor shall, upon the Sponsor's request, grant to the Sponsor, any Replacement Contractor and each Subcontractor, at the Contractor's expense, the right to continue to use any and all intellectual property including patented and/or proprietary information that the Sponsor deems necessary to complete the Facility in accordance herewith;
(c) at the Sponsor's request and at the <Contractor's> <{Sponsor's}> expense, the Contractor shall supply any Equipment needed for the completion and operation of the Facility in accordance herewith;
(d) the Contractor shall, upon the Sponsor's request, assign the Contractor's rights under all Subcontracts and Permits to the Sponsor (or, at the Sponsor's direction, any Replacement Contractor or Sponsor Person [or designee of the Financing Parties]) and the Contractor will execute any documents required to evidence or comply with such assignments[{; provided that, if the Sponsor does accept any assignment of any Subcontract, it will also assume all obligations of the Contractor thereunder to the extent they relate to the Work and have not been breached by the Contractor prior to the date of such assignment and do not include making payments to such Subcontractor with respect to Work for which the Sponsor has already made payment to the Contractor}];
(e) the Sponsor has taken possession of any construction equipment which is in the Contractor's possession [and located at the <Route> <Site> on the date of such termination] for the purpose of completing the Work, and is finishing the Work, or causing a Replacement Contractor to finish the Work, and the Contractor shall not be entitled to receive any further payments hereunder unless required by Article 7.7;
(f) at the Sponsor's request and at the Contractor's expense, the Contractor shall diligently assist the Sponsor in preparing an inventory of all Equipment in use or in storage at the Site;

(g) at the Contractor's expense, the Contractor shall remove from the <Route> <Site> all such Equipment, waste and rubbish as the Sponsor may request; and

(h) upon the Sponsor's request and at the Contractor's expense, the Contractor shall deliver to the Sponsor all design and other information, including any Computer Programs, as may be requested by the Sponsor for the completion and/or operation of the Facility.

17.2.2   The remedies of the Sponsor under Article 17.2.1 are cumulative, and the Sponsor may elect one or more thereof without prejudice to any other right or remedy that the Sponsor may have.

17.2.3   If the Sponsor terminates the Sponsor's obligations under this Contract pursuant to Article 17.2.1(a), the Sponsor shall, as soon as practicable, determine the total expenses incurred and accrued in completing the Work, including all amounts charged by any Replacement Contractor to finish the Work and additional overhead at <<*insert number*>> percent of direct cost incurred by the Sponsor to effect such takeover and complete the Work. If the total expenses incurred by the Sponsor in completing the Facility in accordance herewith (including meeting all Deadlines) exceed the balance of the Contract Price unpaid at the time of the Contractor Event of Default, the Contractor shall pay to the Sponsor the amount of such excess within <<*insert number*>> Days following receipt of the Sponsor's demand for such payment. Under no circumstances shall the Sponsor be required to pay additional amounts to the Contractor unless the Sponsor is able to complete the Facility in accordance with the Deadlines for less than the Contract Price, in which case, upon completion of the Facility, the Sponsor shall refund to the Contractor the amount by which the Contract Price exceeds the costs of the Sponsor referred to in the first two sentences of this Article 17.2.3.

17.3 Sponsor Events of Default[247]

Unless a Contractor Event of Default shall be the cause thereof, the following shall be events of default with respect to the Sponsor (each, a "Sponsor Event of Default") [248]:

(a) the Sponsor does not pay the Contractor any Milestone Payments in accordance with this Contract within <<*insert number*>> Days after the relevant due date established therefor by this Contract and an additional <<*insert number*>> Days have passed after the Sponsor has received written notice thereof from the Contractor;

[(b) {due to an Optional Suspension which is not the result of a proceeding concerning a Permit,[249] all of the Work is stopped for more than <<*two continuous periods of ____insert number*>> consecutive Days plus any period of time during which a Force Majeure Excused Event shall have affected performance of the Work not exceeding <<*insert number*>> Days in total;}

(c) [{the Sponsor fails to maintain the insurance policy that it is required to obtain under Article 15;}] or

(d) {an event or circumstance not caused or attributable to Contractor or any Contractor Person or proceeding concerning a Permit prevents any Work at the Site from being conducted for a period of *<<eight hundred (800 >>* consecutive Days}.

17.4 Contractor's Remedies

If any Sponsor Event of Default has occurred [and is continuing,] the Contractor may stop the Work, in which case the Contractor:

(a) shall take [all] steps [necessary] to preserve, protect and care for all Work, Equipment and materials at the Site, in transit thereto, or at storage areas under its responsibility; and

(b) may make an Adjustment Claim at any time [until *<<insert number>>* Days after the cessation of the Sponsor Event of Default].

If the Work should be stopped for a period of *<<insert number>>* Days in accordance with the preceding sentence, the Contractor may, upon *<<insert number>>* Business Days' written notice to the Sponsor and unless such Sponsor Event of Default is cured before the end of such *<<insert number>>* Business Day period, terminate this Contract and recover from the Sponsor the amount provided in Article 17.3.2 [but hereby waives any and all other claims for damages on account thereof that the Contractor might have against the Sponsor]. Except as expressly provided in the previous sentence, under no circumstance whatsoever shall the Contractor be entitled to a termination of this Contract.

ARTICLE 18. DAMAGES AND LIABILITIES

18.1 Sponsor Solely Liable

The Contractor acknowledges and agrees that in the event of the Sponsor's breach hereunder, the Contractor shall have no recourse and waives any claim(s) that may be made against any other Sponsor Person besides the Sponsor, and the Contractor's sole recourse therefor shall be against the Sponsor. The foregoing acknowledgment and agreement is made expressly for the benefit of all Sponsor Persons, all of whom are intended third-party beneficiaries of this Article 18.1.

18.2 Contractor's Liability[250]

The Contractor's total liability to the Sponsor in respect of this Contract {and the Work howsoever arising (including negligence)} shall not exceed [*<<insert number>>* percent of] the Contract Price [provided that if the Contractor has incurred liability under Article 16, such liability shall not be applied to the total liability limit of the Contractor established under this Article 18.2]. The Contractor shall be liable for any penalties or fines imposed on any Sponsor Person as a result of the Work, such as penalties or fines resulting from the Contractor's Permit reporting requirements, and such liability for penalties or fines shall not be limited by this Article 18.2 and shall not be applied to the total liability limit of the Contractor established in the previous sentence. The required insurance coverage set forth in Article 15 shall in no way affect nor limit the Contractor's liability with respect to its performance of the Work. The Sponsor's approvals or issuance of a Certificate of [Facility] [Unit] Commissioning Completion, a Certificate of [Facility] [Unit] Provisional Acceptance or Certificate of Facility Final Acceptance shall not in any way modify or alter the Contractor's obligations hereunder, and neither the inspection, approval or payment, including the making of final payment, under this Contract shall:

(a) be construed to be an acceptance of defective or deficient Work;
(b) be an admission of the Contractor's satisfactory performance of the Work; or
(c) relieve the Contractor of any of its obligations under this Contract.

18.3 Damage Limitations

In no event, <{whether}> as a result of breach of contract, [{tort liability (including negligence), strict liability or otherwise}] shall either Party be liable to the other Party for the special, [{incidental},] indirect or consequential damages of such Party of any nature whatsoever, including loss of use, loss of bonding capacity, loss of efficiency or loss of profit; provided that any damages (other than contractual damages) whatsoever paid by a Party to another Person (other than a Party) shall not be considered consequential damages. For the avoidance of doubt, the provisions of this Article 18.3 shall not limit the Contractor's obligations to achieve Facility Final Acceptance and pay Liquidated Damages [and/or National Content Actual Damages].

18.4 Limitations on Delay Damages [and Shutdown Liquidated Damages] [and] Loss of Availability] Damages[251]

18.4.1   The Sponsor shall not be entitled to any remedies for failure of the Contractor to meet the Completion Guarantees (unless the Delay Damages to be assessed hereunder are not enforceable by the Sponsor for any reason whatsoever in which case the Sponsor shall be entitled to its actual damages [{up to a maximum of the amount which would have been recovered if the Delay Damages had been enforceable}]) except:

(a) the assessment of Delay Damages under this Contract; or
(b) the remedies listed in Article 17.2 which shall be available for a Contractor Event of Default occurring under Article 17.1(k) [{(provided that the remedies listed in Article 17.2 based on Contractor Event of Default under Article 17.1(k) shall not be available to the Sponsor while the Contractor continues to pay Delay Damages) despite the limit imposed by Article 18.4.2}].

18.4.2   The aggregate amount of Delay Damages [plus Shutdown Liquidated Damages] [plus Loss of Availability Liquidated Damages] that the Contractor shall be required to pay shall not exceed <<*insert number*>> percent of the Contract Price as such may change by Price Adjustment from time to time (the "Delay Damages

Cap"). [In no event shall the Contractor's liability for Loss of Availability Damages exceed <<*insert number*>> percent of the Contract Price as such may change by Price Adjustment from time to time.]

18.4.3   The Contractor's liability with respect to a failure of the Facility to achieve the Facility Minimum Performance Levels or to meet the requirements of applicable Laws and achieve Facility Provisional Acceptance shall not be limited in any way.

18.5 Nature of Liquidated Damages[252]

Each Party acknowledges and agrees that:

(a) the Sponsor will suffer financial loss in case of any situation for which Liquidated Damages have been provided as a remedy herein and such financial loss will be difficult to calculate precisely; and
(b) all Liquidated Damages provided for herein are a fair and reasonable estimation of the Sponsor's expected financial loss, are not meant to serve as penalties designed to deter breach hereof and they reflect the Parties' assessment and estimate of such financial loss referred to above.

In light of the foregoing and the equal bargaining power of the Parties, the Contractor accepts and agrees to the Liquidated Damages as liquidated damages.

18.6 No Excuse of Performance

Notwithstanding the Contractor's payment of Liquidated Damages in accordance with Article 9, the Contractor shall complete the Work and achieve Facility Final Acceptance.

18.7 [Joint and Several Liability]

Any reference to the "Contractor" shall be a reference to each of [Choose name of Consortium Member] and [Choose name of Consortium Member], acting jointly and severally and each of [Choose name of Consortium Member] and [Choose name of Consortium Member] shall be responsible for all the obligations of the Contractor as if it were the only Person comprising the Contractor.

ARTICLE 19. ADJUSTMENTS

19.1 Adjustments Generally[253, 254]

19.1.1   If, pursuant to any provision in this Contract, the Contractor is expressly entitled to an Adjustment, the Parties shall in good faith attempt to agree upon an Adjustment following the procedures set forth herein.

19.1.2   No Adjustment Claim shall be allowed to be submitted for any reason on or after the Facility Provisional Acceptance Date if the incident giving rise to the Adjustment Claim therefor occurred [more than <<*insert number*>> Days] prior to the Facility Provisional Acceptance Date.

19.1.3   In no event shall the Contractor be entitled to:

(a) recover any damages or additional compensation from the Sponsor (or seek to recover such from any Sponsor Person or any of their counterparties); or
(b) any extension of time or relief for its performance hereunder; on account of any delay, obstruction or interference attributable to any Person or cause (including doctrines of so-called "owner delay" or "constructive acceleration"), unless such delay, obstruction or interference is expressly stated herein as grounds for an Adjustment, in which case the Contractor shall only be entitled to such relief as is specified in such Adjustment and the Contractor hereby expressly waives for the consideration of <<*insert currency and amount*>> any and all rights, on its own behalf as well as on the behalf of all Contract Persons, to make any claim against any Sponsor Person for damages or additional compensation or extension of time or excuse for performance as a result thereof under any legal or equitable theory including common law principles of force majeure, mistake, and frustration.

19.1.4   If the Contractor intends to submit to the Sponsor an Adjustment Claim, the Contractor must do so within the time period specified therefor in the relevant provision of this Contract. [No Adjustment shall be made if the Contractor fails to submit such Adjustment Claim to the Sponsor within the relevant period therefor specified in this Contract.] All Adjustment Claims shall be substantially in the form of Exhibit 1A and shall contain:

(a) all changes, if any, as may be required to the Project Variables; and
(b) such additional information or documentation as shall be necessary or helpful in considering such Adjustment Claim.

19.1.5   The Parties shall attempt to agree upon an Adjustment within <<*insert number*>> Days of the Contractor's submission of an Adjustment Claim to the Sponsor; provided always that while any Adjustment Claim is pending, the Contractor shall continue to perform the Work for the Contract Price, in strict adherence to the Baseline Level 3 Schedule, and otherwise in accordance with the provisions of this Contract.

19.2 Price Adjustments Specifically

19.2.1   Any Adjustment Claim submitted by the Contractor for a Price Adjustment shall be accompanied by appropriate supporting documentation containing an itemized breakdown of all elements constituting the basis for such Adjustment Claim, including the following:

(a) estimated engineering man hours at hourly billing rates in accordance with the rates set forth in Exhibit 1A;
(b) equipment quantities and costs thereof with no overhead or profit charge or other markup in excess of those as set forth in Exhibit 1A;
(c) labor costs (identified with each specific operation to be performed); and
(d) construction equipment required with no overhead or profit charge or other markup in excess of those as set forth in Exhibit 1A.

19.3 Time Adjustments Specifically

19.3.1   Any Adjustment Claim submitted by the Contractor for a Time Adjustment shall be accompanied by appropriate supporting documentation containing a detailed explanation in accordance with critical path logic of why each affected Deadline should be postponed and for how long, and such supporting documentation shall include updated critical path schedules satisfying the requirements of Article 4.6 and demonstrating, by comparison of unaffected and affected schedule versions, the relationship between the initiating event and both the existence and extent of the Contractor's need for schedule relief. [No Time Adjustment shall exceed the number of Days that the Work in question was delayed or preserve all or any portion of the Float.][255]

19.3.2   No Time Adjustment shall be permitted unless:

(a) the delay in question:
    (i) is unforeseeable at the Contract Date;
    (ii) is not caused by any Contractor Person; and
    (iii) affects a Critical Path Item; and
(b) the Contractor's performance which is affected by the delay referred to in clause (a) above would not have been simultaneously delayed or interrupted by any other circumstances caused by any Contractor Person.

19.4 Change Orders

When an Adjustment has been determined and agreed upon pursuant to this Article 19, the Parties shall execute a Change Order reflecting such Adjustment. The Contractor may not vary or alter any part of the Work, except in accordance with a Change Order agreed upon by the Parties or, if no agreement is reached, as directed by the Sponsor pursuant to Article 3.6. Once the Parties execute a Change Order with respect to any matter, the Contractor shall not be entitled to any other or further Adjustment for any item that was included in such Change Order, nor will any subsequent Change Order be construed to alter a prior Change Order except to the extent expressly set forth in such subsequent Change Order. Changes in Project Variables shall be made only to the extent expressly set forth in a Change Order executed by the Parties.[256]

19.5 Adjustment Dispute Resolution[257]

If the Parties fail to agree upon any Adjustment Claim in accordance with the procedures set forth herein within <<*insert number*>> Days of the Contractor's submission of such Adjustment Claim, either Party may, by written notice to the other Party, elect to resolve the dispute pursuant to Article 20.

ARTICLE 20. DISPUTE RESOLUTION AND GOVERNING LAW

20.1 *Mutual Discussions*[258, 259]

20.1.1   If a dispute or difference of any kind whatsoever (a "Dispute") shall arise in connection with, relating to, or arising out of this Contract, including any question regarding its existence, validity, interpretation, performance, nonperformance, or termination, the Parties shall attempt to resolve such Dispute by discussions between <their executive vice presidents> <the presidents of each of their <<*insert name of*>> ultimate parents of the Parties>.[260] Such discussions shall take place no later than <<*insert number*>> Days after either Party has notified the other Party of the Dispute in writing (a "Dispute Notice") and shall last no longer than <<*insert number*>> Days. During that period, the Parties' designated representatives may meet as many times as may be appropriate.

20.2 *[Arbitration*[261]

If such Dispute has not been resolved as prescribed above by such officers within <<*insert number*>> Days after the delivery of the Dispute Notice, it shall be finally settled by arbitration conducted under the <<Rules of Arbitration of the International Chamber of Commerce>> <<Rules of the American Arbitration Association>> <<*insert other rules if desired*>> (the "Rules") then in effect. The arbitration shall be heard by three arbitrators appointed [in accordance with the Rules as and if modified by this Article 20.2 (an <"Arbitration")>]. [To the extent the Rules are inconsistent with the provisions hereof, the provisions hereof shall prevail.] The Sponsor shall nominate one arbitrator. The Contractor [and the Parent Guarantor] shall jointly nominate another arbitrator. Each Party shall nominate its arbitrator no later than <<*insert number of days*>> after the request for arbitration has been filed. The third arbitrator shall be appointed by the two Party-nominated arbitrators within <<*insert number*>> of days of the date the second Party-nominated arbitrator is appointed. If either Party fails to nominate an arbitrator, or the two Party-nominated arbitrators fail to appoint a third arbitrator, the arbitrator(s) shall be appointed by the <<American Arbitration Association>> <<International Chamber of Commerce>> <<*insert other body if desired*>>. The tribunal shall have the authority to determine any question or challenge to its jurisdiction, including without limitations, challenges to the existence of a valid arbitration agreement. The seat of arbitration shall be <<*insert location of city*>>. Hearings may be held in any other suitable venue as determined by the arbitral tribunal after having heard the Parties. The language of the arbitration shall be English. The Parties agree to consolidate any [proceeding] [arbitration] hereunder with any [proceeding] [arbitration] commenced under the Parent Guarantee.][262] In order to facilitate the comprehensive resolution of related disputes, and upon request by either Party, the arbitration panel shall at any time [before the first oral hearing of evidence] consolidate the arbitration proceeding with any other arbitration proceeding concerning the Parties in a manner so that the panel first constituted will hear all disputes.

[The arbitrators shall have the power to order, by interlocutory award or otherwise, consolidation of proceedings, joinder of parties, issuance of writs of attachment, foreclosure of mechanics liens, enforcement of stop notices and other provisional remedies, including temporary and permanent restraining orders and injunctions]. The decision of the arbitrators shall be final and binding but may not disregard any express provision hereof and may be entered as a judgment in any court of competent jurisdiction.]

20.3 [Dispute Review Board][263]

20.3.1   The Parties shall impanel a dispute review board (the "Board") to assist in the resolution of Disputes. The Board shall fairly and impartially consider all Disputes brought before it in accordance with the procedures set forth herein and shall provide written recommendations to the Sponsor and Contractor to assist in the resolution thereof. Recommendations of the Board shall not be binding on either the Sponsor or the Contractor.

20.3.2   The Board shall be established as follows.

Within <<insert number>> Days of the Notice to Proceed, the Sponsor and Contractor shall present the other with a list of three Board member candidates. The Parties will make their best endeavors to propose candidates who, in addition to possessing recognized professional credentials or considerable practical experience, are familiar with the subject matter of the Contract, including the preparation, interpretation, and execution of contracts similar to the Contract and the resolution of disputes arising therefrom. The lists of candidates shall include detailed biographical information of each candidate, disclosures of any known or potential conflicts of interest or other disqualifying considerations, and member charge rates inclusive of travel and other incidental costs. The Board will consist of one member. The Sponsor and the Contractor shall each have <<insert number>> Days in which to strike the names of any unacceptable candidates and number the remaining names in order of preference from the other Party's list of candidates. In the event that both Parties' lists of candidates contain the same candidate, that candidate will be the Board member. Otherwise, the Parties shall negotiate a mutually agreeable Board member. The Board will consist of three members, one nominated by the Sponsor and approved by the Contractor, one nominated by the Contractor and approved by the Sponsor, and a third member nominated by the first two members and approved by both the Sponsor and the Contractor. Within <<insert number>> Days of receiving the list of proposed members, each of the Sponsor and the Contractor shall select a member proposed by the other Party and that member shall be appointed to the Board. In the event that either Party is unable to select a member from the other's list, the other Party shall propose a new list of member candidates for consideration. The objecting Party shall not be required to identify reasons for non-acceptance of the other Party's original list of proposed members. Within <<insert number>> Days of receiving the second list, the objecting Party shall select a member proposed by the other Party and that member shall be appointed to the Board. In the event that either Party is unable to select a member from the other's list, the objecting Party must provide a reasonable justification for its refusal on the basis of the qualification or conflict criteria outlined herein. In the event that all proposed members are unacceptable to either Party, the other Party shall propose a new list of member candidates for consideration. Within <<*insert number*>> Days of selection of two (2) acceptable members, those members shall jointly select a third member from those candidates remaining as without objection on either of the Parties' lists of proposed members. The third member shall serve as Chairman of the Board (the "Chairman"). To the extent practicable, the two members shall endeavor to select a third member whose qualifications complement the experience of the first two and who can competently manage the leadership and administrative functions of the Board. In the event that no acceptable candidates remain, the Parties shall generate new lists of proposed members as above. In the event of an impasse in selection of the third member from nominees of the first two members, the third member shall be selected by mutual agreement of the Sponsor and the Contractor. All activities of the Board shall be directed by, and all communications with the Board shall be made through the Chairman. The Chairman will be responsible for directing the course of the hearing and for keeping the official records of the hearing.[264]

20.3.3 [Immediately following selection of the Board, the Sponsor and Contractor shall select an additional non-active member (the "Back-up Member") to serve as a replacement in the event that any Board member is unable to continue service. The Back-up Member shall be included as a recipient of all routine progress reports and contract meeting minutes, but, until such time as the Back-up Member may be called upon to serve on the Board, shall not receive copies of documentation related to a specific Dispute.]

20.3.4 The Board shall at all times remain neutral, act impartially, and be free of any actual or apparent conflict of interest in regard to the project or its participants. Board members may not have:

(a) an ownership interest in any entity involved with the Contract;
(b) a financial interest in the Contract, except for payment for services as a member of the Board;
(c) been previously employed by, or have any financial ties to, any Party, including fee based consulting services, [within a period of <<*insert number*>> years prior to award of the Contract, except with the express written approval of both Parties];
(d) a business or personal relationship with any key members of any Party; or
(e) prior involvement in the Facility of a nature that could compromise that member's ability to participate impartially in the Board's activities.

20.3.5 Board members shall not engage in any ex parte communication with any Party to the Contract, including consultants, attorneys, subcontractors, suppliers or others on any current or pending matter which may come before the Board.

20.3.6 Board members have a continuing duty to disclose to the Parties any circumstance likely to have an actual or apparent effect on their impartiality. Notwithstanding any time limits identified in the member selection process in Article 20.2.2, upon learning of any new information which would disqualify a current Board member, any Party may object to that member's continued participation on the Board. Except in the case of a conflict as outlined in Article 20.2.2(a) or (b) above, any such objection shall only be made where the disqualifying information was not or, with reasonable diligence on the part of the objecting Party could not have been, known to the Party at the time of the member's selection.

20.3.7   If for any reason a Board member is unable to perform his duties, the Backup Member shall replace such member and, if the Parties deem it appropriate, the Parties shall select a new Backup Member in the same manner as the original Backup Member candidate selection process as outlined in Article 20.2.2.

20.3.8   <u>Dispute Review Board Operation</u>. The Board shall operate as follows:
    Upon its being established, the Board, in consultation with the Sponsor and the Contractor, will establish a procedure to provide the Board with <<monthly>> progress reports under the Contract. The Board members will use these reports to keep informed of construction developments and the progress of the Work.

20.3.9   Every <<*three (3)*>> months, the Board will visit the <Route> <Site> to examine the progress of the Work and meet with representatives of the Sponsor and the Contractor. Each meeting shall begin with an informal meeting between the Board and both Parties, followed where appropriate by a <Site> <Route> inspection accompanied by authorized representatives of both the Sponsor and Contractor and, if necessary, a follow up meeting to discuss any ongoing or developing disputes. All inspections and meetings shall be attended by authorized representatives of the Sponsor and Contractor, at which time the Board may engage in and facilitate among and between the Parties informal discussions in order to resolve any pending issues that may become Disputes. At no time during the <Site> <Route> inspections or review meetings shall they engage in ex parte discussions, formal or informal, with either Party. Within two weeks after the <Site> <Route> visit, the Chairman will distribute minutes of the <Site> <Route> visit meeting to the Contractor and the Sponsor.

20.3.10   The Parties shall not indiscriminately refer matters to the Board for resolution. In the presence of a Dispute, either the Sponsor or Contractor may refer a Dispute to the Board if the Parties have exhausted all reasonable good faith efforts to amicably and fairly settle their differences on their own and either Party believes that bilateral negotiations are not likely to succeed or have reached an impasse.

20.3.11   Requests for Board Review of a Dispute (a "<u>Request for Review</u>") shall be submitted by the complaining party (the "<u>Plaintiff</u>") in writing to the Board, with simultaneous complete copy to the opposing Party (the "<u>Respondent</u>"), setting forth the nature of the dispute, the factual and contractual basis of the Dispute, all remedies sought and all documents that support each element of the claim. Within <<*insert number*>> Days after a Request for Review has been received by both the Board and the Respondent, the Respondent shall submit to the Board, with a simultaneous complete copy to the Plaintiff, a written response (the "<u>Response</u>"), setting forth the factual and contractual basis of any defense and all documents that support each element of the defense. If the Respondent wishes to counterclaim, the Respondent shall indicate so in its Response, and within <<*insert number*>> Days after submitting its Response, submit to the Board, with a simultaneous complete copy to the Plaintiff, a written Counterclaim setting forth the factual and contractual basis of the counterclaim, all remedies sought and all documents that support each element of the Counterclaim. Within <<*insert number*>> Days after the Counterclaim has been received by both the Board and the Plaintiff, the original Plaintiff shall submit to the Board, with a simultaneous complete copy to the Respondent, a written response (the "Counterclaim Response"), setting forth the factual and contractual basis of any defense and all documents that support each element of the defense.

20.3.12   The Board process shall not relieve either Party of its contractual obligations, and both Parties shall proceed diligently with the Work and comply with all Contract provisions while any Dispute is pending or before or otherwise under consideration by the Board.

20.3.13   Within <<*insert number*>> Days receipt of the later of a Response to a Request for Review or Counterclaim Response, the Board will, in consultation with the Sponsor and the Contractor, establish dates for any additional pre-hearing submissions and schedule a hearing date. The hearing will generally be conducted at the time of the next regularly scheduled <Site> <Route> visit. The Board may also convene a preliminary hearing by conference call for the purpose of addressing information exchange, the order of proceedings at the hearing, consolidation of disputes, and any other matters that the Board believes may expedite the hearing process.

20.3.14   Hearings shall be held at the <<Site>>.

20.3.15   The Board shall establish rules and procedures for the informal, efficient, fair and expeditious hearing of disputes consistent with the terms of the Contract. To the extent that changes in established procedures do not unfairly impact either Party, the Board shall have the authority to modify those rules and procedures and conduct the hearing as it deems most appropriate, consistent with the Contract.

20.3.16   Within <<*insert number*>> Days receipt of the later of a Counterclaim Response, the Board will notify the Sponsor and Contractor of the dates for any additional pre-hearing submissions and the hearing date. Except in circumstances that in the opinion of the Board require immediate action, the hearing will be scheduled during a regularly scheduled <Site> <Route> visit.

20.3.17   Subject to any contrary direction from the Board, at least <<*insert number*>> Days in advance of any scheduled hearing, each Party shall prepare and submit via hand-delivery or overnight mail to the Board, with a complete copy to the opposing Party, a brief statement of the Dispute describing as concisely as practicable:

(a) the nature of the Dispute, including where applicable the claim amount in time and in money;

(b) the basis for the Party's belief of its entitlement, including clearly defined contractual justification (with copies of relevant contract excerpts) for the stated position;

(c) the Party's reasoning as to why the opposing Party's position is not contractually correct and appropriate; and

(d) all exhibits referred to in the statement of position or to be utilized in planned hearing testimony and any other reference material necessary to present the Party's position.

20.3.18   Subsequent to submission of statements to the Board and opposing Party, unless approved otherwise by the Board, the Parties shall not send the Board or opposing Party any further exhibits or correspondence regarding the Dispute.

20.3.19   The Board will not be bound by any judicial rules of evidence or burden of proof. The Board may limit the presentation of documents or oral statements when it deems them to be irrelevant or redundant or when it determines it has sufficient understanding of the facts underlying a claim or defense to make its recommendations.

20.3.20   Subject to the Board's general authority to control the progress of the hearing and specific determination that all aspects of the Dispute have been adequately addressed, each Party shall make an initial presentation of its position with respect to the Dispute or defense and one or more rebuttals to any assertion by another Party. The Board may limit the presentation of documents or oral statements when it considers them to be irrelevant or redundant. Board members shall control the hearing and may guide the discussions of issues by asking questions of the Parties in order to obtain all information the Board considers necessary to make its recommendations. The Board may permit the questioning of one Party by another Party only if it would facilitate the presentation or clarification of an issue in the Dispute.

20.3.21   The Sponsor and Contractor shall be represented at all hearings by authorized representatives. At least <<*insert number*>> Days prior to a scheduled hearing date, the Parties shall exchange and submit simultaneously to the Board a list of their respective witnesses and representatives planning to attend the meeting, including each individual's name, title, professional affiliation, and a brief summary of the matters that the person will address. If a Party intends to be accompanied at the hearing by counsel or by an independent expert, and has not notified the others within the appropriate time period that it intends to be accompanied by counsel or by an independent expert, but receives another Party's notice that it will be so accompanied, the recipient Party may then elect to be accompanied by counsel or independent expert provided it notifies the other Party and the Board in writing at least <<*insert number*>> Days prior to the scheduled meeting date. Unless the Board determines that it would assist resolution of the Dispute, counsel may not (a) examine directly or by cross examination any witness, (b) object to questions or factual statements made or related during the meeting, or (c) make motions or offer arguments.

20.3.22   The Board will notify all affected Parties in writing if it decides to consolidate Disputes arising out of the same circumstances or involving similar factual or legal issues, or if it decides to bifurcate a Dispute into separate determinations as to merit and costs. The Parties shall conform their submission of documents and presentations at the hearing to be consistent with such consolidation or bifurcation as directed by the Board.

20.3.23   Where any Dispute includes a Subcontractor claim, the Contractor shall provide all relevant Subcontractor documents and clearly identify in all submissions any portion of the Dispute that involves a Subcontractor claim and the identity of the related Subcontractor. When any Board meeting or hearing is convened to consider a Dispute that includes one or more Subcontractor claims, the Board may require, and the Contractor shall be responsible to <{request}> <ensure>, that an authorized representative of each Subcontractor involved in the Dispute with direct and actual knowledge of the facts underlying the Subcontractor claim attend to assist in presenting the Subcontractor claim and to answer questions raised by the Board members or the Sponsor's representatives. The Contractor's failure to secure any Subcontractor's participation may be considered by the Board in making its recommendation.

20.3.24   A Party's failure to comply with the Board's meeting rules and procedures and other requirements, or use of such rules to obstruct the process, may be considered by the Board in making its recommendations.

20.3.25   After the close of any hearing, the Board will meet to formulate recommendations for resolution of the Dispute. All Board deliberations will be conducted in private and will be confidential.

20.3.26   Within <<*insert number*>> Days after the close of a hearing, the Board will forward its recommendation for resolution of the Dispute in writing to both the Sponsor and the Contractor. Within <<*insert number*>> Days of receiving the Board's recommendation, or such other time specified by the Board, both the Sponsor and the Contractor shall provide written notice to the other and to the Board stating their acceptance or rejection of the Board's recommendation. The failure of either Party to respond within the specified period shall be deemed an acceptance of the Board's recommendation. If, with the aid of the Board's recommendation, the Sponsor and the Contractor are able to resolve their Dispute, the Parties will promptly implement any required modifications to the Contract.

20.3.27   Should the Dispute remain unresolved because of a bona fide lack of clear understanding of the Board's recommendation, either Party may request, and the board at its discretion may clarify, specified portions of the recommendation.

20.3.28   *If new information becomes available subsequent to the hearing, either Party may request* that the Board consider such and the Board shall do so.

20.3.29   If the Board's recommendation does not resolve the Dispute and become binding on the Parties because a Party has objected to such recommendation in writing delivered to the other Party within ten (10) Days of its receipt thereof, the written recommendation, including any dissenting opinion, will [not] be admissible as evidence [to the extent permitted by Law] in any subsequent Dispute resolution proceeding. The Board will maintain copies of all party submissions, hearing relating documents (notes, minutes, etc.) and the Board's written recommendation for the duration of its term, at the end of which it shall convey all such documents in its possession to the Sponsor. The Sponsor shall maintain any such documents for a period of <<*insert number*>> years following Facility Provisional Acceptance.

20.3.30   All costs associated with the assembly and function of the Board ("Board Costs") shall be shared equally by the Parties. Board Costs shall include member compensation in accordance with the individual charge rates identified at time of election in accordance with Article 20.2.2, including all incidental travel and related site inspection, meeting and hearing costs.

20.3.31   Where required, and with the mutual agreement of the Parties, the Board may procure as Board Costs additional technical, legal, accounting, research or other specialty services necessary to the Board's function.

20.3.32   Any expenses associated with the presentation of witnesses, experts, or other representatives of the Sponsor and the Contractor shall be exclusively to the account of the Party responsible for procuring such additional assistance.

20.3.33   Upon approval by both Parties of all invoices for Board Costs, the Contractor shall pay the full amount of such invoices and submit evidence of the same to the Sponsor for reimbursement of half thereof, whereupon the Sponsor shall execute a Change Order for that amount.

20.3.34   The Board will be dissolved as of the Facility Final Acceptance Date or, should any Disputes be pending as of that date, the date on which the Board issues its recommendations regarding those Disputes.

20.4 Pendency of Dispute[265]

During the existence of any Dispute under this Contract (including the pendency of any arbitration or other proceedings related thereto):

(a) each Party shall (unless expressly permitted otherwise by a provision of this Contract) continue to perform its obligations under this Contract; and
(b) neither Party shall exercise any other remedy or termination right under this Contract arising by virtue of the matters in dispute.

Where any Dispute relates to the nature of the Parties' obligations under the Contract, the Sponsor shall give the Contractor any instructions as may be necessary for proper performance of the Work and to prevent any delay of the Work pending settlement of the Dispute. The Contractor shall comply immediately and fully with those instructions, it being understood that, by so doing, the Contractor will not prejudice its contractual or legal position in a Dispute. [{If it is subsequently determined that the instructions were contrary to the Contract requirements, the Sponsor shall pay the costs incurred by the Contractor in carrying out those instructions beyond what the Contract correctly understood and interpreted would have required}.] If the Contractor has failed to continue to perform the Work diligently in strict adherence to the Baseline Level 3 Schedule and otherwise in accordance with the provisions of this Contract in connection with a matter involving an Adjustment Claim, the Contractor hereby waives its right to pursue an Adjustment for the matter in question.

20.5 Judgment Currency

If, for the purpose of obtaining an award or judgment before any arbitral tribunal or any court with respect to the enforcement of any award made by any arbitral tribunal or court, it shall be necessary to convert into any other currency any amount due thereunder and a change shall occur between the rate of exchange applied in making such conversion and the rate of exchange prevailing on the date of payment of such award or judgment, each Party agrees to pay such additional amounts (if any) as may be necessary to ensure that the amount paid on the date of payment to the other Party is the amount in such other currency that, when converted in accordance with normal banking procedures, will result in receipt by such other Party of the full amount then due. Any amount due from either Party under this Article 20.5 shall be due as a separate debt and shall not be affected by or merged into any award or judgment being obtained for any other sum due under or in respect of this Contract.[266]

20.6 Costs Reimbursed

The prevailing Party in any Dispute shall [not] be entitled to have all costs that it incurred in connection therewith reimbursed upon demand by the other Party including arbitrator fees, fees to the arbitration organization, and attorneys' and experts' fees and costs. The arbitration panel shall include in its final ruling a determination as to which Party is the prevailing Party for purposes of implementing this provision.

20.7 Non-Waiver

Neither Party shall be deemed to have waived any right under this Contract unless such Party shall have delivered to the other Party a written waiver signed by such waiving Party. No failure or successive failure by either Party to enforce any covenant or agreement, and no waiver or successive waivers by either Party of any condition of this Contract, shall operate as a discharge of such covenant, agreement or condition, or render the same invalid, or impair such Party's right to enforce the same in the event of any subsequent breach thereof by the other Party.

20.8 Severability

[{1}]If any of the terms, covenants or conditions hereof or the application of any such term, covenant or condition shall be held invalid or unenforceable as to either

Party or as to any circumstance by any court or arbitrator having jurisdiction, the remainder of such terms, covenants or conditions shall not be affected thereby, shall remain in full force and effect and shall continue to be valid and enforceable in any other jurisdiction. In such event, the Parties shall negotiate in good faith to substitute a term, covenant or condition in this Contract to replace the one held invalid or unenforceable by a mutually agreed amendment to this Contract with a view toward achieving a valid and enforceable legal and economic effect as similar as is then reasonably possible to that originally provided for in this Contract.

[{(2) Survival of Provisions. In order that the Parties may fully exercise their rights and perform their obligations hereunder arising from the performance of the Work, such provisions of this Contract that are required to insure such exercise or performance shall survive the termination of this Contract for any cause whatsoever. No action shall be brought by either Party based on any Dispute arising out of or related to this Contract, or any breach thereof, more than <<*two (2)*>> years after accrual of the cause of action}.][267]

20.9 Entire Agreement

This Contract constitutes the entire agreement and contains all of the understandings and agreements of whatsoever kind and nature existing between the Parties, and supersedes all prior written or oral agreements, commitments, [representations, communications][268] and understandings between the Parties and all of their affiliates including the Letter of Intent dated <<*insert date*>> between <<*insert names*>> [, for all of which affiliates the Parties are duly authorized by power of attorney to act in this regard].[269,270]

20.10 Amendment

No amendment, waiver or consent relating to this Contract shall be effective unless it is in writing and signed by the Parties.

20.11 Successors and Assigns; Third-Party Beneficiaries[271]

All of the terms and provisions of this Contract shall be binding upon and inure to the benefit of the Parties and their respective successors and assignees expressly permitted hereby. This Contract is for the sole benefit of the Parties and each Person who is expressly stated herein as being an intended third-party beneficiary hereof, and is not for the benefit of any other Person.

20.12 Descriptive Headings

Descriptive headings herein are for convenience only and shall not control or affect the meaning or construction of any provision of this Contract.

20.13 Remedies Non-Exclusive[272]

Except as expressly provided herein,[273] all remedies provided in this Contract shall be deemed cumulative and not in lieu of, or exclusive of, each other or of any other remedy available to either Party under this Contract, at law or in equity, and the exercise of any remedy, or the existence herein of other remedies, shall not prevent the exercise of any other remedy, provided that the Contractor shall not have any right to suspend its performance or terminate this Contract except as expressly provided in Article 17.4.

20.14 Governing Law[274]

[This Contract shall be governed by the laws of England.] [AS PERMITTED BY GENERAL OBLIGATION LAW § 5–1401 OF NEW YORK STATE,][275] THIS CONTRACT SHALL BE GOVERNED BY THE LAWS OF THE STATE OF NEW YORK. [276][ANY LEGAL ACTION OR PROCEEDING WITH RESPECT TO THIS AGREEMENT [MAY] [MUST] BE BROUGHT IN THE COURTS OF THE STATE OF NEW YORK OR OF THE UNITED STATES OF AMERICA FOR THE SOUTHERN DISTRICT OF NEW YORK AND, BY EXECUTION AND DELIVERY OF THIS AGREEMENT, EACH OF THE PARTIES HEREBY ACCEPTS FOR ITSELF AND IN RESPECT OF ITS PROPERTY, GENERALLY AND UNCONDITIONALLY, THE [EXCLUSIVE] JURISDICTION OF THE AFORESAID COURTS].[277] EACH OF THE PARTIES HEREBY IRREVOCABLY CONSENTS TO THE SERVICE OF PROCESS OF ANY OF THE AFOREMENTIONED COURTS IN ANY SUCH ACTION OR PROCEEDING BY THE MAILING OF COPIES THEREOF TO IT BY REGISTERED OR CERTIFIED MAIL, POSTAGE PREPAID, RETURN RECEIPT REQUESTED, TO EACH OF THE PARTIES AT ITS ADDRESS SET FORTH IN ARTICLE 20.18.[278] EACH OF THE PARTIES HEREBY IRREVOCABLY WAIVES ANY OBJECTION WHICH IT MAY NOW OR HEREAFTER HAVE, INCLUDING THAT OF INCONVENIENT FORUM, TO THE LAYING OF VENUE OF ANY SUCH ACTION OR PROCEEDING IN SUCH JURISDICTIONS. TO THE EXTENT PERMITTED BY APPLICABLE LAW, EACH OF THE PARTIES HEREBY IRREVOCABLY WAIVES ALL RIGHT OF TRIAL BY JURY IN ANY ACTION, PROCEEDING OR COUNTERCLAIM ARISING OUT OF OR IN CONNECTION WITH THIS AGREEMENT OR ANY MATTER ARISING HEREUNDER. THE PARTIES HEREBY EXPRESSLY AGREE TO EXCLUDE AND DISCLAIM THE APPLICATION OF THE PROVISIONS OF THE UNITED NATIONS CONVENTION ON CONTRACTS FOR THE INTERNATIONAL SALE OF GOODS, AND ANY SUCCESSOR CONVENTION OR LEGISLATION, TO THIS CONTRACT.

20.15 Language[279]

The official text of this Contract shall be English, regardless of any course of dealings or any translation that may be made for the convenience of the Parties. All correspondence, information, literature, data, manuals, definitive documents, notices, waivers and all other communication, written or otherwise, between the Parties in connection with this Contract shall be in English and no document in any other language may be used to construe any ambiguity determined by a court or arbitral panel to exist herein.

20.16 Agent for Service of Legal Process

Each Party irrevocably appoints <<*insert name*>> of<<*insert location*>> as its agent to receive and acknowledge on its behalf service of any writ, summons, order, judgment or other notice of legal process in [<<*insert jurisdiction*>>] or for purposes of a Dispute. Each Party agrees that any such legal process shall be sufficiently served on it if delivered to such agent for service at its address for the time being in the [<<*insert jurisdiction*>>] whether or not such agent gives notice thereof to the concerned Party.[280]

20.17 Rules of Construction

In the interpretation of this Contract, unless the context otherwise expressly states:

(a) the term "or" is not exclusive;
(b) the terms "including", "include" and "includes" shall be deemed to be followed by the words "without limitation";
(c) any reference to any gender includes the other gender;
(d) [unless referred to in the definition of the Sponsor Information],[281] any reference to any agreement, instrument, contract or other document shall be a reference to such agreement, instrument, contract or other document as amended, supplemented, modified, suspended, replaced, restated or novated from time to time;
(e) any reference to any Law shall include all statutory and administrative provisions consolidating, amending or replacing such Law, and shall include all rules and regulations promulgated thereunder;
(f) any reference to any Person includes its successors and assigns;
(g) all obligations under this Contract of any Party are to be construed as continuing obligations throughout the term hereof;
(h) the fact that counsel to any Party shall have drafted this Contract shall not affect the interpretation of any provision of this Contract in a manner adverse to such Party or otherwise prejudice or impair the rights of such Party; and the preparation of this Contract has been a joint effort of the Parties and the resulting document shall not be construed more severely against one of the Parties than against the other;[282]
(i) the terms "hereof", "herein", "hereunder", "this Contract" and words of similar import shall be deemed to be references to this Contract as a whole, including all Exhibits, and not to any particular Article, Exhibit, or portion of this Contract;
(j) wherever the provisions of this Contract require, provide for or permit an approval, agreement or consent by either Party of or to any action, person, document, or other matter contemplated by this Contract, the following provisions shall apply:
   (i) [unless expressly stated otherwise in the provision in question,] any approval or consent to be given by the Sponsor must be done so expressly and only in writing, and may be withheld whimsically, capriciously, for any reason whatsoever, or no reason whatsoever;
   (ii) such approval, agreement or consent shall not be unreasonably or arbitrarily withheld, conditioned or delayed (unless otherwise specifically provided herein);
   (iii) the Party whose consent is being sought (the "<u>Consenting Party</u>") shall, within the relevant time period set forth herein or, if no time period is specified, within <<*insert number*>> Days, advise the other Party by notice

either that it consents, agrees or approves or that it withholds its consent, agreement or approval, in which latter case it shall set forth, in reasonable detail, its reasons for withholding its consent, agreement or approval; <u>provided</u> that, if the Consenting Party shall fail to give the other Party the notice contemplated in this Article 20.17, the relevant approval, agreement and consent shall be deemed consented to, agreed to or approved by the Consenting Party with no further action;

(iv) if the responding notice mentioned in Article 20.17 indicates that the Consenting Party does not approve, agree or consent, the other Party may take whatever steps may be necessary to satisfy the objections of the Consenting Party set out in the responding notice and, thereupon, may resubmit such request for approval, agreement or consent from time to time and the provisions of this Article 20.17 shall again apply until such time as the approval or consent of the Consenting Party is finally obtained; and

(v) if the disapproval or withholding of consent, agreement or approval mentioned in Article 20.17 is subsequently determined to have been improperly withheld, conditioned or delayed by the Consenting Party, such approval, agreement or consent shall be deemed to have been given on the date on which such approval, agreement or consent was originally required;

(k) if an index or similar reference referred to in this Contract is changed or no longer published or reported by the Person (or such Person's successor) who, at the date hereof, publishes or reports such index or reference, then the Parties shall use their best efforts to replace such index with the most appropriate substitute for the changed or no longer published index or reference; and

(l) conflicts between any provision of Articles 1 through 21 inclusive and any provision of the Exhibits shall be resolved in favor of the provision of Articles 1 through 21 inclusive [{and conflicts between Exhibits shall be resolved by the Sponsor}].

20.18 Notices

[No notice which is required or permitted to be given by the Contractor shall be effective or considered properly given unless it has been executed by <<*choose name of a Consortium Member if the Contractor is a consortium*>>. Any notice which is required or permitted to be given by the Sponsor to the Contractor hereunder shall be effective and considered properly given if it has been directed to <<*choose name of a Consortium Member if Contractor is a consortium*>>.] [In the case of the Contractor's appointment of an arbitrator under Article 20.2, <<*choose name of a Consortium Member if Contractor is a consortium*>> shall be responsible for the choice and notifying the relevant parties at the [International Chamber of Commerce] of such.] All notices to be given herein shall be effective upon receipt and shall be in writing and delivered to the Parties by reputed overnight courier service at the following address (or such other address as may hereafter be designated in writing by a respective Party in accordance with this Article 20.18) with a facsimile thereof sent to such Party with written confirmation thereof by the sending Party's facsimile machine kept by the sending Party.

[if to the Sponsor's Project Manager:

> his office designated in writing in accordance with this Article 20.18 after his appointment;][283]

if to the Sponsor:
[address]
Telecopy: _____
Telephone: _____
Attention: General Counsel

if to the Contractor:
[address]
Telephone: _____
Facsimile: _____
Attention: Contractor's Project Director

with a copy to:
Telephone: _____
Facsimile: _____
Attention: General Counsel

20.19 Cooperation and Consolidation of Disputes

The Contractor agrees at the request of the Sponsor:

(a) to consolidate any Dispute with any dispute or proceeding;
  (i) under [either] [the] Parent Guaranty;
  (ii) with a dispute or proceeding with any other Contractor Person; and
(b) at no cost, to assist and testify in any dispute or proceeding regarding any Sponsor Person, any Contractor Person or the Facility and whether or not any of the foregoing involves this Contract or any Contractor Person.

ARTICLE 21. MISCELLANEOUS

21.1 Assignment[284]

Except as expressly provided in this Article 21.1, [the Contractor] [no Consortium Member] [shall] [may not] assign any of its rights or obligations hereunder, directly or indirectly, whether by pledge, assignment, sale of assets or sale or merger (statutory or otherwise), without the prior written consent of the Sponsor. The subcontracting of portions of the Work in compliance with the requirements of this Contract shall not be deemed an assignment for purposes of the foregoing.

The Sponsor may assign this Contract or its rights or obligations thereunder to any subsidiary or affiliate of the Sponsor without the consent of the Contractor.[285]

21.2 [Financing][286]

21.2.1   The Contractor acknowledges and agrees that, without the consent of the Contractor, the Sponsor may assign or create a security interest in its rights and

interests under or pursuant to this Contract, the Facility, the Site, any movable property of the Sponsor or any other rights or assets of the Sponsor in favor of any of
the Financing Parties. <Each Consortium Member> <The Contractor> agrees within
<<*insert number*>> Business Days after notice from the Sponsor:[287]

(a) to execute and perform the Lender Consent and Agreement in the form of
    Exhibit 21.2.1;
(b) to execute and perform an Assignment and Assumption Agreement substantially
    in the form of Exhibit 21.2.1(b) and for each Vendor Agreement listed in Exhibit
    1B (but updated to reflect the state of affairs at the time of execution); and
(c) if requested by the Sponsor, to execute such agreements as may be reasonably
    requested by the Financing Parties [{in a project financing}] and to provide an
    opinion of [in-house] counsel[288] of each of [the Contractor] [each Consortium
    Member] and [the] [each] Parent Guarantor regarding the enforceability of
    (i) this Contract, (ii) the Lender Consent and Agreement, and (iii) the Parent
    Guarantee[s] addressed to the Sponsor and the Financing Parties, including in
    the form set forth in Exhibit 21.2.1(c) if requested by the Sponsor.

21.2.2   The Contractor shall assist the Sponsor to facilitate obtaining [project]
financing from [international] Financing Parties on a non-recourse basis. The
Contractor acknowledges that the Financing Parties will review this Contract and may
require changes hereto as a condition of providing financing and/or insurance, and
the Contractor agrees, in good faith, to consider amending this Contract to incorporate any such requirements and otherwise to cooperate with the Financing Parties.
The Contractor shall, without undue delay, respond to reasonable requests and
requirements of any Financing Party for documentation and information related to:

(a) the qualifications, experience, past performance and financial condition of
    Contractor;
(b) any loan distribution under any financing of all or any part of the Work; and
(c) other matters pertaining to the Contractor's participation hereunder and in
    the Work.

21.2.3 The Contractor shall provide such documents and other technical assistance
as the Sponsor may request in connection with obtaining financing for the Facility.

21.2.4   The holder of any security interest in this Contract shall not be prevented or
impeded by the Contractor from enforcing such security interest.

[21.2.5   <The Contractor shall execute all consents to assignment and/or acknowledgments of any security interest as are requested by the Sponsor to give effect to the
foregoing, and shall provide such certificates and opinions of counsel addressed to
the Sponsor and the Financing Parties as may be requested in connection with any
financing of the Facility. The Contractor agrees that such consents and acknowledgments may contain:

(a) an agreement by the Contractor to allow the holder of such security interest to
    cure defaults by the Sponsor;

(b) an acknowledgment by the Contractor that the Sponsor is not in default under this Contract if such is the case;

(c) a prohibition against amending, assigning or terminating this Contract without the prior written consent of the holder of such security interest;

(d) a consent by the Contractor to allow the assignment of this Contract to the successors in interest of the holder of such security interest after foreclosure thereon; and

(e) other provisions that are conventional and reasonable in financing facilities such as the Facility on a non-recourse basis.

21.2.6   If so required by any Financing Party, the Contractor shall agree not to terminate this Contract on account of a Sponsor Event of Default until it has provided the Financing Party with written notice of such Sponsor Event of Default at least <<*insert number*>> Days prior to any intended termination. If the Contractor fails to give such notice, any termination of this Contract shall not be effective until <<*insert number*>> Days after such notice is given to such Financing Party. The Contractor shall not terminate this Contract as a result of any Sponsor Event of Default if:

(a) such Sponsor Event of Default has been cured prior to the effective date of termination; or

(b) any Financing Party has instituted and is diligently pursuing corrective action to cure such Sponsor Event of Default or has caused the initiation of, and is diligently pursuing, proceedings to take possession of the Facility; provided, that in either case above the Contractor shall be entitled to make an Adjustment Claim if it is affected thereby.]

21.3 Confidential and Proprietary Information[289]

21.3.1   The Contractor agrees that it shall and it shall cause all Contractor Persons to:

(a) keep confidential this Contract and all information, documentation, data or know-how regarding the business affairs, finances, technology or processes that are disclosed by the Sponsor (collectively, "Confidential Information"); and

(b) not disclose to any Person or otherwise use any Confidential Information or any part thereof.

21.3.2   The restrictions in Article 21.3.1 shall not apply, or shall cease to apply, to any part of the Confidential Information that:

(a) is in the public domain other than by reason of a breach of Article 21.3.1;

(b) was in the possession of Contractor Person at the time of the disclosure;

(c) was obtained by the Contractor in good faith from a third party entitled to disclose it;

(d) is legally required to be disclosed, in which case such disclosure shall be made, if at all, in accordance with Article 21.3.3;

(e) was disclosed to Contractor Persons involved in the Sponsor's development of the Facility whose duties reasonably required such disclosure[; provided that, other than in the case of disclosure to legal counsel, such Person shall first have agreed in writing not to disclose the relevant Confidential Information to any other Person for any purposes whatsoever]; or

(f) was otherwise disclosed as expressly contemplated hereby.

21.3.3   If the Contractor is required by law to disclose Confidential Information or if disclosure thereof is required by the Contractor in connection with the assertion of any claim or defense in any judicial or administrative proceedings involving it, the Contractor may make disclosure thereof, notwithstanding the provisions of this Article 21, <u>provided</u> that the Contractor shall immediately notify the Sponsor of the requirement and the terms thereof prior to such disclosure and shall use its best efforts to obtain proprietary or confidential treatment of such Confidential Information by the third party to whom the Confidential Information is being disclosed, and shall, to the extent such remedies are available, seek protective orders limiting the dissemination and use of the Confidential Information. This Contract does not restrict the rights of the Sponsor to challenge any law requiring the disclosure of any Confidential Information.

21.4 Publicity[290]

Without the prior written approval of the Sponsor, [{such approval not to be unreasonably withheld,}] neither the Contractor nor any Contractor Person shall issue any public statement, press release, publicity handout, photograph or other material about the scope, extent or value of the Work, any details as to materials and equipment to be used or installed or any other matter relating to this Contract. No Contractor Person shall invite or permit any reporter, photographer, television camera crew, commercial radio broadcaster or any other such Person to enter the <Site> <Route>. No Contractor Person shall use any Sponsor Person's name, photograph, logo, trademark, or other identifying characteristics, without the Sponsor's express prior written approval. No Contractor Person shall display, install, erect or maintain any advertising or other signage at the <Route> <Site> without the Sponsor's prior written approval, except as may be required by Law.

21.5 Compliance with Certain Laws[291]

The Contractor represents, warrants and covenants that all Contractor Persons have complied and will comply with all applicable Laws dealing with illegal payments, gifts or gratuities. [Without limiting the foregoing, the Contractor recognizes that the Sponsor and its affiliates desire to and are obligated to comply with all provisions of the United States Foreign Corrupt Practices Act (the "FCPA")].[292] The Contractor hereby agrees to cooperate with and assist the Sponsor and its affiliates with respect to such compliance, and will do such acts, and refrain from doing such acts, in connection with the development of the Facility and the performance of the Work as will enable the Sponsor and such affiliates to comply with [the FCPA] [such laws]. The Contractor represents and warrants to, and covenants with, the Sponsor that in the performance of the Contractor's obligations under this Contract, [{neither}] the Contractor [{nor any Contractor Person}] has not made and will not make, whether on its own behalf or on behalf of any other party, any offer of payment of or promise to pay, or gift of or promise to give, any money or anything of value, directly and indirectly, to:

(a) any officer, official or employee of any government or any department, agency, subdivision or instrumentality thereof;

(b) any political party, party official or candidate for political office; or

(c) any person, while knowing or having reason to know that all or a portion of such money or thing of value will be offered, given or promised, directly or indirectly, to any official, to any political party or official thereof, or to any candidate for political office.

21.6 [Counterparts]

[This Contract may be executed in one or more counterparts, each of which shall be deemed to be an original and all such counterparts shall together constitute one and the same contract.][293]

21.7 Set-Off

Contractor agrees that the Sponsor may set-off against any payment due Contractor hereunder any amount owed by Contractor to the Sponsor or any of its affiliates under this or any other contract.

IN WITNESS WHEREOF, the Parties have caused this Contract to be executed as of the date first above written.

<<[*Name of <Contractor> <Sponsor>*]>>
By:_____
Name:_____
Title:_____

<<*Name of <Contractor> <Consortium Member>*>>
By:_____
Name:_____
Title:_____

<<*Name of <Contractor> <Consortium Member>*>>
By:_____
Name:_____
Title:_____

EXHIBIT 1A

<u>FORM OF CLAIM FOR ADJUSTMENT</u>[294]

CLAIM FOR ADJUSTMENT

CLAIM NO:_____   AREA:_____

REVISION:_____

DATE:_____

PAGE __ OF __

Description of Event that Caused Claim:

_____

_____

_____

_____

Description of Change:

_____

_____

_____

_____

_____

_____

_____

References:

_____

_____

LUMP-SUM BREAKDOWN>

Cost Codes	Description	Quantity	Unit	Mhrs	Material <<*insert currency*>>	Labor <<*insert currency*>>	Sub-Contract <<*insert currency*>>	Total <<*insert currency*>>
401-407	Demolition/Site Work							
411-413	Concrete/Masonry							
414	Structural Steel							
415-419	Architectural							
421-425	HVAC/Plumbing/ Building							
431-439	Major Equipment							
441-449	Process Piping							
451-458	Electrical							
461	DCS System							
462-465	Instrumentation							
471	Insulation							
472	Painting							
473-474	Fireproofing/ Coatings							
Subtotal - Direct Field Costs								

LUMP-SUM BREAKDOWN (CONTINUED)

Cost Codes	Description	Total <<*insert currency*>>	Cost Codes	Description	Total <<*insert currency*>>
481	Proratables		493	Start-up Services	
482	Export Charges		494	Start-up Spares	
482	Ocean Freight		495	Chemicals	
483	Taxes		496	Initial Fuel	
484	Bonds		Sub-total	Start-up	
484	Insurance				
485	Building Permits		120	Engineering	
487	Construction Equipment		516	Initial Spare Parts	
488	Construction Management				
488	Outside Services			Contingency	
489	Temporary Buildings			Escalation	
489	Temporary Construction			Fee	
Sub-total	Construction Indirect Cost		TOTAL	PRICE	

SCHEDULE OF VALUES

Cost Code	AREA	SYST	Description	Quantity	Unit	Material <<insert currency>>	Install <<insert currency>>	Total <<insert currency>>
40X			SITE WORK AND DEMOLITION					
402			Demolition/Salvage/ Relocation					
403			Hazardous Material Removal/ Decontamination					
404			Site Preparation					
405			Piling and Caissons					
406			Site Upgrading and Improvements					
407			Site Utility Piping and Sewers					
41X			CONCRETE/ STRUCTURAL/ARCH					
411			Foundations (List by Foundation):					
412			Concrete Superstructures (List by Structure):					
413			Masonry					
414			Structural Steel (List by Structure):					
415			Wood and Plastics					
416			Thermal & Moisture Protection					
417			Doors and Windows					
418			Interior Finishes					
419			Specialties and Furnishings					

SCHEDULE OF VALUES (CONTINUED)

42X	HVAC/PLUMBING/ BUILDING SERVICES
421	Building Elevators / Conveyors
422	Building Plumbing / Under Slab Sewers
423	Building Fire Protection
424	HVAC
425	Miscellaneous Services
43X	MAJOR EQUIPMENT (List by Item):
44X	PROCESS PIPING/ DUCTING
441	Carbon Steel (List by System):
442	Stainless Steel (List by System):
443	Special Piping (List by System):
444	Instrument Piping & Tubing
445	Process Duct (List by System):
446	Special Piping Requirements (List):
447	Miscellaneous Piping Materials
448	Testing and Cleaning

SCHEDULE OF VALUES (CONTINUED)

45X	ELECTRICAL
451	Electrical Substation (List items):
452	Auxiliary Power System (Swgr/MCC) (List items):
453	Power Utilization (Motor Runs) (List):
454	Lighting Systems
455	Instrument Power and Control Systems
456	Grounding and Cathodic Protection
457	Communication, Alarms, Security Systems
458	Special Electrical Installation (List):
459	Electrical Interconnects (Host/Utility)
46X	INSTRUMENTATION & CONTROLS
461	Process Control System
462	Inst Controllers (Non-microprocessor)
463	Locally Mounted Instruments (List):
464	In Line Instruments (List):
465	Calibration and Loop Test

47X	INSUL / PAINT / FIRE PROT
471	Insulation:
	– Equipment Insulation
	– Process Piping Insulation
472	Painting:
	– Equipment Painting
	– Process Piping Painting
	– Structural Steel Painting
473	Fireproofing:
474	Specialty Coatings

DESCRIPTION OF EFFECT OF ADJUSTMENT

Effect of this adjustment in the Work on the Facility and Baseline Level 3 Schedule (if none, so state):

_____
_____
_____

Effect of this adjustment in the Work on the Completion Guarantees (if none, so state):

_____
_____
_____

Effect of this adjustment in the Work on the Performance Guarantees (if none, so state):

_____
_____
_____

Effect of this adjustment in Work on the Functional Specifications (if none, so state):

_____
_____
_____

CLAIM FOR ADJUSTMENT

CLAIM NO:_____   AREA:_____
REVISION:_____
DATE:_____
PAGE __ OF __

DESCRIPTION OF EFFECT OF ADJUSTMENT (CONT'D)

For this Adjustment, the payment criteria and schedule are:

_____
_____
_____
_____
_____

ISSUED BY CONTRACTOR
<<insert date>>
By:_____
Name:
Title:

ACCEPTED/REJECTED BY SPONSOR

<<*insert date*>>

By:_____

Name:

Title:

CLAIM FOR ADJUSTMENT
NOTES TO CLAIM FOR ADJUSTMENT

Direct Field Costs (Cost Codes 401–474)

Material:

All materials, supplies, and equipment which are for permanent installation in the Facility. The cost of equipment and material includes invoice cost, inland transportation, insurance, handling costs, permits, licenses and permitted markups per Appendix 1 for overhead and profit. Applicable export charges, ocean freight, and taxes shall be included as line items under "Construction Indirect Cost." Cost for Contractor supplied spare parts prior to Facility Provisional Acceptance are not included here but under cost code 516, "Initial Spare Parts."

Costs shall be based upon the Commercial Market Prices for materials or products. "Commercial Market Prices" shall mean, for products and services offered for sale by Contractor, the lowest price for such items offered by any affiliate of Contractor in the ordinary course of business to customers with similar volumes of sales, and for products and services offered for sale by Subcontractors of Vendors, the lowest price that Contractor can obtain after using best efforts to obtain the lowest price for such.

Labor:

The hourly salaries and/or wages for skilled, semiskilled, or unskilled labor up to and including the foreman who install materials. The total cost includes bare payroll, payroll burdens, fringe benefits in accordance with the applicable collective bargaining agreements, and permitted markups per Appendix 1. The overhead includes, but is not limited to, the following:

(a) Scaffolding
(b) Weather protection
(c) Cleanup
(d) General site dewatering
(e) Small tools and consumables
(f) Supplies and sundries
(g) Construction utilities
(h) Recruitment and indoctrination of work force
(i) Safety training
(j) Welder qualification

Subcontractors:
The cost of work performed by Subcontractors is the invoice cost and includes the cost of labor, material, equipment, services, construction equipment, tools, supplies, utilities, transportation and permits provided by the Subcontractor in the performance of the Work. All terms and conditions applicable to Contractor are also applicable to Subcontractors. Applicable permitted markups per Appendix 1 for overhead and profit are also included. Applicable taxes shall be included as a line item under "Construction Indirect Costs."

Construction Indirect Costs (Cost Codes 481–489)

481 Proratables:
Includes materials and labor for the following specific items: Drug testing, fire watch, and work stoppages due to events which may endanger the work force.

487 Construction Equipment and 489 Temporary Buildings (Facilities):
Any major construction equipment or tool which has a purchase value of $1,000 or more or temporary facility (e.g., trailers and warehouses), either purchased or rented, are chargeable only in the following circumstances:

(a) They are brought on-site specifically to perform a Change Order;
(b) They are already on-site and it can be demonstrated that the schedule duration has been extended by a Change Order which caused additional cost to Contractor.

The base cost for third-party rentals shall be the invoice cost as permitted by Appendix 1, which shall include maintenance, fuel, insurance, lubricants, supplies, mobilization, demobilization and fees. The total third-party rental cost is the base cost plus the applicable taxes on the base cost (no markup for overhead and profit). There shall be no additional markup or taxes added to the rental rates for Contractor-owned construction equipment or temporary facilities since they are included in the rental rates shown in Appendix 1.

488 Construction Management:
The hourly salaries or wages for all supervisors above the working foreman, such as general foreman and area foreman, also include all levels of supervision management directly assigned to supervise the Work at the Site. The total cost includes base salaries, payroll burdens, and fringe benefits per Appendix 1. Cost also includes permitted markup per Appendix 1 for overhead and profit. The overhead includes the following:

(a) Field office supplies and consumables.
(b) Field office equipment and furniture.
(c) Costs of field office secretaries and clerks.
(d) Field office reproductions, communications and computer services.

(e) Field office travel and living expenses, including per diems, relocation (move in and move out) and any required business travel away from the construction job site.
(f) Field office utilities.
(g) Field office janitorial, water and coffee services and the like.

488 Outside Services:

The base cost of outside services including guard service, construction debris removal, survey parties, testing laboratories, inspection services and the like. Permitted markups per Appendix 1 for subcontracted services to cover overhead and profit shall apply. Applicable taxes shall be included as a line item under "Construction Indirect Costs."

489 Temporary Construction:

Any temporary construction (e.g. roads, fencing, parking, rail spurs, etc.) are chargeable as an Adjustment only if they must be changed to perform a Change Order.

Start-up Costs (Cost Codes 493–496)

493 Startup Services:

Descriptions for "Labor" and "Construction Management" above shall also apply to startup services. Startup engineers shall be classified as Construction Management personnel while assigned to the Site.

494 Startup Spares and 495 Chemicals:

Spare parts, consumables (gaskets, filters, etc.), and chemicals required for startup. Permitted markups for permanent materials shall also apply to startup spares and chemicals.

493 Initial Fuel:

Permitted markups for permanent materials shall also apply to initial fuel.

Home Office Engineering (Cost Code 120)

The hourly salaries and/or wages for technical personnel, including field technical support. The total cost includes base payroll, payroll burdens, fringe benefits and permitted markups per Appendix 1 for overhead and profit. The overhead includes the following:

(a) All salaries, benefits, overheads, expenses, and costs with respect to Contractor's officers and employees of its staff functions, such as contract, legal, tax, sales, insurance, administrative, accounting, auditing, construction, personnel, public relations, and research and development departments, as well as managers and department heads of such line functions as engineering and purchasing.

(b) Salaries of nontechnical employees such as secretaries, stenographers, typists, clerks, and receptionists, whether or not they work on the project. This includes those employees preparing reports or correspondence on word processing systems and those operating any computers, such as those for CAD services.

(c) All computer usage, including engineering calculations, CAD drafting and that not directly related to the Work, such as for accounting and payroll.

(d) Postage, rent, utilities, communications (including long distance and fax) and similar services.

(e) All reproduction expenses.

(f) Travel, living and relocation expenses for home office personnel.

(g) Furnishings, equipment and supplies for a rearrangement and maintenance of Contractor's offices.

(h) Preparation of Contractor's standards and procedures.

(i) Training programs for Contractor's officers and employees.

(j) Contractor library and reference documents, including governmental and industry standards, codes, regulations and specifications.

(k) All insurance or bonds required by Contractor not specified in the Agreement as being required by the Sponsor.

(l) All permits, licenses, and fees required to be in Contractor's name necessary for the Work and not specified in the Agreement as being required by the Sponsor.

(m) Interest on funds borrowed by Contractor.

(n) All materials, equipment, tools, supplies, taxes and services necessary for the Work not specified elsewhere in the Agreement.

(o) Contractor's income tax.

Initial Spare Parts (Cost Code 516)

Contractor's cost for capital spare parts required up to Facility Provisional Acceptance and spare parts for operation thereafter. Permitted markups for permanent materials shall also apply to capital spares.

Backup Information

Contractor shall complete a Lump Sum Breakdown and Schedule of Values for each Claim. Contractor shall also provide backup cost calculations for Construction Indirect Costs (481–489), Startup (493–496), Engineering (120), and Initial Spare Parts (516).

Payment Schedule

The Contractor is to propose a payment schedule that is independent from the Milestone Payment Schedule.

APPENDIX I TO EXHIBIT IA

CLAIM FOR ADJUSTMENT

RATES AND MARKUPS

1.0   <u>Markup</u>. Contractor shall be entitled to the following markups, except as specifically addressed in Appendix 1 of this Exhibit for all Adjustments, which markups shall compensate Contractor and its Subcontractors for costs specified in the notes attached hereto.

	Contractor Markup	Applied to
For labor and services performed by Subcontractors		Subcontractor invoice
For labor and services performed by own forces		Base wages plus burden and fringes
For material supplied by Subcontractors, at actual cost of the material (less all discounts)		Invoiced cost, point of origin
For material supplied by Contractor at actual cost of the material (less all discounts)		Invoiced cost, point of origin
For Work performed by Subcontractors on a lump sum or fixed unit price basis (which has been accepted by the Sponsor)		Lump Sum or fixed unit prices

On Adjustment Claims covering both increases and decreases in the Contract Price, the markups will be computed on the net change only. On Adjustment Claims for decreases in the amount of the Contract Price, the markups will be included on the decrease in addition to the decrease in direct cost.

When Contractor submits an Adjustment Claim involving the use of Subcontractors the cost of such subcontracted Work shall not exceed the cost of such Work as if it were accomplished entirely by Contractor's own forces.

Contractor agrees to incorporate provisions in the contracts with the Subcontractors for this Facility limiting such Subcontractors to the markups addressed in this Article for all Work performed pursuant Adjustments.

2.0   <u>Price Adjustments</u>. No Price Adjustment shall include markups exceeding those specified below:

1. for labor performed by any Subcontractor on an hourly basis, the Contractor shall add to such Subcontractor's price a markup of no more than [_____ percent];

2. for labor performed by any Subcontractor on a lump sum or fixed unit price basis (acceptable to the Sponsor), the Contractor shall add to such Subcontractor's price a markup of no more than [_____ percent];

3. for labor performed by the Contractor's employees, the Contractor shall add to the actual cost of such labor a markup of no more than [_____ percent]; and

4. for material supplied by the Contractor or any Subcontractor, the Contractor shall add to the actual cost of such material (less all discounts), a markup of no more than [_____ percent].

Rates are firm for the duration of the Work. Markups shall compensate Contractor for costs specified in the notes attached hereto.

Category	Base Pay [<<*insert currency*>>]/HR	Overtime Premium [<<*insert currency*>>]/HR
Site Manager		included in base
Site Supervisor		included in base

Payroll burdens and fringes at [_____ percent (__ percent)] of base pay.
Payroll burdens and fringes at [_____ percent (__ percent)] of overtime premium.
Overhead and profit at [_____ percent (__ percent)] of base pay.
No markup for overhead and profit is allowed on the premium portions of overtime.

3.0 <u>Home Office Engineering Personnel</u>. Contractor shall provide wage rates for all types, classifications and skill levels of personnel that will be chargeable on the Work. Rates are firm for the duration of the Work. Markups shall compensate Contractor for costs specified in the notes attached hereto.

Category	Base Pay [<<*insert currency*>>]/HR	Overtime Premium [<<*insert currency*>>]/HR
Project Manager		included in base
Manager		included in base
Lead Engineer		included in base
Discipline Engineer		included in base

Payroll burdens and fringes included in base pay.
Payroll burdens and fringes included in overtime premium.
Overhead and profit included in base pay.

4.0 <u>Unit Pricing</u>. Unit Pricing shall apply to Adjustment Claims. Sponsor, at its sole discretion, may request lump sum pricing in lieu of unit pricing.

FORM OF CHANGE ORDER

CHANGE ORDER

Change Order No.:   **Change Order Date: <<*insert date*>>**

Reference is made to the Fixed Price, Lump Sum Turnkey Engineering, Procurement and Construction Contract between <<*insert name of Contractor*>> (the "<u>Contractor</u>") and <<*insert name of Sponsor*>> ("<u>Sponsor</u>"), dated <<*insert date*>> (the "<u>Contract</u>"). Capitalized terms not otherwise defined in this Change Order No. __, have the same meanings as specified in this Contract.

Account Code:_____

Vendor/Contractor/Subcontract: _____

Items	Description	Costs
	Contract Price Increase:	
	Reference Documents:	
	Original Contract Price	
	Previous approved Change Orders	
	Previous Contract Price	
	Amount of this Change Order	
	Contract Price	

The Contract, except as specifically amended by this Change Order No. __, remains in full force and effect. No amendment, waiver or consent relating to this Change Order No. __ shall be effective unless it is in writing and signed by the Contractor [the agent for the Financing Parties][295] and Sponsor.

<<*insert name of Contractor*>>   <<*insert name of Sponsor*>>

By: **By:**

Name: ___   **Name:** ___

Title   **Title**

[EXHIBIT 1B]

[LIST OF ASSIGNED VENDOR AGREEMENTS]
<<*insert list of Assigned Vendor Agreements*>>

EXHIBIT 1C

BASELINE SCHEDULE

EXHIBIT 1D

LIST OF INDUSTRY CODES
<<*insert list of Industry Codes*>>

EXHIBIT 1E

LIST OF CONTRACTOR PERMITS
<<*insert list of Permits Contractor will obtain*>>

EXHIBIT 1G

ENVIRONMENTAL REPORT

EXHIBIT 1H

LIST OF ENVIRONMENTAL STANDARDS
<<*insert list of Environmental Standards*>>

EXHIBIT 1J

SPONSOR INFORMATION
<<*insert information that Contactor must take into account*>>

EXHIBIT IK

FUNCTIONAL SPECIFICATIONS

EXHIBIT IL

FORM OF WARRANTY LC[296]
IRREVOCABLE STANDBY LETTER OF CREDIT NO. _____
<<insert date>>
CREDIT NO. _____
<<insert date>>
<<insert Sponsor name and address>>

Ladies and Gentlemen:

1. <u>Stated Amount</u>. We hereby establish in favor *<<insert name of Sponsor>>* (the "<u>Sponsor</u>") an irrevocable Standby letter of credit (the "<u>Credit</u>") whereby you are authorized to draw on [*<< insert name of Acceptable Issuer Bank>>*] (the "<u>Issuing Bank</u>"), Credit No. _____, issued for the account of [*insert name of Contractor*] (the "<u>Contractor</u>"), an aggregate amount not exceeding *<<insert amount of currency in words>>* (*<<insert amount of currency in words and numerals>>*) (the "<u>Stated Amount</u>"). This Credit is issued pursuant to the Turnkey Engineering, Procurement and Construction Contract, dated *<<insert date>>*, between the Contractor and the Sponsor (as amended, restated, supplemented or otherwise modified from time to time, the "<u>Contract</u>").
2. <u>Term</u>. This Credit is effective immediately and expires at the close of banking business at our office in [<New York City> <London> <Hong Kong> <Tokyo>][297] on [*insert date*].
3. <u>Drawing</u>. Funds under this Credit are available to you against your presentation to us at our offices above of a drawing certificate in the form of Annex A hereto. Such certificate shall have all blanks appropriately filled in and shall be signed by your authorized officer, and such certificate shall be on your letterhead. You may make multiple and/or partial drawings hereunder.
4. <u>Drawing Procedures</u>. If a drawing is made by you hereunder, payment shall be made of the amount specified in immediately available funds on the <next> <{second}> succeeding Business Day; provided that such drawing and the documents presented in connection therewith conform to the terms and conditions hereof. All payments hereunder shall be made free and clear of, and without deduction for or on account of, any present or future taxes, duties, charges, fees, deductions or withholdings of any nature and by whomsoever imposed. As used in this Credit, the term "<u>Business Day</u>" shall mean any day other than a Saturday or Sunday or legal holiday or a day on which banking institutions in which our offices listed above are authorized or required by law or executive order to close.
5. <u>Assignment</u>. This Credit may be assigned by you upon written notice thereof to the Issuing Bank.

6. <u>Governing Law</u>. This Credit is subject to the Uniform Customs and Practice for Documentary Credits (2007 Revision), International Chamber of Commerce Publication No. 600 (the "<u>UCP</u>"). This Credit shall be deemed to be a contract made under the laws of <New York> <England> and shall, as to matters not governed by the UCP, be governed and construed in accordance with the laws of <New York> <England>.

7. <u>Notices</u>. Communications with respect to this Credit shall be in writing and shall be addressed to the Issuing Bank at *<insert address>* (or otherwise as designated by the Issuing Bank to the Sponsor by written communication) specifically referring thereon to "[<<*insert name of Issuing Bank*>>] Irrevocable Standby Letter of Credit No. _____."

8. <u>Integration</u>. This Credit sets forth in full our undertaking, and such undertaking shall not in any way be modified, amended, amplified or limited by reference to any document, instrument or agreement referred to herein (including, without limitation, the Contract), except only the drawing certificate referred to herein; and any such reference shall not be deemed to incorporate herein by reference any document, instrument or agreement, except for such drawing certificates.

Very truly yours,
[*insert name of Issuing Bank*]
By:_____
Authorized Officer

ANNEX A to EXHIBIT L

DRAWING CERTIFICATE
[<<*insert date*>>]
[<<*insert name of Issuing Bank*>>]
[<<*insert address of Issuing Bank*>>]

Re: Irrevocable Standby Letter of Credit No. _____

Ladies and Gentlemen:

1. <<*insert name of Sponsor*>> (the "<u>Sponsor</u>") hereby certifies to [<<*insert name of Issuing Bank*>>], with reference to Irrevocable Standby Letter of Credit No. _____ (the "<u>Credit</u>"), that:

2. <u>Reason for Drawing</u>.
   The Contractor has failed to fulfill one or more of the conditions of, or failed to perform one or more of its obligations under, the Contract, and now in accordance with the Contract, the Sponsor is entitled to the amount of <<*insert amount in words and numerals also*>> to attempt to remedy such failure to perform and/or
   This Credit is scheduled to expire within thirty (30) Days after the date hereof, and the Contractor has not posted in our favor a substitute for this Credit meeting the requirements of the Contract.

3.   [{Prior Notice to the Contractor. At least five (5) Business Days prior to the date hereof, the Sponsor gave the Contractor written notice of (a) the foregoing and (b) the Sponsor's intention to make a drawing under the Credit.}]

4.   Demand for Payment. The Sponsor is making a demand for payment under the Credit in the amount of <<insert amount in words and numerals also>>.

5.   Terms. Capitalized terms used herein and not otherwise defined shall have the meanings ascribed thereto in the Turnkey Engineering, Procurement and Construction Contract, <<insert date>> (as amended, restated, supplemented or otherwise modified from time to time, the "Contract"), between the Sponsor and the Contractor.

6.   Payment Instructions. Payment of the amount demanded hereby shall be made by wire transfer to the following account: _____.

IN WITNESS WHEREOF, the Sponsor has executed and delivered this Certificate as of <<insert date>>.

<<insert Sponsor name>>

By:_____

Name:_____

Title:_____

EXHIBIT 1M

OUTLINE OF CONTRACTOR PROVIDED MATERIALS

EXHIBIT 4.1.1

MAJOR ITEMS OF WORK

Item
Pressure Parts:
  - Steam Drum
  - Major piping
Distributed Control System
Turbine
Generator
Boiler Feed Pump
District Heating Pumps
Main Fans—FD/ID/PA
Electrostatic Precipitators
Water Treatment Plant
Civil Works
Main Transformers
High Voltage Switchgear

EXHIBIT 4.2

<u>LIST OF APPROVED SUBCONTRACTORS</u>[298]

Item of Work   Subcontractor

1. BOILER CODE WORK
    1.1 Steam Drum
    1.2 Boiler Pressure Parts
    1.3 Boiler Code Piping
    1.4 Auxiliary Piping
    1.5 Other Pressure Parts

2. STEEL STRUCTURE AND PLATE WORK
    2.1 Steel structure
    2.2 Wall Claddings
    2.3 Plate Work

3. SYSTEMS, EQUIPMENT AND COMPONENTS
    3.1 Boiler Valves and Fittings
    3.1.1 Safety Valves
    3.1.2 HP Valves
    3.2 Refractory
    3.3 Insulation and Lagging
    3.4 Conveyors/Ash Handling Equipment
    3.4.1 Mechanical Fuel/Ash Conveyors, Ash Cooling Screws
    3.4.2 Pneumatic Ash Conveyors
    3.4.3 Fuel Crushers
    3.5 Start Up Burners
    3.6 Sootblowers
    3.7 Flue Gas Air Preheaters
    3.8 Steam Coil Air Preheaters
    3.9 Fans and Blowers
    3.9.1 Fans
    3.9.2 Blowers
    3.10 Electrostatic Precipitator
    3.11 Pumps
    3.12 Compressors
    3.13 Lifts and Hoists
    3.14 Civil Works
    3.15 HVAC Contractors

4. ELECTRICAL, INSTRUMENTATION AND AUTOMATION
    4.1 Electrical

5. SERVICE
    5.1 Erection Work
    5.2 Chemical Cleaning

6. ENGINEERING
    6.1 Layout, Basic Design and other Mechanical Engineering
    6.2 Electrical and Instrumentation Engineering
    6.3 Civil Engineering
    6.4 HVAC Engineering

7. STEAM TURBINE

8. BALANCE OF PLANT EQUIPMENT
    8.1 Pumps
    8.2 Valves
    8.3 Valve Actuators
    8.4 Heat Exchangers
    8.5 Water treatment plant
    8.6 Cooling Water Towers

9. INSTRUMENTATION & CONTROL
    9.1 Digital Automation System

10. ELECTRICAL SYSTEMS
    10.1 Generators
    10.2 110 kV Circuit Breakers
    10.3 110 kV Disconnectors
    10.4 110 kV Current Transformers
    10.5 110 kV Voltage Transformers
    10.6 110 kV Surge Arresters
    10.7 Generator Transformer
    10.8 Auxiliary Transformer
    10.9 Protection Relays
    10.10 Synchronizers
    10.11 6 kV Switchgears
    10.12 Distribution Transformers
    10.13 400 V Switchgears and MCC's
    10.14 Motors
    10.15 Inverters
    10.16 UPS and DC Equipment
    10.17 Cables
    10.18 Electrical Installations

11. ERECTION AND INSTALLATION WORK

12. CIVIL WORKS AND HVAC
    12.1 Civil Works Generally
    12.2 Steel structures
    12.3 Insulation
    12.4 Civil

13. ELCHO OFF SITE ELECTRICAL INTERCONNECTION AND DISTRIBUTION FACILITIES

14. RAILWAY SYSTEM

15. COAL STORAGE YARD

16. APPROVED SUBCONTRACTORS—ALTERNATE

Equipment	Approved Subcontractor(s)
CTG	
STG	
HRSG	
Duct Burners	
Power Transformers (Main)	
Power Transformers (Aux.)	
Switchyard	
Distributed Control System	
Bridge Cranes	
Structural Steel Supply	
Elec. Bulk Materials	
Anchor Bolts	
Rebar	
Fixators	
Fans	
High Resistance Grounding Equipment	
HVAC Systems	
Motion Detectors	
Motors	
Neutral Grounding Resistor	
Panel Fabricators, large	
Panel Fabricators, small	
PLCs	
Power Distribution Centers	
Auxiliary Boilers	
Chemical Feed Systems	
Surface Condensers	
Air Dryers	
Deaerators	
Feedwater Heaters	
Freeze Protection Systems	
Gas Compressors	
Inlet Air Filters	
Boiler Feed Water Pumps	
Chemical Feed Pumps	
Circulating Water Pumps	
Condensate Pumps	
Pumps—General	
Fire Water Pumps	
Pumps—Rotary Gear & Screw	
Pumps—Submersible	
Pumps—Sump	
Pumps—Sanitary Lift Stations	
SCRs	
Shop Fabricated Vessel & Tank	
CPI—Oily Water Separator	
Dampers	
Shop Fabricated Piping	
Strainers	
Tubing, Fittings	
Valves & Manifolds Instruments	
Valves, Plug	
Valves: Cast HP≥2 1/2"	
Valves: Cast LP≥2 1/2"	

Equipment	Approved Subcontractor(s)
Valves: Diaphragm	
Valves: Forged ≤2"	
Valves: Ball	
Valves: Large Butterfly	
Valves: Safety	
Control Valves	
Valves: Motor Operated	
Valves: Steam Conditioning	
Expansion Joints Metallic	
Expansion Joints Rubber	
Piping—FRP (CW)	
Piping Shoes—Clamp On	
Pipe Supports	
Piping Bulk Materials	
Cathodic Protection	
ISO Phase Bus Duct	
High Voltage Cable	
Medium and Low Voltage Cable	
Cable Tray	
Dry Type Transformers	
Generator Circuit Breakers	
MCCs (480V and 4160V)	
Medium and Low Voltage Switchgear	
Station Batteries	
UPS and Battery Chargers	
CEMs	
Air Compressors; Rotary	
Air Compressors; Rotary Screw	
Air Compressors; Centrifugal	
Air Compressors; Reciprocating	
Fuel Gas Compressors	
Fuel Gas Pressure Reducing Skids	
Fuel Gas Filter Separators	
Diesel Generators	
Closed Loop Cooling Water Exchangers	
Heat Exchangers (Shell & Tube)	
Electric Heaters	
Bus Ducts (non seg)	
Actuators, Electric	
Actuators, Pneumatic	
Air Regulators	
Annunciators	
Chlorination Equipment	
Combustibles Meter	
Control Drives	
Controllers, Single Loop	
Data Loggers	
DP Gauges	
Drum Level, Conductance	
Enclosures	
Flow Elements (Orifices, Venturi)	
Flow Meters, Displacement	
Flow Meters, Magnetic	
Flow Meters, Mass	

Equipment	Approved Subcontractor(s)
Flow Meters, Sonic	
Flow Meters, Turbine	
Flow Meters, Vortex	
Fuel Safety Systems	
Gauge Glasses	
I/Ps	
Indicators, Bar Graph	
Instrument Stands	
Instrument Enclosures	
Instrument Tube Bundles	
Level Probes, Capacitance	
Level Switches	
Level, Tank	
Level, Ultrasound	
Lighting Fixtures	
Limit Switches	
Nuclear Devices	
O_2 Meters	
Pilot Tubes	
Positioners	
Pressure Gauges	
Pressure Switches	
Process Transmitters	
Recorders	
Regulators, Liquids	
Regulators, Pneumatic	
Relays	
Lighting Transformers and Panels	
RTDs	
Signal Conditioners	
Solenoid Valves	
Temperature Indicators	
Temperature Switches	
Temperature Transmitters	
Thermocouples	
Transducers—Electrical	
Transmitters, Nuclear	
Transmitters, Pressure, Dp	
Vibration Monitors	
Water Quality Probes	
Concrete (Foundations)	
Piling	
Civil/Soil/Concrete Testing	
Geotechnical Investigation	
Cooling Towers	
Field Fabricated Tanks	
Water Treatment Systems	
NDE Services & Welder Testing	
Fire Protection	
Commissioning Services	
Electrical Inspection & Testing	
Painting	
Finish Grading & Paving	
General Site Testing Services	

Equipment	Approved Subcontractor(s)
Modular Buildings (Trailers)	
Specialty Pipe & Equipment Cleaning	
Scaffolding, Forming, Shoring & Rebar	
Fogging Systems	
Insulation & Heat Tracing	
Site Surveying	
Centerline Installation (Power Island)	
Site Preparation	
Structural Steel Erection	
Buildings	
Electrical & Instrumentation Installation	
Piping Installation	
Mechanical Erection	
HRSG Erection	
Instrument Calibration	
Sequence of Events (SOE)	
River Water Intake System	
Rigging & Heavy Haul	
Structural Steel Erection	
Transformer Dress out/Testing	
Electrical Field Testing	

EXHIBIT 5.1(a)(iv)

[SITE][ROUTE] ASSESSMENT [SCOPE][299]

[EXHIBIT 5.1(j)]

[LIST OF EXPERTS]

EXHIBIT 5.1(qq)(A)

FORM OF CONTRACTOR'S INDEMNITY[300]

THIS INDEMNITY AGREEMENT (the "Agreement") is executed by [_____], a [_____] ("Contractor Indemnitor"), [insert Parent Guarantor] (hereinafter collectively referred to as "Indemnitors").

With reference to the following facts:

(a) Certain works of improvement have been or will be commenced upon the land, as defined in Article 2.1.

(b) Contractor Indemnitor has an interest in the works of improvement on the land, or thereon to be constructed, as contractor, and desires that Title Company issue its policies of title insurance without mechanic's lien exclusions, insuring marketability of title and/or priority of encumbrances upon the terms and conditions hereafter set forth.

NOW THEREFORE, Indemnitors jointly and severally agree as follows:

1. <u>COVENANTS</u>

1.1   Indemnitors will hold harmless, protect and indemnify Title Company from and against any and all liabilities, losses, damages, expenses, and charges including attorney's fees and expenses of litigation, which Title Company may sustain under any policy of insurance respecting the land resulting directly from any Mechanic's Lien or claim thereof filed by Contractor Indemnitor and/or any of Contractor Indemnitor's subcontractors or suppliers at any tier and which arise from Indemnitor's failure to comply with the requirement of 1.1.2 herein.

1.1.1   Indemnitors will pay, or cause to be paid, all bills, charges, or expenses of any works of improvements on the land, which Indemnitors, in their reasonable judgment, determine to be valid bills, charges, or expenses which Indemnitors are obligated to pay.

1.1.2   If a Mechanic's Lien affecting the land for which a policy of title insurance has been issued in reliance on this Agreement is filed, a Priority Lien established, or an action to foreclose a Priority Lien commenced, Indemnitors shall, in the case of Mechanic's Lien filed by Contractor Indemnitor's subcontractors and suppliers, cause such lien to be released of record and/or such action to be dismissed with prejudice; and in the case of a Mechanic's Lien, filed by Contractor shall execute and deliver such instruments as may be necessary to subordinate such Mechanic's Lien to the lien insured by the policy of title insurance.

1.1.3   Title Company shall have the right, at any time after thirty (30) days written notice to Indemnitors, to pay, discharge, satisfy, or remove from the title to the land any Mechanic's Lien or claim thereon; and Indemnitors covenant and agree to pay to Title Company on demand all amounts expended by Title Company to pay, discharge, satisfy, or remove any Mechanic's Lien or claim thereof resulting from a valid (as determined by Indemnitors' reasonable judgment) bill, charge, or expense which Indemnitors are obligated to pay.

1.2   If a Mechanic's Lien affecting the land for which a policy of title insurance has been issued in reliance on this Agreement is filed, a Priority Lien established, or an action to foreclose a Priority Lien commenced, Contractor Indemnitor shall, upon request of Title Company, promptly furnish Title Company with copies of all receipted bills or other evidence of payment or set-off for works of improvement upon the land, but only to the extent that any such bills or other evidence of payment or set-off relate to the lien or claim of lien in question.

1.3   Title Company is hereby granted the right to rely upon this Agreement whether or not an indemnitor is the person ordering the policy of title insurance, regardless of any change in ownership, title, or interest in the land or the works of improvement thereon, or any change in Contractor Indemnitor's interest therein. Said right shall extend to subsequent policies issued with respect to the land. However, nothing contained herein shall be construed so as to obligate Title Company to issue any policies of title insurance in the form above desired, but, should Title Company issue any such policies of title insurance, it will do so in reliance upon the undertakings of the undersigned and in consideration thereof.

2.   <u>ADDITIONAL PROVISIONS</u>

2.1   As used herein, the term "land" includes the real property described on Exhibit "A", and any part, parcel, or subdivision of the described real property and any part, parcel, or subdivision of the legal or equitable interest in said real property.

2.2   As used herein, the term "Mechanic's Lien" shall be deemed to refer to the applicable statutes of the state in which the land is situated that enable mechanics, materialmen, artisans, and laborers to have a lien and/or enforce such lien against the land any improvements constructed thereon for the value of labor bestowed thereon and/or materials furnished thereto; provided that such term as used herein shall exclude any liens actually of record as of the date of issuance of the policies of title insurance which Indemnitors are not obligated to pay.

2.3   The term "policy of title insurance" includes such policies as are customarily issued by title companies insuring priority of liens and marketability of title, and all endorsements thereon. The term also includes other documents and reports customarily issued by title companies concerning the state of title, ownership, or interest in real property.

2.4   The term "Priority Lien" means a Mechanic's Lien asserting priority over the lien of the mortgage or deed of trust insured by the policy of title insurance.

2.5   The term "Title Company" means _____ or any other title company which issues the policy of title insurance referred herein.
    INDEMNITORS EXECUTE THIS AGREEMENT BECAUSE OF THE BENEFITS DIRECTLY AND INDIRECTLY ACCRUING TO INDEMNITORS BY REASON OF THE ISSUANCE OF SAID POLICIES.
    IN WITNESS WHEREOF, the undersigned have executed this Agreement this __ day of <<*insert date*>>.

<<_____>>
By:_____
Title:_____

<<_____>>
By:_____
Title:_____

EXHIBIT 5.1(qq)B

FORM OF CONTRACTOR PRIORITY AGREEMENT

WHEREAS, _____ (the "<u>Sponsor</u>"), the owner of the land described in Schedule A attached hereto (the "<u>Premises</u>"), proposes to erect thereon certain improvements (the "Project") and for the purpose of raising necessary funds has applied to a group of banks led by _____, as agent (hereinafter collectively, the "Lender") for a mortgage loan to be secured by a first priority lien [mortgage] [deed of trust] (hereinafter referred to as the "Mortgage") on such Premises; and

WHEREAS, the Lender has applied to Title Company (as defined below) for a policy of title insurance to be issued to the Lender insuring the first priority lien of the Mortgage;

WHEREAS, "Title Company" means _____ or any other title company which issues the policy of title insurance referred to herein;

WHEREAS, the undersigned (the "Contractor") and Sponsor entered into a Turnkey Engineering, Procurement and Construction Agreement dated <<*insert date*>>;

WHEREAS, the undersigned has been employed to furnish materials or to perform labor or both thereof incident to the said improvements, for which the undersigned may have a statutory or constitutional right of lien; and

WHEREAS, the undersigned is desirous of said mortgage loan being consummated and said title insurance being issued.

NOW THEREFORE, in consideration of the premises and an inducement to the Lender to enter into a loan agreement with the owner and to the Title Company to issue the said policy of title insurance, respectively, the undersigned does hereby agree that the lien of the proposed Mortgage shall be, and at all times remain, prior, paramount and superior to any statutory or constitutional right of lien that the undersigned may now have or hereafter acquire, whether for materials furnished or labor performed or for both thereof.

IN WITNESS WHEREOF, the undersigned has executed this Contractor Priority Agreement this ____ day of _____, 20__.

[_____]

By:_____

Title:_____

EXHIBIT 5.1(rr)

FORM OF SUBCONTRACTOR PRIORITY AGREEMENT

WHEREAS, _____ (the "Sponsor"), the owner of the land described on the attached Exhibit "A" (the "Premises"), proposes to erect thereon certain improvements, and for the purpose of raising necessary funds has applied to a group of banks led by _____, as agent (hereinafter collectively, the "Lender") for a mortgage loan to be secured by a first priority lien [mortgage] [deed of trust] (hereinafter referred to as the "Mortgage") on such Premises; and

WHEREAS, the Lender has applied to Title Company for a policy of title insurance to be issued to the Lender insuring the first priority lien of the Mortgage;

WHEREAS, "Title Company" means _____ or any other title company which issues the policy of title insurance referred to herein;

WHEREAS, the undersigned has been employed by the contractor, [_____], to furnish materials or to perform labor or both thereof incident to the said improvements, for which the undersigned may have a statutory or constitutional right of lien; and

WHEREAS, the undersigned is desirous of said mortgage loan being consummated and said title insurance being issued.

NOW, THEREFORE, in consideration of the premises and as an inducement to the Lender to enter into a loan agreement with the Sponsor and to the Title Company to issue the said policy of title insurance, respectively, the undersigned does hereby agree that the lien of the proposed Mortgage shall be, and at all times remain, prior, paramount and superior to any statutory or constitutional right of lien that the undersigned may now have or hereafter acquire, whether for materials furnished or labor performed or for both thereof.

IN WITNESS WHEREOF, the undersigned has executed this Priority Agreement this ___ day of _____, 20__.

_____

Subcontractor

By:_____

Title:_____

EXHIBIT 5.1(w)

TRAINING PROGRAM

EXHIBIT 5.1(z)

LIST OF DOCUMENTATION TO BE PROVIDED BY CONTRACTOR

EXHIBIT 5.4(f)

QUALITY CONTROL PROGRAM

EXHIBIT 5.6.5

FORM OF PROGRESS REPORT

1 Scheduling Methods

1.1 Contractor will maintain the project master schedule. The master schedule will be of sufficient detail as to allow any user of *<<insert program name>>* program to easily identify the single "CRITICAL PATH" and overall project "PERCENTAGES COMPLETE" for main areas of effort and the entire project.

1.2 Contractor shall provide the Sponsor with both a hard copy and an electronic copy of the updated schedule monthly with the copies of the monthly Progress Report.

2 Monthly Progress Report

The monthly Progress Report, covering each calendar month from the Work Commencement Date until the Facility Final Acceptance will consist of a tabular report with five (5) copies to be delivered to the Sponsor [at the <Route><Site>]. The monthly Progress Report will consist of, at a minimum, the following contributing reports:

2.1. Executive SummaryProject status consisting of a brief report which represents overall Project progress as compared to planned progress and accompanied by an overall Project "S curve" which shows overall Project completion percentage as compared to planned overall Project completion percentage. The Contractor will represent in definitive terms whether the Project is on schedule and on budget. The contractor, in this summary, will provide information regarding any recovery plans either being contemplated or in progress.

2.2   Shipping Status

(a) The Contractor will maintain and present a detailed shipping and transportation report for all Project materials. This report, in a spreadsheet format, will indicate actual shipping of project materials against planned shipping of materials. This information will include dates, ports of dispatch and vessel names.
(b) In addition to the spreadsheet format report mentioned above, the shipping status will include a narrative Article dealing with shipping problems either actual or anticipated and any actual or anticipated recovery plans associated with these problems.

2.3   Engineering and Design Status

(a) The engineering and design progress and status of the Project will be tracked by using actual drawing submissions to the Sponsor as compared to the planned drawing submissions schedule (this information will be furnished in spreadsheet format and will be broken down into civil, electrical, instrument & control, mechanical, and main flow diagrams, etc.) Individual "S curves" for each of these main areas of engineering will be developed from the spreadsheet data and will show planned vs. actual progress on a monthly basis.
(b) Details of tracking engineering open items and issues should be included in this Article of the monthly Progress Report. This will be accomplished using a spreadsheet format which will clearly summarize the agreed upon status of open engineering issues.

2.4   Project Schedule Status

(a) The Project Schedule will be updated monthly and made part of the monthly Progress Report.
(b) Milestone status for milestones as per the Contract will be presented in a spreadsheet format clearly showing planned dates of milestones and the actual dates that milestones have been achieved. The data contained in this report shall be consistent with the Project Schedule.
(c) Look Ahead Schedule: A three month look ahead construction schedule and report will summarize activities that are planned for the three months subsequent to the month being reported on. This schedule and report shall be based on the report function of the <<*insert name*>> scheduling software and be made part of the monthly Progress Report.

2.5   Milestone Payment StatusMilestone Payment status will be represented in a spreadsheet format which clearly shows amounts and dates payments are received and amounts and dates payments are due. (Information regarding submission of invoices should also be included in this report.)

2.6   Procurement Status

(a) Procurement status will consist of a statement summarizing procurement activities and which would incorporate areas of concern and recovery plans as appropriate for procurement which is behind schedule.

(b) Monthly Equipment Procurement progress and status will be tracked by using actual purchase order submissions as compared to the planned purchase order submissions schedule (this information will be furnished in a spreadsheet format and will be broken down into civil, electrical, instrument & control, mechanical, and main flow diagrams, etc.) Individual "S curves" for each of these main areas of engineering will be developed from the spreadsheet data and will show planned vs. actual progress on a monthly basis.

(c) Major Equipment fabrication status will be presented in a bar chart format showing the monthly progress of fabrication and manufacture for the following major equipment:

    (i)   Gas turbines
   (ii)   Gas turbine starting equipment
  (iii)   HRSG's (steam drums and prefabricated tube panels)
  (iv)   Steam turbine
   (v)   Condenser
  (vi)   Fuel gas compressors
 (vii)   Boiler HP/IP and LP feedwater pumps
(viii)   Generators
  (ix)   Main transformers
   (x)   Distributed control system

2.7   Construction Status

(a) Construction status will consist of a narrative summary of the construction activities incorporating areas of concern and recovery plans as appropriate for construction activities which are behind schedule. Each monthly report should logically progress month by month from start of Work until completion of the Work for those activities being reported on.

  (i)   Civil (BOQ's and "S curves" showing planned against actual)
 (ii)   Mechanical (BOQ's and "S curves" showing planned against actual)
(iii)   Electrical (BOQ's and "S curves" showing planned against actual)
(iv)   Instrument & Controls (BOQ's and "S curves" showing planned against actual)

2.8   Safety and Manpower Status

(a) Safety statistics will be presented by an agreed-upon format.
(b) Manpower statistics will be represented by an agreed-upon format.

2.9   Startup and Commissioning StatusStartup progress will be presented in a summarization narrative which clearly indicates current or anticipated problems concerning system turnover and system commissioning.

2.10   System turnover will be presented in a spreadsheet format and will clearly show planned against actual progress for each system which is to be turned over to the Sponsor startup and commissioning activities.

2.11   Attachments

(a) Progress Photos

2.12   Environmental Status

(a) The environmental status of the project will be presented in a narrative summary of all environmentally related activities which had occurred during the previous month or are occurring on an on-going basis.
(b) Current Schedule

3   Other Reports

From time to time and normally in response to Contractor activities which are reported to be or appear to be behind planned schedule, the Sponsor will require special reports to be submitted by Contractor.

Note: The overall construction progress report will combine monthly progress reports from any other contractors engaged by the Sponsor whose work may impact that of the Contractor.

EXHIBIT 5.6.7

FORM FOR CURRENT SCHEDULE

EXHIBIT 5.10

LIST OF KEY CONTRACTOR PERSONNEL

1. Project Director
2. Safety Coordinator
3. Security Coordinator
4. QA/QC Manager
5. construction manager
6. alternate construction manager
7. project engineer
8. resident engineer
9. purchasing agent
10. start-up engineer
11. Site superintendent
12. lead structural
13. mechanical
14. chemical
15. electrical
16. instrumentation

17. each of control, civil, cost, schedule, procurement, construction and training supervisors.

<<insert any other positions desired>>

EXHIBIT 5.10.1(d)A

FORM OF AFFIDAVIT OF PAYMENT AND PARTIAL RELEASE OF CLAIMS FOR PAYMENT

In consideration of the sum of [*<<insert amount of Milestone Payment>>*], paid by [*<<insert name of Sponsor>>*], the receipt of which is hereby acknowledged, the undersigned hereby releases, waives, and discharges any rights or claims for payment, including, but not limited to, lien rights, the undersigned now has against the Sponsor or any Sponsor Person assign arising out of or relating to the Facility on account of labor performed, work done and/or material furnished thereon.

The undersigned hereby represents, warrants and covenants to the Sponsor that:

(a) all of its Subcontractors, suppliers, laborers, materialmen, and any other person or entity that has provided labor, material or other services for the Facility have been paid in full or will be paid from the proceeds of the sums being paid by the Sponsor in accordance with the terms of their Subcontracts except as noted herein;

(b) all social security taxes, withholding taxes, sales and use taxes, works taxes and insurance premiums that have accrued in connection with the Facility have been fully paid and discharged; and

(c) all payrolls, payroll taxes, liens, charges for payment, claims, demands for payment, judgments, security interests, bills for equipment and other indebtedness connected with the Facility have been paid or released, as the case may be, except as noted in the list of exceptions below.

List of Exceptions: <None> *<list exceptions and describe in detail>*.

The releases and certification made herein by the undersigned are for labor performed, work done and/or material furnished by or on behalf of the undersigned through the period ending on [*insert date of Contractor requisition*], *<<insert calendar date>>*.

Capitalized terms used and not otherwise defined herein have the meanings given them in the Fixed Price, Lump Sum, Turnkey, Engineering, Procurement and Construction Contract, dated as of *<<insert calendar date>>*, between the Sponsor and the Contractor.

IN WITNESS WHEREOF, the undersigned has executed these presents this *<<insert calendar date>>*

> [*insert name of Contractor*]
> By:_____
> Name:_____
> Title:_____

EXHIBIT 5.10.1(d)B

FORM OF AFFIDAVIT OF PAYMENT AND FINAL RELEASE

In consideration of the sum of [*insert amount of Contract Price plus all Price Adjustments*], paid by [*insert name of Sponsor*], the receipt of which is hereby acknowledged, the undersigned hereby releases, waives and discharges any rights and claims that the undersigned now has or may have against the Sponsor Person or any Sponsor or Indemnitee arising out of or relating to the Facility or Work on account of labor performed, work done and/or material furnished.

The undersigned represents and warrants to the Sponsor that:

(a) it has not assigned any of the above referenced rights or claims;

(b) attached hereto is a list of all Subcontractors that were or are still involved in the Work;

(c) all of its Subcontractors have waived and released any of their rights to file or assert any Lien against the Facility or the Work;

(d) all of its Subcontractors, suppliers, laborers, materialmen, and any other person or entity that has provided labor, material or other services for the Facility have, in accordance with the terms of their Subcontracts, been paid in full [{or will be immediately paid from the proceeds of the sums being paid by the undersigned}];

(e) all social security taxes, withholding taxes, sales and use taxes, works taxes and insurance premiums that have accrued in connection with the Work and Facility have been fully paid and discharged; and

(f) all payrolls, payroll taxes, liens, charges, claims, demands, judgments, security interests, bills for equipment and other indebtedness of the Contractor or any Subcontractor connected with the Facility or the Work have been paid in full or released, as the case may be.

Capitalized terms used and not otherwise defined herein have the meanings given them in the Fixed Price, Lump Sum, Turnkey Engineering, Procurement and Construction Contract, dated as of <<*insert calendar date*>>, between the Sponsor and the Contractor.

IN WITNESS WHEREOF, the undersigned has executed these presents <<*insert calendar date*>>.

[*insert name of Contractor*]
By:_____
Name:_____
Title:_____

EXHIBIT 5.12

CONTEMPLATED SOURCING
<<*insert list of contemplated sources*>>

EXHIBIT 5.12.1

NATIONAL CONTENT ACTUAL DAMAGES
<<insert list of national content actual damages>>

EXHIBIT 5.29.1

LIST OF SPARE PARTS [AND STRATEGIC SPARES]
<<insert list of spare parts [and strategic spares]>>

EXHIBIT 6.1.4

LIST OF WITNESS/HOLD POINTS
<<insert list of witness/hold points>>

EXHIBIT 7.1.1

LIST OF SPONSOR PERMITS
<<insert list of Permits, if any, Sponsor will obtain>>

EXHIBIT 7.7.2

OPTIONAL TERMINATION SCHEDULE[301]

Before the Work Commencement Date, only the <<Site Assessment *<<insert other items if any>>* cost>> shall be chargeable to the Sponsor. After the Work Commencement Date, the following schedule shall apply:

Days from Work Commencement Date	Termination Payment as percent of Contract Price

OPTIONAL TERMINATION SCHEDULE

Month	Cancellation fee as % of Contract Price
1	_____%
2	_____%
3	_____%
4	_____%
5	_____%
...	...
...	...

Month	Cancellation fee as % of Contract Price
32	_____%
33	_____%
34	_____%
35	_____%
36	_____%

EXHIBIT 7.10

ADDITIONAL SCOPE OPTION AND ADDITIONAL FACILITIES PRICE

EXHIBIT 8.1

PERFORMANCE TESTS

EXHIBIT 9.2.1

PERFORMANCE GUARANTEES

Performance, Availability and Emissions Guarantee Levels and Buy Down Amount
Reference Conditions

2.1   Site specific reference conditions are:

[Turbine Inlet Air Conditions	
Compressor inlet]	
Temperature (dry bulb)	[__]°F
Relative Humidity	[__] percent
Temperature (wet bulb)	[__]°F
Elevation	[__] meters AMSL
[Inlet Loss	
Exhaust Loss]	
Fuel Specifications	Set out in Exhibit 1K
[Generator]	__ kV
[Power Factor]	__
[Evaporative Cooler]	[on/off]

1. <u>Emission Warranty</u>. Contractor represents, covenants and warrants that when operating the combustion turbine and the HRSG supplemental firing in accordance with manufacturers' requirements, that the hourly average Facility air emissions will not exceed the values set forth in the Permits [and _____].
2. <u>Water Discharge Warranty</u>. Contractor represents, covenants and warrants that the Facility's water discharge during operation shall meet the requirements of Permits [and Exhibit ___].
3. <u>Noise Pollution Warranty</u>. Contractor represents, covenants and warrants that during operation the Facility shall meet the applicable noise emission requirements as shown in the Permits [and in Exhibit _____].
4. <u>Offtaker Tests.</u>
5. "<u>Stack Emissions</u>" shall mean the emissions requirements set forth in Article 12._____ as measured during the Performance Tests used to determine compliance with the Performance Guarantees.

6. "Wastewater Quality" shall mean the Wastewater requirements set forth in Article [_____] as measured during the Performance Tests used to determine compliance with the Performance Guarantees.

EMISSIONS

	Full Condensing	<<insert number>> MW$_{th}$	<<insert number>> MW$_{th}$
Sulphur Dioxide Concentration, 6% O$_2$, drymg/Nm^3			
NO$_x$ Concentration, 6% O$_2$, dry mg/Nm^3			
Particulate Concentration, 6% O$_2$, dry mg/Nm^3			
Carbon Monoxide Concentration, 6% O$_2$, dry mg/Nm^3			

The Contractor shall guarantee that the NOx emissions as measured at the exit plane of the exhaust stack shall not exceed <<*insert number*>> ppmvd @ 15% 02 when firing Fuel [over the gas turbine load range specified below:
From____% to 100% G.T. Load
and [___] ppmvd @ 15% 02 when firing on distillate fuel over the gas turbine load range specified below:
From _____% to 100% G.T. Load]

NOISE LEVELS

Boundary noise limit as measured at the nearest receptor on the boundary of the Facility, excluding background noise from the Existing Plant. Day time (between 06.00 and 22.00) Night time (between 22.00 and 06.00)	__ dbA during the worst 8 hour reference period __ dbA during the worst 1 hour reference period
General work areas	__dbA during the worst 8 hour reference period. This guarantee does not apply inside sound insulated room but does apply in areas where operation and maintenance require extended periods of attendance.
Control Room and office areas	__dbA during the worst 8 hour reference period

1. **Continuous Noise** The Contractor shall guarantee the Facility sound power level at maximum power production shall not exceed [_____] dB(A) at any point on the site boundary.
2. **Infrequent Noise Sources** The Contractor shall guarantee that any noise source that is planned to operate less than [_____] times a year and for less than [_____] minutes at a time shall not exceed the continuous noise limits by more than [_____] dB(A).
3. **Specific Areas**. The Contractor shall guarantee:
   (i) Noise levels in general work areas shall not exceed an SPL of [_____] dB(A)
   (ii) Noise levels in control rooms and offices shall not exceed an SPL of [_____] dB(A)

"Minimum Performance Guaranty" is defined in Section [__].

"Optional Facility Performance Guarantee Cure Period" shall have the meaning ascribed thereto in Article 9<<__>>.

[**"Optional Unit One Performance Guarantee Cure Period"** is defined in Article 9<<__>>.]

[**"Optional Unit Two Performance Guarantee Cure Period"** is defined in Article 9<<__>>.]

"Reference Conditions" shall mean an ambient Site temperature of 95°F in the case of the 84° Test and 28°F in the case of the 32° Test (both dry bulb), 40% relative humidity, atmospheric pressure of 14.7 pounds per square inch, a generator power factor of <<*insert coefficient*>> lagging. Performance Test results shall be adjusted from ambient conditions occurring during the time of such tests to the Reference Conditions using <<Correction Curves>>.

[**"Unit Buy Down Amounts"** shall be the Unit Buy Down Amounts described in Article 9<<__>>.]

[**"Unit Minimum Performance Level"** shall be met when the Unit in question is capable of achieving all of the following:

(a) <<insert minimum acceptable levels>>; and
(b) achieving the Emission Guarantee.]

"Unit Performance Guarantees" are each of the guarantees of performance for a Unit in Article 7.2.

2.2   The Contractor guarantees that the New Facility will be capable of simultaneously achieving all of the following (the "Emissions Guarantee"):

(a) the emissions from the New Facility shall not exceed the emissions delineated in Appendix 1 of this Exhibit U when either the New Facility or a Unit is operated at any load point above 40% (whether on Gas or Oil);
(b) not exceeding noise levels as set forth in Appendix 1 of this Exhibit U;
(c) wastewater quality other than sanitary waste shall not be worse than that set forth in Appendix 1 of this Exhibit U;
(d) the New Facility shall be designed such that clean rain water runoff will meet all limitations set forth in applicable Laws for effluent emissions; and
(e) all emissions rates from the New Facility shall not exceed the emission standards of the relevant Governmental Authorities and shall fully comply with all applicable Laws.

"Wastewater" shall mean cooling tower blowdown, intermittent steam boiler blowdown, washdowns, equipment drips, contaminated rainwater runoff and neutralized demineralization plant wastes.

2.3   Cure Period for Failure to Achieve Performance Guarantees. Once the Contractor has achieved the Facility Minimum Performance Level, the Contractor shall have a period from the date of such achievement until one (1) month thereafter to try to achieve the Performance Guarantees and if the Contractor is successful, all Impaired Performance Damages will be promptly refunded to the extent of the improvements.

2.4   <u>Minimization</u>. The Contractor and Project Company shall use their best efforts to minimize the economic impact on Project Company during periods of repairs or Performance Tests carried out during the above cure period(s) and the Contractor shall be responsible for any damage or degradation to the New Facility it causes during such period, if any.

(a) [<u>Other Performance Guarantees</u>. The Contractor guarantees that the New Facility will pass the 172-hour reliability test described in the Functional Specifications].

(b) _____

[EXHIBIT 9.2.2

<u>PERFORMANCE GUARANTEES</u> RELATING TO AVAILABILITY]

[**"Availability Period"** for a Unit shall commence <no earlier than << *insert number*>> Days after> <on> the Facility Provisional Acceptance Date <or the period commencing on the last day of the First Maintenance Period for such Unit, whichever occurs first (but in no event earlier than << *insert number*>> Days after the Facility Provisional Acceptance Date)>, and end on {the earlier of} the last day of the Warranty Notification Period [for such Unit] {or <<*insert number*>> Days after the commencement of such period}.]

[**"Availability Period"** for a Unit shall commence [on _____] no earlier than [six months after the Facility Substantial Completion Date or the period commencing on the last day of the First Maintenance Period for such Unit, whichever occurs first (but in no event earlier than six months after the Facility Substantial Completion Date)], and end on [on _____] [the earlier of the last day of the Warranty Notification Period for such Unit or 18 months after the commencement of such period,] [provided that in all cases in which the Project Company's failure to comply with the Availability Stipulations has affected the Unit Availability of either Unit, the duration of such effects shall not be included in the Availability Period].][302, 303]

[**"Availability Stipulations"** shall mean that the Project Company has:

(a) on a consistent basis, operated the Facility on feedstocks with the qualities outlined in the Functional Specifications;
(b) kept reasonable records of disturbances, repairs, fuel quantities and qualities as available and operational data; and
(c) operated, inspected and maintained [the Facility] [the Units] generally in accordance with the Contractor's written manuals.][304]

[{**"Availability Stipulations"** shall mean that the Sponsor has [generally]:

(a) operated the Facility on feedstocks with the qualities outlined in the Functional Specifications;
(b) kept reasonable records of extraordinary disturbances, major repairs, fuel quantities and qualities as available and operational data; and

(c) operated, inspected and maintained the Facility generally in accordance with the Contractor's written manuals so long as such manuals do not call for practices beyond those practices which would be employed if <{Good}> <Prudent> <Industry> <{Utility}> Practices are being observed[305}].]

["**First Maintenance Period**" shall be the first period during which the Sponsor carries out the annual maintenance and related inspections for a Unit.]

["**Forced Outage**" shall mean an interruption or reduction of the <<*generating/throughput*>> capacity of the Facility that is <not><the result of failure of:

(a) Equipment;
(b) Facility Systems; or
(c) Computer Programs;

and is> caused by a Force Majeure Excused Event or a Person other than a Contractor Person [{or the Sponsor's failure to keep available <at<<*insert warehouse name and location*>> > [or] <on-Site> any spare part listed in Exhibit 5.29}].]

"**Forced Outage Hours**" means the duration, expressed in hours, of a Forced Outage, but excluding any periods of Force Majeure occurring during such Forced Outage.

["**Loss of Availability Damages**" is defined in _____.]

["**Maintenance Outage Hours**" means the duration, expressed in hours, of a Maintenance Outage, but excluding any periods of Force Majeure occurring during such Maintenance Outage.]

["**Planned Outage Hours**" means the duration, expressed in hours, of a Planned Outage, but excluding any periods of Force Majeure occurring during such Planned Outage.]

["**Unit Availability**" for a Unit shall be determined by the following formula:

$$(APH - APOH - \sum_i^n FOH_1) \div APH$$

where:

$$FOH_i = H_i \left[1 - (R_i \div F)\right]$$

F = [*insert MW guarantee from Article 9.1*], or, if the Contractor has made payment in full of all required Unit Buy Down Amounts for the Unit in question, the Unit Net Electrical Output of the relevant Unit which formed the basis for the Unit Buy Down Amount at << *insert number* >> MW;

FOH_i = the number of hours during which the reduction or elimination of the relevant Unit's electrical generating capacity is in effect during the particular Forced Outage;

APOH = AMH (APH ÷ 8760);

AMH = the actual number of hours a Unit is out of service as the result of the annual scheduled maintenance during the Availability Period but in no event greater than <<500>>;

APH = the number of hours in the Availability Period;

R_i =the generating capacity of the relevant Unit during the particular Forced Outage i;

i = each period of Forced Outage affecting the relevant Unit during the Availability Period; and

n = total number of periods of Forced Outage affecting the relevant Unit during the Availability Period {excluding the Period of any Forced Outage which is a result of the failure of the Sponsor or its operator to comply with the Availability Stipulations}.

["**Unit One Availability Guarantee**" is defined in Article 9<<__>>.]
["**Unit Two Availability Guarantee**" is defined in Article 9<<__>>.]
["**Unit Availability**" for a Unit shall be determined by the following formula:

$$(APH - APOH - \sum_i^n FOH_1) \div APH$$

where:

$$FOH_i = H_i [1-(R_i \div F)]$$

F = [fill in MW_e guarantee pt]., or, if the Contractor has made payment in full of all required Unit Buy Down Amounts for the Unit in question, the Unit Net Electrical Output of the relevant Unit which formed the basis for the Unit Buy Down Amount at ____ MW_e;

FOH_i = the number of hours during which the reduction or elimination of the relevant Unit's electrical generating capacity is in effect during the particular Forced Outage i;

APOH = AMH (APH $\div$ 8760);

AMH = the actual number of hours a Unit is out of service as the result of the annual scheduled maintenance during the Availability Period but in no event greater than [500];

APH = the number of hours in the Availability Period;

R_i = the generating capacity of the relevant Unit during the particular Forced Outage i;

i = each period of Forced Outage affecting the relevant Unit during the Availability Period; and

n = total number of periods of Forced Outage affecting the relevant Unit during the Availability Period].[306]

   "**Unit One Availability Guarantee**" shall have the meaning ascribed thereto in Article 7.2.10.

   The Contractor guarantees that during [any <<*insert number*>> consecutive months of [its] [the] Availability Period:[307]

(a) Unit One will achieve a Unit Availability of at least <<*insert number*>> percent (the "Unit One Availability Guarantee"); and

(b) Unit Two will achieve a Unit Availability of at least <<*insert number*>> percent (the "Unit Two Availability Guarantee").

If the Unit One Availability Guarantee is not achieved, the Contractor shall pay the Sponsor as liquidated damages <<*insert currency and amount*>> for each full increment of <<*insert number*>> tenths of a percentage point by which <<*insert number*>> percent exceeds the actual Unit Availability of Unit One.

If the Unit Two Availability Guarantee is not achieved, the Contractor shall pay the Sponsor as liquidated damages multiplied by <<*insert currency and amount*>> for each full increment of <<*insert number*>>-tenths of a percentage point by which <<*insert number*>> percent exceeds the actual Unit Availability of Unit Two.

The liquidated damages referred to in Articles 7.2.11 and 7.2.12 shall be referred to as "Loss of Availability Damages." [Any Loss of Availability Damages assessed on a Unit may be reduced if the Unit Availability of the other Unit exceeds <<*insert number*>> percent by adding the amount by which the Unit Availability of such over-performing Unit exceeds <<*insert number*>> percent to the Unit Availability of the under-performing Unit, provided, that, in no case shall the amount added to the Unit Availability of the under-performing Unit exceed <<*insert number*>> percentage points].[308] The Availability Liquidated Damages, if any, shall be paid by the Contractor to the Company within <<*insert number*>> Days after the end of the Availability Period.

[EXHIBIT 9.3

FORM OF ESCROW AGREEMENT]

ESCROW AGREEMENT ("Escrow Agreement"), made as of the <<*insert number*>> day of <<*insert month and year*>><<*insert name of Contractor*>>by and among [ ], a <<*insert type and jurisdiction*>> (the "Contractor") <<*insert name of Sponsor*>>, (the "Sponsor"), a <<insert type and jurisdiction>> and _____ BANK N.A. (the "Bank"), [a National Banking Association], as Escrow Agent (the "Escrow Agent").

WITNESSETH:

Sponsor and Contractor have entered into a Turnkey Engineering, Procurement and Construction Contract, dated as of [_____, 20 __] (the "Contract"). The Contract requires Contractor and/or Sponsor to place in escrow amounts in dispute with respect to Delay Damages (as defined in the Contract) [and certain disputed amounts as described more fully therein]. Sponsor and Contractor wish to appoint the Banks the Escrow Agent for such purposes and provide for the terms and conditions for such Escrow Agent as more particularly set forth below.

NOW THEREFORE, in consideration of the premises and the covenants of the parties herein, the parties hereby agree as follows:

1. Appointment of Escrow Agent. Sponsor and Contractor hereby appoint the Bank as Escrow Agent, and the latter accepts appointment as such, all upon the terms and provisions set forth in this Escrow Agreement.
2. Contract Terms. All terms used herein shall have the same meaning as set forth in the Contract unless the context clearly indicates otherwise.

3. [Sponsor's Agent. Sponsor appoints _____ and _____, and each of them, as Sponsor's Agent, with full power to act in behalf of Sponsor in connection with all matters relating to this Escrow Agreement and to appoint a successor or successors for either or both of them. Contactor and Escrow Agent shall be entitled to rely on any taken act taken, or communication, instrument or document signed by, either of such named persons as Sponsor's Agent (or any person of whom Contractor and the Escrow Agent shall have given notice of the appointment of such person as a successor of Sponsor's Agent). Either or both of such persons (or any successor or successors) may resign as Sponsor's Agent by giving written notice to the Escrow Agent, Contractor and any other person then acting as Sponsor's Agent. Upon the resignation or death of any person acting as Sponsor's Agent, the person, if any, theretofore appointed in writing to become his successor as Sponsor's Agent, shall become such successor, or there is no such appointed successor for such resigning or deceased Sponsor's Agent, and if he shall fail to do so within 30 days after such resignation or death, either the Escrow Agent or Contract must apply to a court of competent jurisdiction for the appointment of such a successor. Each successor Sponsor's Agent shall have the same power and authority as though named herein.]

4. [Contractor's Agent. The Contractor appoints <<*insert agents' names*>>and each of them, as Contractor's Agent, with full power to act in behalf of the Contractor in connection with all matters relating to this Escrow Agreement and to appoint a successor or successors for either or both of them. Sponsor and the Escrow Agent shall be entitled to rely on any act taken, or communication, instrument or document signed by, either of such named persons as Contractor's Agent (or any person of whom Sponsor and the Escrow Agent shall have been given notice of the appointment of such person as successor Contractor's Agent). Either or both of such persons (or any successor or successors) may resign as Contractor's Agent by giving written notice to the Escrow Agent, Sponsor and any other person then acting as Contractor's Agent. Upon the resignation or death of any person acting as Contractor's Agent, the person, if any, theretofore appointed in writing to become his successor as Contractor's Agent shall become such successor, or if there is no such appointed successor, the remaining Contractor's Agent shall appoint a successor for such resigning or deceased Contractor's Agent, and if he shall fail to do so within 30 days after such resignation or death, either the Escrow Agent or Sponsor may apply to a court of competent jurisdiction for the appointment of such successor. Each successor Contractor's Agent shall have the same power and authority as though named herein.]

5. Deposits. No initial deposit is made with the Escrow Agent. Any future cash deposits made by Contractor and any income earned thereon shall be held by the Escrow Agent in an escrow account, hereinafter called the "Contractor Escrow Fund." Any future cash deposits made by Sponsor and any income earned thereon shall be held by the Escrow Agent in an escrow account, hereinafter called the "Sponsor Escrow Fund." Each of the Contractor Escrow Fund and the Sponsor Escrow Fund shall be invested, administered and disbursed in accordance with the terms and provisions of this Escrow Agreement. The Contractor Escrow Fund and the Sponsor Escrow Fund are herein collectively called "Escrow Fund."

6. <u>Procedure for Drawing upon the Escrow Fund</u>. The following procedure shall be used for payment from the Escrow Fund:

   (a) If the [arbitrator] determines that Sponsor, and not the Contractor, is due any amount, Escrow Agent shall pay to Sponsor such amount (including interest thereon from the deposit date to the date of payment thereof at the Applicable Rate) from (not to exceed) the Contractor Escrow Fund, all upon receipt by the Escrow Agent of a joint letter to such effect signed by Contractor['s Agent] and Sponsor['s Agent], and shall also pay to Sponsor the Sponsor Escrow Fund.

   (b) If the [arbitrator] determines that Contractor, and not Sponsor, is due any amount, the Escrow Agent shall pay to Contractor such amount (including interest thereon from the deposit date to the date of payment thereof at the Applicable Rate) from (but not to exceed) the Sponsor Escrow Fund, all upon receipt by Escrow Agent of a joint letter to such effect signed by Contractor['s Agent] and Sponsor['s], and shall also pay to Contractor the Contractor Escrow Fund.

   (c) Immediately following any payment under paragraph (a) or (b) above, or if the [arbitrator] determines that neither Sponsor nor Contractor is due any amount, Escrow Agent shall pay to Sponsor all funds remaining in the Sponsor Escrow's Fund and to Contractor all funds remaining in the Contractor Escrow Fund, all upon receipt by the Escrow Agent of a joint letter to such effect signed by Contractor['s Agent] and Sponsor['s Agent].

7. <u>Investments</u>. At the written direction of (a) the Contractor['s Agent] with respect to the Contractor Escrow Fund and (b) Sponsor['s Agent] with respect to the Sponsor Escrow Fund, the Escrow Agent shall invest and reinvest all cash amounts in the respective portions of the Escrow Fund in (i) short-term debt obligations of the United States Government or any agency thereof, (ii) short-term debt obligations guaranteed by the United States Government, (iii) prime commercial paper rated at least P-1 or the equivalent by Moody's Investors Service, Inc. ("<u>Moody's</u>") or A-1 or the equivalent by Standard and Poor's Corporations ("<u>S&P</u>") at the time of purchase, or (iv) short-term tax exempt obligations of state or municipal governments rated at least AA or the equivalent by S&P or AA or the equivalent by Moody's. At the time of purchase or (iv) short-term tax exempt obligations of state or municipal governments rated at least AA or the equivalent by S&P or AA or the equivalent by Moody's. At the written direction of Contractor['s Agent], as to the Contractor Escrow Fund, and Sponsor['s Agent] as to the Sponsor's Escrow Fund, the Escrow Agent shall release any profits realized from such investments to the Contractor['s Agent], or to Sponsor['s Agent] as the case may be, Contractor as to the Contractor Escrow Fund, and Sponsor as to the Sponsor Escrow Fund, shall upon the direction of the Escrow Agent, add cash to the Contractor Escrow Fund or Sponsor Escrow Fund, as the case may be, because of any diminution of cash as a result of losses realized on investments, and the Escrow Agent is obligated to so direct. The Escrow Agent may sell any investment in its sole direction as may be necessary to raise cash in order to make disbursements from the Escrow Fund.

8. <u>Administration</u>. It is agreed that the Escrow Agent shall have no duties or responsibilities whatsoever hereunder except as specifically provided herein; that the Escrow Agent may look solely to this Agreement with respect to its rights, duties

and obligations with respect to the subject matter hereof; that the Escrow Agent shall be fully protected and incur no liability to anyone in acting upon any notice, request, consent, certificate, document, letter, telegram, order, resolution or other paper believed by it to be genuine and to be signed or sent by the proper persons that the Escrow Agent shall be under no obligation to commence, continue or defend any suit or proceeding in connection with the Escrow Fund or this Escrow Agreement unless requested to do so by either Contractor's Agent or Sponsor's Agent and indemnified to its satisfaction; that the Escrow Agent may from time to time in its discretion, engage legal counsel for advice concerning the subject matter of this Escrow Agreement and anything done or suffered to be done in good faith by the Escrow Agent in accordance with the opinion of counsel shall be conclusive in favor the Escrow Agent against all those now of hereafter interested in the subject matter of this Escrow Agreement; that the Escrow Agent shall be entitled to reimbursement for all expenses, including reasonable counsel fees, incurred by it in carrying out its duties hereunder and reasonable compensation for all services rendered by it in connection with this matter, which expenses and compensation shall be paid equally by Contractor and Sponsor and the source of such expenses and compensation shall, as to each of Contractor and Sponsor, first be paid from any income attributable to it in the Escrow Fund to the extent that such income is sufficient therefor, and Contractor and Sponsor shall each pay directly to the Escrow Agent its share of the balance thereof.

9. Resignation. The Escrow Agent may resign by giving written notice to Contractor['s Agent] and Sponsor['s Agent], in which case Contractor['s Agent] and Sponsor['s Agent] shall appoint a successor Escrow Agent. If they fail to do so within 60 days, the resigning Escrow Agent may apply to a court of competent jurisdiction for the appointment of a successor. The resigning Escrow Agent shall transfer and deliver the Escrow Fund to the successor Escrow Agent after the expenses and compensation of the resigning Escrow Agent shall have been paid to it, and all of the provisions of this Escrow Agreement shall apply to the successor Escrow Agent as though it had been named herein.

10. Notices. Any and all notices or other communications between the parties which are required or permitted under this Agreement shall be in writing and shall be sufficiently given if delivered in person to or if deposited in the United States mail, postage prepaid, registered or certified, return receipt requested, addressed as follows:

To Sponsor [and Sponsor's Agent]:
    [ADDRESS]
    [Telephone: _____]
    [Facsimile: _____]
    [Attention: _____]

with a copy to:
    [ADDRESS]
    [Telephone: _____]
    [Facsimile: _____]
    [Attention: _____]

To Contractor [and Contractor's Agent]:
    [ADDRESS]
    [Telephone: _____]
    [Facsimile: _____]
    [Attention: _____]
with a copy to:
    [ADDRESS]
    [Telephone: _____]
    [Facsimile: _____]
    [Attention: _____]
To Escrow Agent:
    [ADDRESS]

or to such other address as shall be furnished in writing by any party to the other and shall be deemed to be given as of the date so delivered or deposited.

11. <u>Consents and Waivers</u>. No consent, approval or waiver of any provisions of this Escrow Agreement shall be valid unless made in writing and no such waiver shall constitute a waiver of any other provision hereof (whether or not similar) nor shall such waiver constitute a continuing waiver.

12. <u>Governing Law; Binding Effect; Assignment</u>. This Agreement shall be governed by the laws of the State of New York. To the fullest extent permitted by law, this Escrow Agreement shall be binding upon and inure to the benefit of the parties hereto, and their respective heirs, successors and assigns; provided, however, that no assignment of any rights or delegation of any obligations provided for herein may be made by any party hereto without the express written consent of the other party.

13. <u>Amendments or Modifications</u>. No amendment or modification of this Agreement shall be effective unless the same shall be in writing and signed by the parties duly authorized hereto.

14. <u>Captions</u>. The captions used in this Escrow Agreement are for convenience and identification purposes only, are not an integral part of this Escrow Agreement and are not to be considered in the interpretation of any part of this Escrow Agreement.

15. <u>Counterparts</u>. This Agreement may be executed in one or more counterparts, all of which taken together shall constitute but one instrument.

16. <u>Reports</u>. The Escrow Agent shall supply Sponsor['s Agent] and Contractor['s Agent] with monthly statements of the Escrow Fund.

IN WITNESS WHEREOF, the parties hereto, by their duly authorized officers, have caused this Escrow Agreement to be executed on and as of the day and year first above written.

[SPONSOR]

By:_____

Name:

Title:

ATTEST:

Secretary

[Corporate Seal]

[CONTRACTOR]

By:_____

Name:

Title:

ATTEST:

Secretary

[Corporate Seal]

[_____BANK, N.A].

By:_____
___

Name:

Title:

EXHIBIT 10.1.2

<u>FORM OF ADVANCE PAYMENT LC</u>

<<insert date>>

CREDIT NO. _____

<<insert date>>

<<insert Sponsor name and address>>

Ladies and Gentlemen:

1. <u>Stated Amount</u>. We hereby establish in favor *<<insert name of Sponsor>>* (the "<u>Sponsor</u>") an irrevocable Standby letter of credit (the "<u>Credit</u>") whereby you are authorized to draw on [*<< insert name of Acceptable Issuer Bank>>*] (the "<u>Issuing Bank</u>"), Credit No. _____, issued for the account of [*insert name of Contractor*] (the "<u>Contractor</u>"), an aggregate amount not exceeding *<<insert amount of currency in words>>* (*<<insert amount of currency in numerals>>*) (the "<u>Stated Amount</u>"). This Credit is issued pursuant to the Fixed Price, Lump Sum Turnkey Engineering, Procurement and Construction Contract, dated as of *<<insert date>>*, between the Contractor and the Sponsor (as amended, restated, supplemented or otherwise modified from time to time, the "<u>Contract</u>").

2. <u>Term</u>. This Credit is effective immediately and expires at the close of banking business at our office in <New York City> <London> <Hong Kong> <Tokyo> on [*insert date*].

3. <u>Drawing</u>. Funds under this Credit are available to you against your presentation to us at our offices above of a drawing certificate in the form of Annex A hereto. Such certificate shall have all blanks appropriately filled in and shall be signed

by your authorized officer, and such certificate shall be on your letterhead. You may make multiple and/or partial drawings hereunder.

4.   <u>Drawing Procedures</u>. If a drawing is made by you hereunder, payment shall be made of the amount specified in immediately available funds on the <next> <{second}> succeeding Business Day; provided that such drawing and the documents presented in connection therewith conform to the terms and conditions hereof. All payments hereunder shall be made free and clear of, and without deduction for or on account of, any present or future taxes, duties, charges, fees, deductions or withholdings of any nature and by whomsoever imposed. As used in this Credit, the term "<u>Business Day</u>" shall mean any day other than a Saturday or Sunday or legal holiday or a day on which banking institutions in which our offices listed above are authorized or required by law or executive order to close.

5.   <u>Assignment</u>. This Credit may be assigned by you upon written notice thereof to the Issuing Bank.

6.   <u>Governing Law</u>. This Credit is subject to the Uniform Customs and Practice for Documentary Credits (2007 Revision), International Chamber of Commerce Publication No. 600 (the "<u>UCP</u>"). This Credit shall be deemed to be a contract made under the laws of <New York> <England> and shall, as to matters not governed by the UCP, be governed and construed in accordance with the laws of <New York> <England>.

7.   <u>Notices</u>. Communications with respect to this Credit shall be in writing and shall be addressed to the Issuing Bank at <*insert address*> (or otherwise as designated by the Issuing Bank to the Sponsor by written communication) specifically referring thereon to "[<<*insert name of Issuing Bank*>>] Irrevocable Standby Letter of Credit No. _____."

8.   <u>Integration</u>. This Credit sets forth in full our undertaking, and such undertaking shall not in any way be modified, amended, amplified or limited by reference to any document, instrument or agreement referred to herein (including, without limitation, the Contract), except only the drawing certificate referred to herein; and any such reference shall not be deemed to incorporate herein by reference any document, instrument or agreement, except for such drawing certificates.

Very truly yours,
[*insert name of Issuing Bank*]
By:_____
Authorized Officer

ANNEX A to EXHIBIT 10.1.2

<u>DRAWING CERTIFICATE</u>
[<<*insert date*>>]
[<<*insert name of Issuing Bank*>>]
[<<*insert address of Issuing Bank*>>]

Re:Irrevocable Standby Letter of Credit No.

Ladies and Gentlemen:

1. <<*insert name of Sponsor*>> (the "Sponsor") hereby certifies to [<<*insert name of Issuing Bank*>>], with reference to Irrevocable Standby Letter of Credit No. _____ (the "Credit"), that:
    (a) the Contractor has failed to fulfill one or more of the conditions of, or failed to perform one or more of its obligations under, the Contract and now in accordance with the Contract, the Sponsor is entitled to <<*insert currency and amount*>>; and/or
    (b) this Credit is scheduled to expire within <<thirty (30) >> Days after the date hereof and the Contractor has not posted in our favor a substitute for this Credit meeting the requirements of the Contract or refunded an amount of cash equal to available and undrawn amount under this Credit.
2. [{Prior Notice to the Contractor. At least <<*five (5)*>> Business Days prior to the date hereof, the undersigned gave the Contractor written notice of:
    (a) the foregoing; and
    (b) the Sponsor's intention to make a drawing under the Credit.}]
3. Demand for Payment. The Sponsor is making a demand for payment under the Credit in the amount of <<*insert amount in words and numerals also*>>.
4. Terms. Capitalized terms used herein and not otherwise defined shall have the meanings ascribed thereto in the Fixed Price, Lump Sum Turnkey Engineering, Procurement and Construction Contract, dated as of <<*insert date*>> (as amended, restated, supplemented or otherwise modified from time to time, the "Contract"), between the Sponsor and the Contractor.
5. Payment Instructions. Payment of the amount demanded hereby shall be made by wire transfer to the following account: _____.

IN WITNESS WHEREOF, the Sponsor has executed and delivered this Certificate as of <<*insert date*>>.

<<insert Sponsor name>>
By:_____
Authorized Representative
Name: _____
Title: _____

EXHIBIT 10.1.3(c)

CONTRACT PRICE AND COMPONENTS THEREOF[309]

Item of Equipment or Work	Contract Price Component	Country of Origin

TOTAL

EXHIBIT 10.3.1A

MILESTONE PAYMENT SCHEDULE

Milestone	Days from Work Commencement Date	Expected Calendar Date to be Achieved	Percentage of Contract Price	Milestone Description	Supporting Documentation
1	1	____	____ percent	Notice of Proceed and Site Assessment complete	Written Notice to Proceed, Site Assessment, Performance LC posted and effective and delivered to Sponsor
2	____	____	____ percent	Order of all steam and combination turbine generators units	Unpriced Subcontract, technical specification
3	____	____	____ percent	Order of main pressure parts materials including drum and boiler pressure part basic design package	Unpriced Subcontract, submission of boiler drawings, boiler material selection and temperature charts calculation notes
4	____	____	____ percent	Civil design package complete	Piping and instrumentation diagrams and layout drawings, general arrangement drawings and elevational drawings of boiler island
5	____	____	____ percent	Main demolitions works finished	Inspection certificate, first civil design package
6	____	____	____ percent	Start of civil works	Main civil unpriced contract, mobilization on site
7	____	____	____ percent	Basic facility design package complete	Piping and instrumentation diagrams of main processes, basic plant layout drawings and electrical system one line diagrams

Milestone	Days from Work Commencement Date	Expected Calendar Date to be Achieved	Percentage of Contract Price	Milestone Description	Supporting Documentation
8	_____	_____	_____ percent	Order of water treatment plant and main transformers	Unpriced Subcontracts
9	_____	_____	_____ percent	Delivery of steel structures for 1st erection stage, Unit One	Delivery confirmed by Sponsor in writing
10	_____	_____	_____ percent	All turbine building foundations ready	Completion agreed by Sponsor in writing
11	_____	_____	_____ percent	Start of steel structure. erection Unit Two	Written protocol of commencement
12	_____	_____	_____ percent	Arrival of (a) steam turbine castings to shop and (b) combination turbines	Manufacturer certificate
13	_____	_____	_____ percent	Order of DCS	Copy of written purchase order sent to manufacturer
14	_____	_____	_____ percent	First delivery of main pressure parts on Site	Delivery confirmed by Sponsor in writing
15	_____	_____	_____ percent	Delivery of steam drum boiler 1, preliminary training program	Delivery confirmed by Sponsor in writing, preliminary training program document submitted to the Sponsor
16	_____	_____	_____ percent	Steam turbine 1 balancing tests done	Test report by approved Sponsor in writing
17	_____	_____	_____ percent	Delivery of steam drum boiler 2 on Site	Delivery confirmed by Sponsor in writing

Milestone	Days from Work Commencement Date	Expected Calendar Date to be Achieved	Percentage of Contract Price	Milestone Description	Supporting Documentation
18	____	____	____ percent	Steam turbine 1 exworks, delivery of Fuel feed. eq.	Certificate of readiness for shipping
19	____	____	____ percent	Economizer boiler 2 on Site	Delivery confirmed by Sponsor in writing
20	____	____	____ percent	Steam turbine 2 exworks, boiler 1 on Site	Delivery confirmed by Sponsor in writing
21	____	____	____ percent	Hydrostatic test boiler 1	Test certificate
22	____	____	____ percent	Delivery of packages for boiler 2	Delivery confirmed by Sponsor in writing
23	____	____	____ percent	Hydrostatic test boiler 2	Test certificate
24	____	____	____ percent	Transformer of Unit Two installed	Installation inspection certificate
25	____	____	____ percent	Unit Two ready for energizing	Readiness report
26	____	____	____ percent	Start of cold tests of Unit Two	Protocol for commencement
27	____	____	____ percent	First fire of Unit One	Control room log book record
28	____	____	____ percent	First steam to steam turbine 1 of Unit One	Control room log book record
29	____	____	____ percent	First fire of Unit Two	Control room log book record
30	____	____	____ percent	Unit Commissioning Completion Date of Unit One	Certificate of Unit Commissioning Completion for Unit One
31	____	____	____ percent	Start of reliability run, Unit One	Protocol of commencement
32	____	____	____ percent	Unit Commissioning Completion Date of Unit Two	Certificate of Unit Commissioning Completion for Unit Two
33	____	____	____ percent	Unit Provisional Acceptance Date of Unit One	Certificate of Unit Provisional Acceptance for Unit One

Milestone	Days from Work Commencement Date	Expected Calendar Date to be Achieved	Percentage of Contract Price	Milestone Description	Supporting Documentation
34	____	____	____ percent	Start Reliability Run, Unit Two	Protocol on commencement
35	____	____	____ percent	Unit Provisional Acceptance Date of Unit Two	Certificate of Unit Provisional Acceptance for Unit One
36	____	____	____ percent	Facility Provisional Acceptance Date	Certificate of Facility Provisional Acceptance
37	____	____	____ percent	Warranty Notification Period Ends	

EXHIBIT 10.3.1B

FORM OF PAYMENT APPLICATION[310]

Date:<<*insert calendar date*>>

<<[*insert name of Sponsor*]>>

<<[*address*]>>

Ladies and Gentlemen:

The undersigned submit[s] this payment application pursuant to Article __ of the Fixed Price, Lump Sum, Turnkey Engineering, Procurement and Construction Contract between <<*insert Contractor's name*>> (the "Contractor") and <<*insert Sponsor's name*>> (the "Sponsor"), dated as of <<*insert calendar date*>> (the "Contract"). Unless otherwise defined herein, all capitalized terms used herein shall have the meanings assigned to such terms in the Contract. The undersigned acknowledge[s] that the Sponsor [and each Financing Party] will rely hereon. On the basis of the foregoing, and for the purpose of inducing the Sponsor to make payment to [or on behalf of] the undersigned [to [Vendor/ECA]],[311] the undersigned hereby represent[s] and warrant[s] as of the date hereof that:

(a) The following is the current status of the Contract account:

Contract Price (not including Price Adjustments)	<<*insert Contract Price in words*>> <<*insert Contract Price in numerals*>>
All Price Adjustments	<<*insert sum of all Price Adjustments in words*>> <<*insert sum of all Price Adjustment in numerals*>>
Total Milestone Payments made by Sponsor	<<*insert amount of all Milestone Payments made by Sponsor in words*>> <<*insert amount of all Milestone Payments made by Sponsor in numerals*>>

(less Retainage and any other withheld amounts or adjustments)	*<<insert in words all cash Retainage and other cash withholding from payments only>>* *<<insert in numerals only all cash Retainage, and cash payments withheld>>*
Approximate Unpaid Contract Price	*<<insert in words unpaid Contract Price>>* *<<insert in numerals unpaid Contract Price>>*

(b) the information in all documents and materials prepared or signed by the Contractor or any of its officers, agents or employees and submitted to the Sponsor in support hereof or in connection with the Facility is, in all respects, true, correct and complete;

(c) except as has been previously explained to the Sponsor in detail in writing, the progress of the Work has been in accordance with the Baseline Level 3 Schedule and the Contractor believes all the Deadlines will be met except as has been previously explained to the Sponsor in detail in writing;

(d) the portion of the Work, as more particularly set forth in an Annex A hereto and corresponding to milestone number ___ in the Milestone Payment Schedule, was completed during the month of *<<insert applicable month>>*, *<<insert applicable calendar year>>* entitling the Contractor to a Milestone Payment of ___percent of the Contract Price minus the applicable Retainage:

    (i) which corresponds to [*<<insert sum due>>*] [of which the Contractor requests that *<<insert percentage>>* percent thereof be paid in [*insert local currency in Contract allows for payments*]];

    (ii) [[*insert currency*] *insert amount*] as a result of the [*insert currency*] Notice;[312]

(e) all Contractor Permits necessary to permit the foregoing are in full force and effect;

(f) Annex B hereto sets forth in detail the basis of the calculation of the amounts due under paragraph (d) above as determined by Article __ of the Contract;

(g) no materially adverse change in the financial condition of the Contractor has occurred since [*insert date of last application for payment*] except as has been disclosed in writing in detail to the Sponsor;

(h) the Work performed to date has, unless otherwise stated by the Contractor in detail in a writing delivered to the Sponsor, been performed in accordance with all the requirements of the Contract and all Payment Applications have been submitted, including the current Payment Application in accordance with the Contract;

(i) to the Contractor's knowledge, all insurance required of the Contractor under the Contract has been bound and is in place, in full force and effect;

(j) the Contractor has been paid all amounts invoiced by it under the Contract, except as otherwise described in Annex B attached hereto except as has been described in detail in a writing delivered to the Sponsor, and all Subcontractors and Vendors engaged or employed by the Contractor have been paid to the extent that such amounts are due or such payment (or a portion thereof) is subject to a good faith contest which is being diligently prosecuted by the Contractor; and

(k) [{no Contractor Event of Default exists}.]

IN WITNESS WHEREOF, the undersigned have executed this Certificate on the date first above written.

<<insert name of Contractor>>
<<insert name of Consortium Member>>
By:_____
Name:_____
Title:_____
[*<<insert name of Consortium Member>>*
By:_____
Name:_____
Title:_____]

ANNEX A to **EXHIBIT 10.3.1B**

DETAILED DESCRIPTION OF MILESTONES MET

ANNEX B to **EXHIBIT 10.3.1B**

CALCULATIONS

EXHIBIT 10.3.5

FORM OF RETAINAGE LC

IRREVOCABLE STANDBY LETTER OF CREDIT NO. _____

<<insert date>>
CREDIT NO. _____
<<insert date>>
<<insert Sponsor name and address>>

Ladies and Gentlemen:

1. <u>Stated Amount</u>. We hereby establish in favor *<<insert name of Sponsor>>* (the "<u>Sponsor</u>") an irrevocable Standby letter of credit (the "<u>Credit</u>") whereby you are authorized to draw on [*<< insert name of Acceptable Issuer Bank>>*] (the "<u>Issuing Bank</u>"), Credit No. _____, issued for the account of [*insert name of Contractor*] (the "<u>Contractor</u>"), an aggregate amount not exceeding *<<insert amount of currency in words>>* (*<<insert amount of currency in numerals>>*) (the "<u>Stated Amount</u>"). This Credit is issued pursuant to the Fixed Price, Lump Sum Turnkey Engineering, Procurement and Construction Contract, dated as of *<<insert date>>*, between the Contractor and the Sponsor (as amended, restated, supplemented or otherwise modified from time to time, the "<u>Contract</u>").

2. <u>Term</u>. This Credit is effective immediately and expires at the close of banking business at our office in <New York City> <London> <Hong Kong> <Tokyo> on [*insert date*].

3. <u>Drawing</u>. Funds under this Credit are available to you against your presentation to us at our offices above of a drawing certificate in the form of Annex A hereto. Such certificate shall have all blanks appropriately filled in and shall be signed by your authorized officer, and such certificate shall be on your letterhead. You may make multiple and/or partial drawings hereunder.

4. <u>Drawing Procedures</u>. If a drawing is made by you hereunder, payment shall be made of the amount specified in immediately available funds on the <next> <{second}> succeeding Business Day; provided that such drawing and the documents presented in connection therewith conform to the terms and conditions hereof. All payments hereunder shall be made free and clear of, and without deduction for or on account of, any present or future taxes, duties, charges, fees, deductions or withholdings of any nature and by whomsoever imposed. As used in this Credit, the term "<u>Business Day</u>" shall mean any day other than a Saturday or Sunday or legal holiday or a day on which banking institutions in which our offices listed above are authorized or required by law or executive order to close.

5. <u>Assignment</u>. This Credit may be assigned by you upon written notice thereof to the Issuing Bank.

6. <u>Governing Law</u>. This Credit is subject to the Uniform Customs and Practice for Documentary Credits (2007 Revision), International Chamber of Commerce Publication No. 600 (the "<u>UCP</u>"). This Credit shall be deemed to be a contract made under the laws of <New York> <England> and shall, as to matters not governed by the UCP, be governed and construed in accordance with the laws of <New York> <England>.

7. <u>Notices</u>. Communications with respect to this Credit shall be in writing and shall be addressed to the Issuing Bank at <*insert address*> (or otherwise as designated by the Issuing Bank to the Sponsor by written communication) specifically referring thereon to "[<<*insert name of Issuing Bank*>>] Irrevocable Standby Letter of Credit No. _____."

8. <u>Integration</u>. This Credit sets forth in full our undertaking, and such undertaking shall not in any way be modified, amended, amplified or limited by reference to any document, instrument or agreement referred to herein (including, without limitation, the Contract), except only the drawing certificate referred to herein; and any such reference shall not be deemed to incorporate herein by reference any document, instrument or agreement, except for such drawing certificates.

Very truly yours,
[*insert name of Issuing Bank*]
By:_____
Authorized Officer

ANNEX A to EXHIBIT 10.3.5

<u>DRAWING CERTIFICATE</u>

[<<*insert date*>>]
[<<*insert name of Issuing Bank*>>]
[<<*insert address of Issuing Bank*>>]

Re: Irrevocable Standby Letter of Credit No. _____
Ladies and Gentlemen:

1. <<*insert name of Sponsor*>> (the "<u>Sponsor</u>") hereby certifies to [<<*insert name of Issuing Bank*>>], with reference to Irrevocable Standby Letter of Credit No. _____ (the "<u>Credit</u>"), that:
    (a) the Contractor has failed to fulfill one or more of the conditions of, or failed to perform one or more of its obligations under, the Contract and now in accordance with the Contract, the Sponsor is now entitled to apply Retainage in the amount of <<*insert amount*>>; and/or
    (b) this Credit is scheduled to expire within <<*thirty (30)* >> Days after the date hereof and the Contractor has not posted in our favor a substitute for this Credit meeting the requirements of the Contract or refunded an amount of cash equal to available and undrawn amount under this Credit.
2. [{<u>Prior Notice to the Contractor</u>. At least <<*five (5)*>> Business Days prior to the date hereof, the undersigned gave the Contractor written notice of:
    (a) the foregoing; and
    (b) the Sponsor's intention to make a drawing under the Credit}.]
3. <u>Demand for Payment</u>. The Sponsor is making a demand for payment under the Credit in the amount of <<*insert amount in words and numerals also*>>.
4. <u>Terms</u>. Capitalized terms used herein and not otherwise defined shall have the meanings ascribed thereto in the Turnkey Engineering, Procurement and Construction Contract, dated as of <<*insert date*>> (as amended, restated, supplemented or otherwise modified from time to time, the "<u>Contract</u>"), between the Sponsor and the Contractor.
5. <u>Payment Instructions</u>. Payment of the amount demanded hereby shall be made by wire transfer to the following account: _____.

IN WITNESS WHEREOF, the Sponsor has executed and delivered this Certificate as of <<*insert date*>>.

<<*insert Sponsor name*>>
By:_____
Authorized Representative
Name:_____
Title:_____

EXHIBIT 10.16

<u>FORM OF PERFORMANCE STANDBY LC</u>[313]
IRREVOCABLE STANDBY LETTER OF CREDIT NO. _____
<<*insert date*>>
CREDIT NO. _____
<<*insert date*>>
<<*insert Sponsor name and address*>>

Ladies and Gentlemen:

1. <u>Stated Amount</u>. We hereby establish in favor <<*insert name of Sponsor*>> (the "<u>Sponsor</u>") an irrevocable Standby letter of credit (the "<u>Credit</u>") whereby you are authorized to draw on [<< *insert name of Acceptable Issuer*>>] (the "<u>Issuing Bank</u>"), Credit No. _____, issued for the account of [*insert name of Contractor*] (the "<u>Contractor</u>"), an aggregate amount not exceeding <<*insert amount of currency in words*>> (<<*insert amount and currency in numerals*>>) (the "<u>Stated Amount</u>"). This Letter of Credit is issued pursuant to the Fixed Price, Lump Sum Turnkey Engineering, Procurement and Construction Contract, dated <<*insert date*>>, between the Contractor and the Sponsor (as amended, restated, supplemented or otherwise modified from time to time, the "<u>Contract</u>").

2. <u>Term</u>. This Credit is effective immediately and expires at the close of banking business at our office in <New York City> <London> <Hong Kong> <Tokyo> on [*insert date*].

3. <u>Drawing</u>. Funds under this Credit are available to you against your presentation to us at our offices above of a drawing certificate in the form of Annex A hereto. Such certificate shall have all blanks appropriately filled in and shall be signed by your authorized officer, and such certificate shall be on your letterhead. You may make multiple and/or partial drawings hereunder.

4. <u>Drawing Procedures</u>. If a drawing is made by you hereunder, payment shall be made of the amount specified in immediately available funds on the <next> <{second}> succeeding Business Day; provided that such drawing and the documents presented in connection therewith conform to the terms and conditions hereof. All payments hereunder shall be made free and clear of, and without deduction for or on account of, any present or future taxes, duties, charges, fees, deductions or withholdings of any nature and by whomsoever imposed. As used in this Credit, the term "<u>Business Day</u>" shall mean any day other than a Saturday or Sunday or legal holiday or a day on which banking institutions in Finland are authorized or required by law or executive order to close.

5. <u>Assignment</u>. This Credit may be assigned by you upon written notice thereof to the Issuing Bank.

6. <u>Governing Law</u>. This Credit is subject to the Uniform Customs and Practice for Documentary Credits (2007 Revision), International Chamber of Commerce Publication No. 600 (the "<u>UCP</u>"). This Credit shall be deemed to be a contract made under the laws of <New York> <England> and shall, as to matters not governed by the UCP, be governed and construed in accordance with the laws of <New York> <England>.

7. <u>Notices</u>. Communications with respect to this Credit shall be in writing and shall be addressed to the Issuing Bank at <*insert address*> (or otherwise as designated by the Issuing Bank to the Sponsor by written communication) specifically referring thereon to "[<<*insert name of Issuing Bank*>>] Irrevocable Standby Letter of Credit No. _____."

8. <u>Integration</u>. This Credit sets forth in full our undertaking, and such undertaking shall not in any way be modified, amended, amplified or limited by reference to any document, instrument or agreement referred to herein (including, without limitation, the Contract), except only the drawing certificate referred to herein; and any such reference shall not be deemed to incorporate herein

by reference any document, instrument or agreement, except for such drawing certificates.

Very truly yours,
[*insert name of Issuing Bank*]
By:_____
Authorized Officer

ANNEX A to EXHIBIT 10.16.

DRAWING CERTIFICATE
[<<*insert date*>>]
[<<*insert name of Issuing Bank*>>]
[<<*insert address of Issuing Bank*>>]

Re:Irrevocable Standby Letter of Credit No. _____

Ladies and Gentlemen:

1. <u>Defined Terms</u>. Capitalized terms used herein and not otherwise defined shall have the meanings ascribed thereto in the Fixed Price, Lump Sum Turnkey Engineering, Procurement and Construction Contract, dated as of <<*insert date*>>, (as amended, restated, supplemented or otherwise modified from time to time, the "<u>Contract</u>"), between the Sponsor and the Contractor.

2. <<*insert name of Sponsor*>> (the "<u>Sponsor</u>") hereby certifies to [<<*insert name of Issuing Bank*>>], with reference to Irrevocable Standby Letter of Credit No. _____ (the "<u>Credit</u>"), that:

   (a) [the Contractor has not paid in full the amount of _____ <<*insert currency*>> (_____) as Liquidated Damages that are now due and payable pursuant to Article 9 of the Contract;]

   (b) [the Contractor has not paid in full the amount of _____ <<*insert currency*>> (_____) in respect of a <breach> <Contractor Event of Default>, which amount is now due and payable pursuant to the Contract;]

   (c) [the Contractor has not paid in full the amount of _____ <<*insert currency*>> (_____) that is now due and payable pursuant to Article _____ of the Contract;]

   (d) [a Lien exists on the Facility, the Site or the Equipment as a result of the Contractor's breach of the Contract and, in order to remove or bond such Lien, the Sponsor will need _____ <<*insert currency*>> (_____);]

   (e) [the Contractor has failed to fulfill one or more of the conditions of, or failed to perform one or more of its obligations under, the Contract in accordance with the terms of the Contract;] [and/or]

   (f) [the Credit is scheduled to expire within <<*thirty (30)*>> Days after the date hereof, and the Contractor has not had posted in our favor a substitute for this Credit meeting the requirements of the Contract].

3. [{<u>Prior Notice to the Contractor</u>. At least <<*five (5)*>> Business Days prior to the date hereof, the undersigned gave the Contractor written notice of:

   (a) the foregoing; and

   (b) the Sponsor's intention to make a drawing under the Credit}.]

4. <u>Demand for Payment</u>. The Sponsor is making a demand for payment under the Credit in the amount of <<*insert amount in words and numerals also*>>.

5. <u>Payment Instructions</u>. Payment of the amount demanded hereby shall be made by wire transfer to the following account: <<*insert account details*>>.

IN WITNESS WHEREOF, the Sponsor has executed and delivered this Certificate as of <<*insert date*>>.

<<*insert Sponsor name*>>
By:_____
Name:_____
Title:_____

EXHIBIT 10.17

<u>FORM OF PARENT GUARANTEE</u>[314, 315]

GUARANTEE, dated as of <<*insert date*>> (this "<u>Parent Guarantee</u>"), is made by [*insert name of Parent Guarantor*], a [*insert type of entity*] (the "<u>Parent Guarantor</u>"), in favor of [*insert name of Sponsor*], a [*insert type of entity*] (the "<u>Sponsor</u>").

<u>WITNESSETH</u>:

WHEREAS, [*insert name of Contractor*], a [*insert type of entity*] (the "<u>Contractor</u>"), and the Sponsor are parties to the Fixed Price, Lump Sum Turnkey Engineering, Procurement and Construction Contract, dated <<*insert date*>>, between the Contractor and the Sponsor (as amended, restated, supplemented or otherwise modified from time to time, the "<u>Contract</u>"); and

WHEREAS, the Parent Guarantor currently owns <<*insert percentage*>>percent of the outstanding stock of the Contractor[316] and will derive substantial direct and indirect benefit from the Contractor's execution and performance of the Contract; and

WHEREAS, the Sponsor has agreed to enter into the Contract only if this Parent Guarantee is executed by the Parent Guarantor and delivered to the Sponsor.

NOW, THEREFORE, in order to induce the Sponsor to enter into the Contract and to consummate the transactions contemplated thereby, and for other good and valuable consideration including the sum of <<*insert amount and currency*>>, receipt and sufficiency of all of which are hereby acknowledged, the Parent Guarantor hereby agrees as follows:

1. <u>Definitions</u>. All capitalized terms used herein that are not defined herein shall have the meanings set forth in the Contract. In addition, as used herein, the following terms shall have the following meanings:

"**<u>Bankruptcy Code</u>**" shall mean <<*insert citation for applicable jurisdiction*>>.

"**Change of Control**" of the Parent Guarantor shall be deemed to have occurred at such time as any Person, together with any affiliates of such Person, is or becomes the beneficial Owner, directly or indirectly, of shares of capital stock of the Parent Guarantor entitling such Person acting by itself or with any of its affiliates or contractual counterparties to exercise more than <<fifty (50)>> percent of the total voting power of any class of stock of the Parent Guarantor [{entitled to vote in elections of directors of the Parent Guarantor}].

"**Consolidated Net Worth**" means, with respect to any Person, as of any date, the consolidated stockholders' equity of such Person determined on a <non> <{consolidated}> basis in accordance with <<*insert accounting standard desired*>>.

"**Guaranteed Obligations**" shall mean:

(a) all obligations of the Contractor under the Contract; and
(b) any and all claims that the Sponsor may have against the Contractor that arise out of the performance or breach by the Contractor of any provision of the Contract.

2. <u>Parent Guarantee</u>. The Parent Guarantor hereby acknowledges receipt of the Contract, and the Parent Guarantor hereby irrevocably and unconditionally, under any and all circumstances whatsoever, guarantees to the Sponsor and its successors, transferees and assigns the due and punctual payment and performance by the Contractor of all of the Guaranteed Obligations.

3. <u>Parent Guarantor Indemnification</u>. The Parent Guarantor hereby indemnifies the Sponsor for all the Sponsor's costs incurred with the [{successful}] enforcement hereof [and {successful}] enforcement of the Contract].

4. <u>Obligations Absolute</u>. The guaranty obligations of the Parent Guarantor are absolute, unconditional and irrevocable and shall constitute a guarantee of payment, performance and discharge and not merely of collection. Such guaranty obligations shall not be subject to any counterclaim, set-off, deduction, diminution, abatement, recoupment, suspension, deferment, reduction or defense for any reason whatsoever, and the Parent Guarantor shall have no right to terminate this Parent Guarantee or to be released, relieved or discharged from its obligations hereunder for any reason whatsoever (whether or not the Parent Guarantor or the Contractor shall have any knowledge or notice thereof), including, without limitation:
   (a) any:
       (i) amendment, modification, addition, deletion or supplement to or other change in the Contract or any other instrument or agreement applicable to any of the parties to the Contract,
       (ii) assignment, subcontracting or transfer of any thereof or of any interest therein, or
       (iii) furnishing or acceptance of additional security or an additional guarantee, or any release, substitution or variation of any security or an additional guarantee, for the obligations of the Contractor or any other party to the Contract;
   (b) any failure, omission or delay on the part of the Contractor or the Sponsor to perform or comply with any term of the Contract;

(c) any waiver, consent, extension, indulgence, compromise, release or other action or inaction under or in respect of the Contract or any obligation or liability of the Contractor or the Sponsor, or any exercise or non-exercise of any right, remedy, power or privilege under or in respect of any such instrument or agreement or any such obligation or liability;

(d) any bankruptcy, insolvency, reorganization, arrangement, readjustment, composition, liquidation or similar proceeding with respect to the Parent Guarantor, the Contractor or the Sponsor or any other Person or any of their respective properties or creditors, or any action taken by any trustee or receiver or by any court in any such proceeding;

(e) any discharge, termination, cancellation, frustration, irregularity, invalidity, unenforceability, illegality or impossibility of performance, in whole or in part, of the Contract;

(f) any dissolution, merger or consolidation (whether permitted or otherwise) of the Contractor or the Parent Guarantor into or with any other Person or any sale, lease or transfer of any of the assets of the Contractor or the Parent Guarantor to any other Person;

(g) any change in the ownership of the Contractor;

(h) any payment by the Parent Guarantor to the Contractor or the Sponsor pursuant to an agreement other than this Parent Guarantee; or

(i) any other occurrence or circumstance whatsoever, whether similar or dissimilar to the foregoing, that might otherwise constitute a legal or equitable defense or discharge of the liabilities of a guarantor or surety or that might otherwise limit recourse against the Parent Guarantor.

5. Waivers by the Parent Guarantor. To the extent permitted by applicable Law, the Parent Guarantor hereby unconditionally waives and agrees to waive at any future time any and all rights that the Parent Guarantor may have or that now or at any time hereafter may be conferred upon it, by applicable Law or otherwise, to terminate, cancel, quit or surrender this Parent Guarantee. Without limiting the generality of the foregoing, it is agreed that, at any time or from time to time, the occurrence or existence of any one or more of the following shall not release, relieve or discharge the Parent Guarantor from liability hereunder, and the Parent Guarantor hereby unconditionally and irrevocably waives, to the extent permitted by applicable Law:

(a) notice of any of the matters referred to in Article 3;

(b) all notices that may be required by applicable Law or otherwise, now or hereafter in effect, to preserve intact any rights against the Parent Guarantor including, without limitation, any demand, presentment and protest, proof of notice of nonpayment under the Contract, and notice of any default or failure on the part of the Contractor to perform and comply with any covenant, agreement, term or condition of the Contract;

(c) the enforcement, assertion or exercise against the Contractor of any right, power, privilege or remedy conferred in the Contract or otherwise;

(d) any requirement of diligence on the part of any Person;

(e) any requirement to proceed against the Contractor prior to proceeding against the Parent Guarantor or any other guarantee or to utilize or exhaust any remedies;

(f) acceptance of this Parent Guarantee by the Sponsor; or

(g) any other occurrence or circumstance whatsoever, whether similar or dis-similar to the foregoing, that might otherwise constitute a legal or equitable discharge, release or defense of a guarantor or surety, or that might other-wise limit recourse against the Parent Guarantor.

6. <u>Parent Guarantee Terms</u>. The Parent Guarantor hereby agrees that a separate action may be brought against it whether or not an action is commenced against the Contractor with respect to any of the Guaranteed Obligations. It is the inten-tion hereof that the Parent Guarantor shall remain liable as principal until the full amount of all sums payable with respect to the Guaranteed Obligations have been fully paid and the Guaranteed Obligations have been fully performed.

7. <u>Reinstatement of Parent Guarantee</u>. This Parent Guarantee shall continue to be effective, or be reinstated, as the case may be, if at any time payment, or any part thereof, of any of the Guaranteed Obligations is rescinded or must otherwise be restored or returned by the recipient thereof upon the insolv-ency, bankruptcy, dissolution, liquidation or reorganization of the Contractor, or upon or as a result of the appointment of a custodian, receiver, intervenor or conservator of, or trustee or similar officer for, the Contractor or any sub-stantial part of its property, or upon any settlement or compromise of any claim effected by the Sponsor in connection with any such insolvency, bankruptcy, dis-solution, liquidation or reorganization with any claimant (including, without limitation, the Contractor) or otherwise, all as though such payments had not been made. If any event specified in the immediately preceding sentence shall occur, and such occurrence shall prevent, delay or otherwise affect the right of the Sponsor to receive any payment in respect of any Parent Guaranteed Obligation, the Parent Guarantor agrees that, for purposes of this Parent Guarantee and its obligations hereunder and notwithstanding the occurrence of any of the foregoing events, the Parent Guarantor shall forthwith pay any such Parent Guaranteed Obligation at such times and in such amounts as are specified in the Contract.

8. <u>No Subrogation</u>. Notwithstanding anything to the contrary in this Parent Guarantee, the Parent Guarantor hereby irrevocably waives <forever> <{until sat-isfaction of the Guaranteed Obligations}> all rights that may have arisen in con-nection with this Parent Guarantee to be subrogated to any of the rights (whether contractual, under the Bankruptcy Code, under common Law or otherwise) of the Sponsor against the Contractor or against any collateral security or guaran-tee held by the Sponsor for the payment of the Guaranteed Obligations. So long as any of the Guaranteed Obligations remain outstanding, if any amount shall be paid by or on behalf of the Contractor to the Parent Guarantor on account of any of the rights herein, such amount shall be held by the Parent Guarantor in trust, segregated from other funds of the Parent Guarantor, and shall, forthwith upon receipt by the Parent Guarantor, be turned over to the Sponsor in the exact form received by the Parent Guarantor (duly endorsed by the Parent Guarantor to the Sponsor, if required), to be applied against the Guaranteed Obligations in such order as the Sponsor may determine. The provisions of this paragraph shall sur-vive the term of this Parent Guarantee and the payment in full of the Guaranteed Obligations.

9. [{<u>Rights of Third Parties</u>. This Parent Guarantee shall not be construed to create any right in any Person other than the Sponsor or to be a contract in whole or in part for the benefit of any Person other than the Sponsor, except to the extent the Sponsor makes an assignment permitted under this Parent Guarantee}.]

10. <u>Representations and Warranties</u>. The Parent Guarantor hereby represents and warrants to the Sponsor as follows:

(a) the Parent Guarantor is a <<*insert type of entity*>> duly organized, validly existing and in good standing under the laws of <<*insert jurisdiction of domicile*>> and is qualified to do business in all jurisdictions in which the nature of the business conducted by it makes such qualification necessary, and has all requisite legal power and authority to execute this Parent Guarantee and to perform the terms, conditions and provisions hereof;

(b) the execution and delivery by the Parent Guarantor of this Parent Guarantee have been duly authorized by all requisite [<<corporate>>] action;

(c) this Parent Guarantee constitutes the legal, valid and binding obligation of the Parent Guarantor, enforceable against the Parent Guarantor in accordance with its terms;

(d) neither the execution, delivery or performance by the Parent Guarantor of this Parent Guarantee, nor the consummation of the transactions contemplated thereby, will result in:

 (i) a violation of, or a conflict with, any provision of the organizational documents of the Parent Guarantor;

 (ii) a contravention or breach of, or a default under, any term or provision of any contract, agreement or instrument to which the Parent Guarantor is a party or by which it or its property may be bound, which contravention, breach or default could be reasonably expected to have a material adverse effect on the ability of the Parent Guarantor to perform its obligations under this Parent Guarantee to consummate the transactions contemplated by this Parent Guarantee; or

 (iii) a violation by the Parent Guarantor of any Law;

(e) the Parent Guarantor is not in violation of any Law promulgated, or judgment entered, by any Governmental Authority[{, which violations, individually or in the aggregate, could reasonably be expected to have a material adverse effect on it or its performance of any obligations under this Parent Guarantee}];

(f) there are no actions, suits or proceedings, now pending or [{(to its best knowledge)}] threatened against the Parent Guarantor before any court or administrative body or arbitral tribunal that might materially adversely affect the ability of the Parent Guarantor to perform its obligations under this Parent Guarantee; and

(g) attached hereto as Appendix I is a true and correct chart setting forth the ownership of the Parent Guarantor and the Parent Guarantor's ownership of the Contractor.

11. Assignment. The Parent Guarantor may not assign or delegate its obligations under this Parent Guarantee, directly or indirectly, whether by pledge, assignment, sale of assets or the sale or merger (statutory or otherwise). The Parent Guarantor acknowledges and agrees that, without the consent of the Parent Guarantor[:

   (a) the Sponsor may assign for collateral and security purposes or create a security interest in its rights and interests under or pursuant to this Parent Guarantee in favor of any of the Financing Parties; and

   (b)]the Sponsor may assign this Parent Guarantee or its rights thereunder to any subsidiary or affiliate of the Sponsor [or to any Financing Party.

[12. Security Interest. The holder of any security interest in this Parent Guarantee shall not be prevented or impeded by the Parent Guarantor from enforcing such security interest. The Parent Guarantor shall execute within <<*five (5)*>> Days all consents to assignment and/or acknowledgments of any security interest or assignment as are requested by the Sponsor to give effect to the foregoing, and shall provide such certificates and opinions of counsel addressed to the Sponsor and the Financing Parties as may be requested in connection with any financing of the Facility any of which may or may not be similar in form to those attached to the Contract or Appendix II. The Parent Guarantor agrees that such consents and acknowledgments may contain:

   (a) an acknowledgment by the Parent Guarantor that the Sponsor is not in default under the Contract[{, if such is the case}];

   (b) representations and warranties by the Parent Guarantor;

   (c) a prohibition against amending, assigning or terminating this Parent Guarantee or the Contract without the prior written consent of the holder of such security interest;

   (d) a consent by the Parent Guarantor to allow the assignment of this Parent Guarantee to the successors in interest of the holder of such security interest after foreclosure thereon; and

   (e) other provisions that are conventional in financing facilities such as the Facility on a non-recourse basis.

13. [The Parent Guarantor:

   (a) acknowledges that the Financing Parties will review this Parent Guarantee and may require changes thereto as a condition of providing financing and/or insurance and/or guarantees; and

   (b) agrees to consider any such requirements in good faith.]

14. Remedies Non-Exclusive. All remedies provided in this Parent Guarantee shall be deemed cumulative and not in lieu of, or exclusive of, each other or of any other remedy available to the Sponsor at law or in equity, and the exercise of any remedy, or the existence herein of other remedies, shall not prevent the exercise of any other remedy.

15. Notices. All notices to be given hereunder shall be effective upon receipt and shall be in writing and delivered by hand, internationally recognized overnight courier service or first class mail (postage prepaid), to the Parent Guarantor or the Sponsor, as the case may be, at the following address (or such other address as may hereafter be designated in writing by a party):

(a) if to the Sponsor:
[*insert name of Sponsor*]
[*insert address*]
Attention: General Counsel
Telecopy:
and

(b) if to the Parent Guarantor:
[*insert name of Parent Guarantor*]
[*insert address*]

_____

_____

Attention: General Counsel
Telecopy:_____

16. <u>Non-Waiver</u>. The Sponsor shall not be deemed to have waived any right under this Parent Guarantee unless the Sponsor shall have delivered to the Parent Guarantor an express written waiver signed by the Sponsor. No failure or successive failure by the Sponsor to enforce any covenant or agreement, and no waiver or successive waivers by the Sponsor of any condition of this Parent Guarantee, shall operate as a discharge of such covenant, agreement or condition, or render the same invalid, or impair the Sponsor's right to enforce the same in the event of any subsequent breach or breaches by the Parent Guarantor.

17. <u>Severability</u>. If any of the terms, covenants or conditions hereof or the application of any such term, covenant or condition shall be held invalid or unenforceable as to the Parent Guarantor or as to any circumstance by any court or arbitrator having jurisdiction, the remainder of such terms, covenants or conditions shall not be affected thereby, shall remain in full force and effect and shall continue to be valid and enforceable in any other jurisdiction, and the Parent Guarantor and the Sponsor shall negotiate in good faith to substitute a term or condition in this Parent Guarantee to replace the one held invalid or unenforceable.

18. <u>Entire Agreement</u>. This Parent Guarantee constitutes the entire agreement and contains all of the understandings and agreements of whatsoever kind and nature existing between the Parent Guarantor and the Sponsor [with respect to the subject matter hereof][317] and the rights, interests, understandings, agreements and obligations of the Parent Guarantor and the Sponsor relating thereto, and supersedes all prior written or oral agreements, commitments, representations, communications and understandings between the Parent Guarantor and the Sponsor [with respect to the subject matter hereof].[318]

19. <u>Amendment</u>. No amendment, waiver or consent relating to this Parent Guarantee shall be effective unless it is in writing and signed by the Parent Guarantor and the Sponsor.

20. <u>Benefits of Parent Guarantee</u>. All of the terms and provisions of this Parent Guarantee shall be binding upon and inure to the benefit of the Parent Guarantor and the Sponsor and their respective successors and assigns expressly permitted hereby.

21. <u>Descriptive Headings</u>. Descriptive headings are for convenience only and shall not control or affect the meaning or construction of any provision of this Parent Guarantee.

22. <u>Rules of Construction</u>. In the interpretation of this Parent Guarantee, unless the context otherwise expressly states:

(a) the singular includes the plural and vice versa and, in particular (but without limiting the generality of the foregoing), any word or expression defined in the singular has the corresponding meaning used in the plural and vice versa;

(b) the term "or" is not exclusive;

(c) the terms "including", "includes" and "include" shall be deemed to be followed by the words "without limitation";

(d) any reference to any gender includes the other gender;

(e) any reference to any agreement, instrument, contract or other document (A) shall include all appendices, exhibits and schedules thereto and (B) shall be a reference to such agreement, instrument, contract or other document as amended, supplemented, modified, suspended, restated or novated from time to time;

(f) any reference to any Law shall include all statutory and administrative provisions consolidating, amending or replacing such Law, and shall include all rules and regulations promulgated thereunder;

(g) any reference to "writing" includes printing, typing, lithography and other means of reproducing words in a visible form;

(h) any reference to any Person other than those as noted in Article 19 includes such Person's successors and assigns;

(i) unless otherwise expressly specified, a reference to an Article is to the Article of this Parent Guarantee;

(j) unless otherwise expressly specified, any right may be exercised at any time and from time to time;

(k) all obligations under this Parent Guarantee are continuing obligations throughout the term hereof; and

(l) the fact that counsel to the Sponsor shall have drafted this Parent Guarantee shall not affect the interpretation of any provision of this Parent Guarantee in a manner adverse to the Sponsor or otherwise prejudice or impair the rights of the Sponsor.

23. <u>GOVERNING LAW, ETC</u>. [AS PERMITTED BY GENERAL OBLIGATION LAW § 5–1401 OF NEW YORK STATE,][319] THIS GUARANTEE SHALL BE GOVERNED BY THE LAWS OF THE STATE OF NEW YORK. ANY LEGAL ACTION OR PROCEEDING WITH RESPECT TO THIS GUARANTEE MAY BE BROUGHT IN THE COURTS OF THE STATE OF NEW YORK OR OF THE UNITED STATES OF AMERICA FOR THE SOUTHERN DISTRICT OF NEW YORK AND, BY EXECUTION AND DELIVERY OF THIS AGREEMENT, EACH OF THE PARTIES HEREBY ACCEPTS FOR ITSELF AND IN RESPECT OF ITS PROPERTY, GENERALLY AND UNCONDITIONALLY, THE JURISDICTION OF THE AFORESAID COURTS AND APPELLATE COURTS. EACH OF THE PARTIES HEREBY IRREVOCABLY CONSENTS TO THE SERVICE OF PROCESS OF ANY OF THE AFOREMENTIONED COURTS IN ANY SUCH ACTION OR PROCEEDING BY THE MAILING OF COPIES THEREOF TO IT BY REGISTERED OR CERTIFIED MAIL, POSTAGE PREPAID, RETURN RECEIPT REQUESTED, TO EACH OF THE PARTIES AT ITS ADDRESS

SET FORTH HEREIN. EACH OF THE PARTIES HEREBY IRREVOCABLY WAIVES ANY OBJECTION WHICH IT MAY NOW OR HEREAFTER HAVE TO THE LAYING OF VENUE OF ANY SUCH ACTION OR PROCEEDING IN SUCH RESPECTIVE JURISDICTIONS. TO THE EXTENT PERMITTED BY APPLICABLE LAW, EACH OF THE PARTIES HEREBY IRREVOCABLY WAIVES ALL RIGHT OF TRIAL BY JURY IN ANY ACTION, PROCEEDING OR COUNTERCLAIM ARISING OUT OF OR IN CONNECTION WITH THIS GUARANTEE OR ANY MATTER ARISING HEREUNDER.[320]

24. <u>Language</u>. The official text of this Parent Guarantee is English, regardless of any translation that may be made for the convenience of the Parent Guarantor or the Sponsor. All definitive documents, notices, waivers and all other communication, written or otherwise, between the Parent Guarantor and the Sponsor in connection with this Parent Guarantee shall be in English, and no document in any other language may be introduced in an attempt to resolve any ambiguity herein.

[{25. <u>Term</u>. This Parent Guarantee and all guarantees, covenants and agreements of the Parent Guarantor contained herein shall terminate and be discharged on the date on which all of the Guaranteed Obligations shall be paid and performed or otherwise discharged in full; provided that this Parent Guarantee shall continue to be effective or reinstated, as the case may be, to the extent provided herein.}][321]

26. <u>Further Assurances</u>. The Parent Guarantor hereby agrees to execute and deliver all such instruments (including without limitation a notice of assignment in the form of Index II promptly upon receipt) and take all such action as the Sponsor may from time to time reasonably request in order to effectuate fully the purposes of this Parent Guarantee.

27. <u>Reliance Presumed</u>. The Parent Guarantor waives notice of reliance upon this Parent Guarantee by the Sponsor. Each of the Guaranteed Obligations (now or hereafter in effect) shall be deemed conclusively to have been created, contracted or incurred in reliance upon this Parent Guarantee.

28. [{<u>Parent Guarantor's Condition</u>. The Parent Guarantor shall not consolidate with or merge with or into any Person or permit any Person to merge with or into it or transfer (by lease, assignment, sale or otherwise) all, substantially all or a substantial portion of its properties and assets, in a single transaction or through a series of related transactions, to another Person or group of Persons or permit any of its subsidiaries to enter into any such transaction or transactions, unless:

(a) the Parent Guarantor shall be the continuing Person, or the Person formed by such consolidation or into which the Parent Guarantor is merged or to which all or substantially all of the properties and assets of the Parent Guarantor are transferred or to which all or substantially all of the assets of the Parent Guarantor and its subsidiaries on a consolidated basis are transferred (the "<u>surviving entity</u>") shall be a Person organized and existing under the laws of <<*insert country*>> and shall expressly assume obligations of the Parent Guarantor hereunder; and

(b) the surviving entity (whether or not the Parent Guarantor) shall, after giving pro forma effect to such transaction (including the assumption by the surviving entity of the obligations hereunder) have a Consolidated Net Worth equal to or greater than the Consolidated Net Worth of the Parent Guarantor immediately preceding <such transaction> <the date hereof>.

If the Parent Guarantor breaches the provision of the above sentence or undergoes a Change of Control, it shall immediately deposit with the Sponsor the sum of <<*insert amount of currency*>> in cash as collateral to secure its obligations hereunder, and the Sponsor shall be entitled to use such sum to satisfy any of the Parent Guarantor's obligations hereunder. Notwithstanding the foregoing, all provisions of this Parent Guarantee shall remain in effect following such deposit].[322]

29. <u>Service of Process</u>. The Parent Guarantor irrevocably appoints <<*insert agent's name*>> of <<*insert address of agent*>> as its agent to receive and acknowledge on its behalf service of any writ, summons, order, judgment or other notice of legal process in <<*insert jurisdiction*>>. The Parent Guarantor agrees that any such legal process shall be sufficiently served on it if delivered to such agent for service at its address for the time being in <<*insert jurisdiction*>> whether or not such agent gives notice thereof to the Guarantor.

30. <u>Proceedings</u>. The Parent Guarantor hereby agrees to the consolidation of any dispute or proceeding hereunder with any dispute or proceeding involving the Contract and/or and other guaranty thereof given by any Person.

IN WITNESS WHEREOF, the Parent Guarantor, intending to be legally bound, has caused this Parent Guarantee to be duly executed and delivered as of the day and year first above written.

<<[*insert name of Parent Guarantor*]>>
By:_____
Name:_____
Title:_____

Accepted:
<<[*insert name of Sponsor*]>>
By:_____
Name:_____
Title:_____

APPENDIX I TO EXHIBIT 10.17

OWNERSHIP CHART

PARENT GUARANTEE[323]

THIS GUARANTEE is made by way of deed the [*insert date*], 2[___]
BETWEEN:
    [   ] a company incorporated in accordance with the laws of [   ] whose registered office is at [   ] (the "<u>Parent Guarantor</u>"); and
    [   ] whose office is at [ ] (the "<u>Sponsor</u>").

WHEREAS:

By an agreement dated [   ] (the "<u>Contract</u>," which term includes all amendments to variations of or supplements to it from time to time in force) the Project Company has agreed to engage [   ] (the "<u>Contractor</u>") for the design, manufacture, delivery, erection testing and completion of a complex to be located <<*insert locator*>>.

It is a condition of the Contract that the Contractor procures the execution and delivery to the Sponsor of a parent company guarantee in the form of this Parent Guarantee.

The Parent Guarantor has agreed to guarantee the due performance of the Contract by the Contractor.

The Parent Guarantor is the owner of [<<*100* percent>>] of the shares of the Contractor.

It is the intention of the parties that this document be executed as a deed.

IT IS AGREED as follows:

1. In consideration of the Sponsor entering into the Contract with the Contractor, the Parent Guarantor:
    (a) as primary obligor and not as surety guarantees to the Sponsor the due and punctual performance by the Contractor of each and all the obligations, warranties, duties and undertakings of the Contractor under and pursuant to the Contract when and if such obligations, duties and undertakings shall become due and performable according to the terms of such Contracts; and
    (b) agrees, in addition to its obligations set out in clause 1(a) above, to indemnify the Sponsor against all losses, damage, costs and/or expenses which the Sponsor may incur by reason of any breach by the Contractor of its obligations, warranties, duties and undertakings under and pursuant to the Contract save that this shall not be construed as imposing greater or different obligations or liabilities on the Parent Guarantor than are imposed on the Contractor under the Contract.

2. The Parent Guarantor agrees that it shall not in any way be released from liability under this Parent Guarantee by any act, omission, matter or other thing whereby (in absence of this provision) the Parent Guarantor would or might be released in whole or in part from liability under this Parent Guarantee including, without limitation, and whether or not known to the Parent Guarantor:
    (a) any arrangement made between the Contractor and the Sponsor including any addendum or amendment to the obligations of Sponsor under the Contract (including, without limitation any adjustment to the amount payable to Contractor under the Contract); or
    (b) any alteration in the obligations undertaken by the Contractor whether by way of any addendum or variation referred to in clause 3 below, any suspension of the Work, extension of the time or otherwise; or
    (c) any waiver or forbearance by the Sponsor whether as to payment, time, performance or otherwise; or
    (d) the taking, variation, renewal or release of, the enforcement or neglect to perfect or enforce any right, guarantee, remedy or security from or against the Contractor or any other person; or

(e) any unenforceability, illegality or invalidity of the Contract or any of the provisions of the Contract or any of the Contractor's obligations under the Contract, so that this Parent Guarantee shall be construed as if there were no such unenforceability, illegality or invalidity; or

(f) the termination of the Contract or the employment of the Contractor under the Contract; or

(g) any breach of the Contract or other default of Sponsor; or

(h) any legal limitation, disability, incapacity or other circumstances relating to the Contractor, or any other person; or

(i) the dissolution, amalgamation, reconstruction, reorganization, change in status, function, control or ownership, insolvency, liquidation or the appointment of an administrator or receiver of the Contractor or any other person; or

(j) any other matter or things (whether similar to the foregoing or otherwise) whereby the obligations of the Parent Guarantor hereunder might under any applicable law be discharged or affected;and the Parent Guarantor hereby waives notice of the foregoing.

3. The Parent Guarantor by this Parent Guarantee authorizes the Contractor and the Sponsor to make any addendum, variation or amendment to the Contract, or the Work, the due and punctual performance of which addendum and variation shall be likewise guaranteed by the Parent Guarantor in accordance with the terms of this Parent Guarantee.

4. This Parent Guarantee shall be a primary obligation of the Parent Guarantor and accordingly the Sponsor shall not be obliged before enforcing this Parent Guarantee to take any action in any court or arbitral proceedings against the Contractor, to make any claim against or any demand of the Contractor, to enforce any other security held by it in respect of the obligations of the Contractor under the Contract or to exercise, levy or enforce any distress, diligence or other process of execution against the Contractor. In the event that the Sponsor brings proceedings against the Contractor, the Parent Guarantor will be bound by any findings of fact, interim or final award or judgment made by an arbitrator or the court in such proceedings.

5. This Parent Guarantee is a continuing guarantee and accordingly shall remain in full force and effect (notwithstanding any intermediate satisfaction by the Contractor, the Parent Guarantor or any other person) until all obligations, warranties, duties and undertakings now or hereafter to be carried out or performed by the Contractor under the Contract shall have been satisfied or performed in full and is not revocable and is in addition to and not in substitution for and shall not merge with any other right, remedy, guarantee or security which the Project Company may at any time hold for the performance of such obligations and may be enforced without first having recourse to any such security. When all obligations, warranties, duties and undertakings now or hereafter to be carried out or performed by Contractor under the Contract shall have been satisfied and performed in full and accepted by the Sponsor, pursuant to the Contract, this Parent Guarantee shall become of no further effect and shall be returned to Parent Guarantor by the Sponsor.

6. Until all amounts which may be or become payable under the Contract or this Parent Guarantee have been irrevocably paid in full, the Parent Guarantor shall not as a result of this Parent Guarantee or any payment or performance under this Parent Guarantee be subrogated to any right or security of the Sponsor or claim or prove in competition with the Project Company against the Contractor or any person or demand or accept repayment of any monies or claim any right of contribution, set-off or indemnity, and any sums received by the Parent Guarantor or the amount of any set-off exercised by the Parent Guarantor in breach of this provision shall be held by the Parent Guarantor in trust for, and shall be promptly paid to, the Project Company.

7. The Parent Guarantor shall not hold any security from the Contractor in respect of this Parent Guarantee and any such security which is held in breach of this provision shall be held by the Parent Guarantor in trust for, and shall promptly be transferred to, the Sponsor.

8. Until all amounts which may be or become payable under the Contract or this Parent Guarantee have been irrevocably paid in full, if (notwithstanding the provisions of sub clauses (1) and (2)) the Parent Guarantor has any rights of subrogation against the Contractor or any rights to prove in a liquidation of the Contractor, the Parent Guarantor agrees to exercise such rights in accordance with the directions of the Project Company.

9. Each payment to be made by the Parent Guarantor under this Parent Guarantee shall be made in <<*insert currency*>>, without any set-off or counterclaim and free and clear of all deductions or withholdings of any kind whatsoever or howsoever arising. If any deduction or withholding must be made by law (including double taxation treaties) the Parent Guarantor will pay that additional amount which is necessary to ensure that the Sponsor receives on the due date a net amount equal to the full amount which it would have received if the payment had been made without the deduction or withholding. The Parent Guarantor shall promptly deliver to the Sponsor any receipts, certificates or other proof evidencing the amounts paid or payable in respect of any such deduction or withholding.

10. The Parent Guarantor shall have <<*five (5)*>> Business Days from the date of demand to make payment in full to the Sponsor of any amount due under this Parent Guarantee. The Parent Guarantor shall pay interest on any amount due under this Parent Guarantee from the date of demand until the date of payment in full (as well after as before any judgment) calculated on a daily basis at the rate of <<   >> per cent per annum above the base rate for the time being of <<   >>.

11. The Parent Guarantor will reimburse the Sponsor for all legal and other costs (including non-recoverable VAT) incurred by the Sponsor in connection with the enforcement of this Parent Guarantee.

12. The Sponsor may appropriate any sum paid by the Contractor, the Parent Guarantor or any other person or recovered or received on account of the obligations the subject of this Parent Guarantee as it sees fit.

13. Any settlement or discharge between the Sponsor and the Contractor and/or the Parent Guarantor shall be conditional upon no settlement with security or payment to the Sponsor by the Contractor or the Parent Guarantor or any other person being avoided or set aside or ordered to be refunded or reduced by virtue

of any provision of enactment relating to bankruptcy, insolvency or liquidation for the time being in force and accordingly (but without limitation the Sponsor's other rights hereunder) the Sponsor shall be entitled to recover from the Parent Guarantor, as if such settlement or discharge had not occurred, the value which the Sponsor has placed upon such settlement or security or the amount of any such payment.

14. The Parent Guarantor warrants that this Parent Guarantee is its legally binding obligation, enforceable in accordance with its terms, and that all necessary governmental and other consents and authorizations for the giving and implementation of this Parent Guarantee have been obtained.

15. The Parent Guarantor warrants and undertakes to the Sponsor that it shall take all necessary action directly or indirectly to perform the obligations expressed to be assumed by it or contemplated by this Parent Guarantee and to implement the provisions of this Parent Guarantee.

16. The Parent Guarantor warrants and confirms to the Sponsor that it has not entered into this Parent Guarantee in reliance upon, nor has it been induced to enter into this Parent Guarantee by any representation, warranty or undertaking made by or on behalf of the Sponsor (whether express or implied and whether pursuant to statute or otherwise) which is not set out in this Parent Guarantee.

17. The Sponsor shall be entitled by notice in writing to the Parent Guarantor to assign the benefit of this Parent Guarantee at any time in connection with an assignment of the Contract in accordance with the provisions of the Contract, to any person without the consent of the Parent Guarantor being required, and any such assignment shall not release the Parent Guarantor from liability under this Parent Guarantee.

18. Any notice to or demand on the Parent Guarantor to be served under this Parent Guarantee may be delivered or sent by first class recorded delivery post or telex or facsimile transmission to the Parent Guarantor at its address appearing in this Parent Guarantee or at such other address as it may have notified to the Sponsor in accordance with this clause.

19. Any such notice or demand shall be deemed to have been served:
    (a) if delivered, at the time of delivery; or
    (b) if posted, at 10.00 a.m. on the second Business Day after it was put into the post; or
    (c) if sent by telex or facsimile process, at the expiration of 2 hours after the time of dispatch, if dispatched before 3.00 p.m. on any Business Day, and in any other case at 10.00 a.m. on the next Business Day.

20. In proving service of a notice or demand it shall be sufficient to prove that delivery was made or that the envelope containing the notice or demand was properly addressed and posted as a prepaid first class recorded delivery letter or that the telex or facsimile message was properly addressed and dispatched, as the case may be.

21. For the purposes of this Parent Guarantee "Business Day" means a day (other than a Saturday or a Sunday) on which banks are generally open in <<*London, England or New York, New York, U.S.A.*>> for normal business.

22. No delay or omission of the Sponsor in exercising any right, power or privilege under this Parent Guarantee shall impair or be construed as a waiver of such

right, power or privilege nor shall any single or partial exercise of any such right, power or privilege preclude any further exercise of such right, power or privilege or the exercise of any other right, power or privilege.

23. A waiver given or consent granted by the Sponsor under this Parent Guarantee will be effective only if given in writing and then only in the instance and for the purposes for which it is given in writing and then only in the instance and for the purpose for which it is given.

24. A waiver by the Sponsor shall not constitute a continuing waiver and shall not prevent the Sponsor from subsequently enforcing any of the provisions of this Parent Guarantee.

25. The invalidity, illegality or unenforceability in whole or in part of any of the provisions of this Parent Guarantee shall not affect the validity, legality and enforceability of the remaining part or provisions of this Parent Guarantee.

26. The Parent Guarantor shall pay all stamp duties and taxes, if any, to which the execution and delivery of this Parent Guarantee may be subject in <<insert place of incorporation of the Parent Guarantor>> and/or in the <<*insert jurisdiction*>> and shall indemnify Sponsor against any and all liabilities with respect to or arising from any delay or omission to pay any such duties and taxes.

27. [The Parent Guarantor recognizes and accepts that Sponsor and <<*insert name of other Parent Guarantor*>> have entered into or will enter into a guarantee pursuant to which <<*insert name of other Parent Guarantor*>> guarantees to the Sponsor as primary obligor and not as surety the due and punctual performance by the Contractor of each and all of the obligations, warranties, duties and undertakings of the Contractor under and pursuant to the Contract and agrees to indemnify the Sponsor against all losses, damage, costs and/or expenses which the Sponsor may incur by reason of any breach by the Contractor if its obligations, warranties, duties and undertakings under and pursuant to the Contract. The Parent Guarantor agrees, in accordance with clause 2(d) hereof, that the obligations of the Parent Guarantor hereunder shall not be affected or discharged by the taking, variation, renewal, release, enforcement or neglect to perfect or enforce the guarantee with <<*insert name of Parent Guarantor*>>. The Parent Guarantor further agrees that the Sponsor may, at its option, make a demand upon either or both the Parent Guarantor and/or <<*insert name of other Parent Guarantor*>> in respect of any liability of the Contractor under the Contract and that the making of or failure to make any such demand on <<*insert name of the other Contractor*>> shall not affect or discharge the liability of the Parent Guarantor hereunder.]

28. Capitalized words and expressions used in this Parent Guarantee shall have the meaning assigned to them in the Contract.

29. The Parent Guarantor submits to the nonexclusive jurisdiction of the English courts for all purposes relating to this Parent Guarantee and agrees that the Sponsor may, at its option, include in any proceedings relating to this Parent Guarantee any claim against <<*insert name of other Parent Guarantor*>> relating to the guarantee entered into or to be entered into between <<*insert name of other Parent Guarantor*>> and the Sponsor referred to in Clause 17 hereof. The Parent Guarantor irrevocably appoints <<*insert name*>> (at its registered office for the time being in England and Wales) to receive for it and on its behalf, service of process issued out of the English Courts in any legal action or proceedings

concerned. Nothing in this paragraph shall affect the right of the Sponsor to join other parties in any proceedings to serve process in any other manner permitted by law, but if the said process agent ceases to exist or have an office in England and Wales, the Parent Guarantor shall forthwith appoint another process agent which shall be a company incorporated in and having its registered office in England and Wales.

30. The parties hereto agree to consolidate any proceeding commenced hereunder with any proceeding commenced under the Contract. This Parent Guarantee shall be governed by the laws of England.

31. Notice shall be given:

    (a) if to the Sponsor:

        [*insert name of Sponsor*]

        [*insert address*]

        Attention: General Counsel

        Telecopy:

        and

    (b) if to the Parent Guarantor:

        [*insert name of Parent Guarantor*]

        [*insert address*]

        _____

        _____

        Attention: General Counsel

        Telecopy:_____

[32. The Parent Guarantor agrees to comply with the terms and provisions of Article <<*insert applicable Article*>> of the Contract to the extent the Parent Guarantor has been given responsibilities thereunder in such Article and the Parent Guarantor and the Sponsor shall be bound by any dispute resolved thereunder.]

33. <u>Assignment</u>. The Parent Guarantor may not assign or delegate its obligations under this Parent Guarantee, directly or indirectly, whether by pledge, assignment, sale of assets or the sale or merger (statutory or otherwise). The Parent Guarantor acknowledges and agrees that, without the consent of the Parent Guarantor[:

    (a) the Sponsor may assign for collateral and security purposes or create a security interest in its rights and interests under or pursuant to this Parent Guarantee in favor of any of the Financing Parties; and

    (b)]the Sponsor may assign this Parent Guarantee or its rights thereunder to any subsidiary or affiliate of the Sponsor [or to any Financing Party.

[34. <u>Security Interest</u>. The holder of any security interest in this Parent Guarantee shall not be prevented or impeded by the Parent Guarantor from enforcing such security interest. The Parent Guarantor shall execute within <<*five (5)*>> Days all consents to assignment and/or acknowledgments of any security interest or assignment as are requested by the Sponsor to give effect to the foregoing, and shall provide such certificates and opinions of counsel addressed to the Sponsor and the Financing Parties as may be requested in connection with any financing of the Facility any of which may or may not be similar in form to those attached to the Contract or Appendix II. The Parent Guarantor agrees that such consents and acknowledgments may contain:

(a) an acknowledgment by the Parent Guarantor that the Sponsor is not in default under the Contract[{, if such is the case}];

(b) representations and warranties by the Parent Guarantor;

(c) a prohibition against amending, assigning or terminating this Parent Guarantee or the Contract without the prior written consent of the holder of such security interest;

(d) a consent by the Parent Guarantor to allow the assignment of this Parent Guarantee to the successors in interest of the holder of such security interest after foreclosure thereon; and

(e) other provisions that are conventional in financing facilities such as the Facility on a non-recourse basis.

IN WITNESS of which this Parent Guarantee has been executed by the Parent Guarantor as a deed and has been delivered on the date which appears thereby first on page 1.

Signed as a deed on behalf of)	
A company incorporated in <<*insert jurisdiction*>>)	_____
by ,)	[Signature 1]
and being persons who,)	
in accordance with the laws of that)	
territory are acting under the authority of)	_____
	[Signature 2]

Witness' Signature:_____

Name:_____

Address:_____

_____   _____

Director   Director/Secretary

APPENDIX I TO EXHIBIT 10.17

OWNERSHIP CHART

APPENDIX II TO EXHIBIT 10.17

NOTICE OF ASSIGNMENT

Re: Fixed Price, Lump Sum Turnkey Engineering, Procurement and Construction Contract (the "<u>Contract</u>") dated [DATE], between <<*insert name of Contractor*>> and <<*insert name of Sponsor*>> (the "<u>Sponsor</u>") and guarantee thereof dated <<*insert date*>> by <<*insert name of Parent Guarantee*>> (the "<u>Parent Guarantee</u>").

Dear Sirs:

This letter constitutes notice to you that the Sponsor has assigned to us all its rights and benefits under the Parent Guarantee. Notwithstanding the assignment, the

Sponsor remains liable to perform all its obligations under the Parent Guarantee, if any, and we shall have no liability in respect of those obligations.

Please pay all moneys payable by you to the Sponsor under the Parent Guarantee to the account number: << >>, sort code: << >>, reference: << >>, attention: << >>, with << >> or such other account number and/or bank as we may otherwise instruct you in a writing which has also been sent to the Sponsor.

Please acknowledge receipt of this Notice of Assignment by signing the duplicate of this Notice of Assignment and returning the same to us for the attention of <<*insert title of person to receive notice*>>.

Yours faithfully,

_____
<<*agent for Financing Parties*>>

We acknowledge receipt of the Notice of Assignment <from ____>> on [<<*insert date*>>.

We confirm that we have not received any prior notice of assignment in relation to the same and agree to pay any monies payable by us to the Sponsor under the Parent Guarantee to the above account and bank or as you shall otherwise instruct us in writing.

_____
for and on behalf of
[Contractor]

SPONSOR

_____   _____
Director   Director/Secretary

IN WITNESS of which this Parent Guarantee has been executed by the Parent Guarantor as a deed and has been delivered on the date which appears first on page 1.

Executed as a deed under seal by	)	
[PARENT GUARANTOR], and signed and	)	
delivered as a deed on its behalf	)	L.S.
by [ ],	)	
in the presence of:	)	

Witness' Signature:_____
Name:_____
Address:_____

_____   _____
Director   Director/Secretary

[The Common Seal of	)	
[ ]	)	
was affixed to this deed in the	)	
presence of:	)	

_____   _____
Director   Director/Secretary

EXHIBIT 12.2(g)

<u>TRANSPORTATION AND TRAFFIC STUDY</u>

[EXHIBIT 15]

<u>[INSURANCE OUTLINE]</u>[324]

A. Erection All Risks

Insured Parties:

1. the Sponsor,
2. all Financing Parties,
3. the Contractor,
4. all Subcontractors,
5. the [Wheeling Company],
6. the Sponsor's operations and maintenance contractor,
7. all Sponsors, and
8. all officers, employees, [suppliers] and consultants of each of the above parties.

Property Insured:
The Facility and all associated property, including all materials, supplies and equipment, spares, consumables and temporary works referred to in "Cover" below and the Existing Plant.

Geographical Limits:
The <Route><Site> and other areas within <<___>> including inland transit and offsite storage.

Cover:
On an "All Risks" basis covering physical loss or damage to the Facility and associated property including risks associated with design, construction and erection, start up and testing of the Facility and all materials, supplies and equipment comprising or to be used by the Facility or intended for installation into the Facility or temporary works and covering all materials, supplies or equipment during transit to or from the <Route><Site> during construction or otherwise (save to the extent covered by the Marine/Inland Transit and Air Cargo insurance specified in paragraph B below).

Period of Cover:
From the commencement of construction activities at the <Route><Site> or the Work Commencement Date, whichever is the earlier, until Facility Final Acceptance Date and for at least <<24>> months in respect of defect guarantees under of the Contract.

Main Extensions:

1. DE3 on buildings and civil works and LEG 3/96 on remainder.
2. Automatic reinstatement of sum insured.
3. 50/50 clause.
4. Escalation in value 25 percent.
5. 72 hour clause.
6. Inland transit in respect of goods sourced in *<<insert country>>*.
7. Extensions of time held covered at terms not exceeding pro rata.
8. Offsite storage (limit *<<insert currency amount>>*).
9. Debris removal (limit *<<insert currency amount>>*).
10. Professional fees.
11. Plans and specifications (limit *<<insert currency amount>>*).
12. Public authorities.
13. Expediting expenses (limit *<<insert currency amount>>* or 25 percent of claim, whichever is the greater).
14. Lenders' clauses.
15. Inflation on incomplete works.

Main Exclusions:

1. Contractor's and Subcontractors' plant and equipment.
2. Wear and tear.
3. Inventory losses.
4. Loss of cash and negotiable instruments.

Sum Insured:
<<insert currency amount>> for the Facility
<<insert currency amount>> for the Existing Plant

Deductibles:

Shall not be more than:

(A) in the case of any claim arising as a result of testing, commissioning, design and maintenance *<<insert currency amount>>*; and
(B) in the case of any other claim *<<insert currency amount>>*.
(C) for the Existing Plant

<<insert currency amount>> in respect of machinery breakdown
<<insert currency amount>> in respect of all other losses

B. Marine/Inland Transit and Air Cargo

The Insured Parties:

1. the Sponsor,
2. all Financing Parties,
3. the Contractor,
4. all Subcontractors.

The Insured Property:

Goods, cargo, plant, machinery, equipment, accessories, materials and/or any interest in connection with or incidental to the Facility.

Period of Cover:

Commencement of first loading or the Work Commencement Date, whichever is the earlier.

Sum Insured:

Not less than the value of the largest single cargo shipment or the value in storage, whichever is the greater (CIF plus 10 percent).

Cover:

Marine/Inland Transit and Air Cargo Insurance covering the insured property against "All Risks" of physical loss or damage while in transit by land, sea or air from country of origin anywhere in the world to the <Route><Site> including loading, or vice versa, from the commencement of the time the insured items are loaded prior to leaving the warehouse or factory for shipment to the <Route><Site>.

Basis of Settlement:
CIF plus 10 percent.

Main Terms/Exclusions:

The policy shall include:

1. 50/50 Clause.
2. Institute Cargo Clause "A."
3. Institute Cargo Clauses (Air Cargo).
4. Institute Strikes Clauses (Cargo).
5. Institute Strikes Clauses (Air Cargo) including loading and unloading prior to and [in] transit per Cargo Clauses.
6. Institute Replacement Clause.
7. Institute Classification Clause.
8. Consolidation/Packers Risk at Loading.
9. Automatic Cover for Transshipment by Barge/Craft.
10. Containers Worthiness Admitted.
11. DIC where property covered elsewhere.
12. Returned Shipments Covered Automatically.
13. General Average.
14. Shipments on/under deck.
15. [200 percent Accumulation Clause to be confirmed].
16. Survey Warranty Key Items.
17. Debris Removal (10 percent of cargo).

18. All Inland Transit Risks except to the extent insured under the Contractor's All Risks policy.
19. 7 days' notice of cancellation war, strikes, riot and civil commotion.
20. Lenders' Clauses.

Deductible:

Not more than <<*insert currency and amount*>> for each loss.

C. Third Party General Liability

The Insured Parties:

1. the Sponsor,
2. all Financing Parties,
3. the Contractor,
4. all Subcontractors,
5. [the Transmission Company,]
6. the Sponsor's operations and maintenance contractor,
7. all Sponsor Persons,
8. all officers, employees and consultants of each of the above parties.

Cover:

Against legal liability to third parties for death, bodily injury or damage to property (excluding property forming part of the Facility) arising out of the construction, testing and commissioning of the Facility.

Period of Cover:

From the commencement of construction activities at the <Route><Site> or the Work Commencement Date, whichever is the earlier, until Facility Final Acceptance Date and [for at least 24 months in respect of defect guarantees under the Contract].

Main Extensions:

1. Contingent Employers Liability.
2. Cross Liabilities.
3. Tools of Trade.
4. Sudden and Accidental Pollution.
5. Worldwide Jurisdiction Excluding USA/Canada.
6. Contingent Motor Liability.
7. Legal Defense Costs.
8. Data Protection Liability.
9. Lenders' Clauses.

Main Exclusions:

1. Loss (other than property damage or bodily injury or death) arising from professional services (which are not within the scope of the Work) rendered for a fee.
2. Own employees.
3. Ownership, possession or use of aircraft, waterborne craft and/or motor vehicles under circumstances where road traffic legislation or equivalent applies.
4. Property in care, custody or control (covered by insurances in A [and B] above).
5. [The Work].

Geographical Limits:

Worldwide in connection with the Work.

Sum Insured:

(A) Up to [<<*insert currency amount*>>] [__] for any one claim other than environmental claims;
(B) For environmental claims, the aggregate claims limit shall be [<<*insert currency amount*>>] [__] during any one year (commencing on the commencement of the Period of Cover); and
(C) Defense costs, charges and expenses.

Deductible:

Not to exceed [<<*insert currency amount*>>] [__] for each claim for damage to property. None for injury to natural persons.

EXHIBIT 15—ALTERNATE

INSURANCE

(a) I. Commercial General Liability insurance in limits not less than:

<<*insert currency amount*>>	Occurrence
<<*insert currency amount*>>	Aggregate
<<*insert currency amount*>>	Products & Completed Operations
<<*insert currency amount*>>	Personal Injury

The coverage should be provided on forms that are in no event less than the coverage that would be provided by the following coverage forms and endorsements:

-	*Commercial General Liability Coverage Form*	CG 00 01 07 98
-	*New York Changes Commercial General Liability Coverage Form*	CG 01 63 09 99
-	*Additional Insured Designated Person or Organization*	

[Naming as Additional Insured for liability arising out of Contractor negligence in the performance of this Contract for or on behalf of the named parties:]

-	*Sponsor Corporation*	CG 20 26 11 85
-	*Waiver of Transfer of Rights of Recovery Against Others to Us*	CG 24 04 10 93
-	*Boats*	CG 24 12 11 85

Project All Risk Insurance:

-	*Builders Risk Coverage Form*	CP 00 20 06 95
-	*Ordinance or Law Coverage*	CP 04 OS 06 95
-	*Ordinance or Law Increase Period of Restoration*	CP 15 31 06 95
-	*Pollutant Clean Up & Removal Additional Aggregate Limit*	CP 04 07 10 91
-	*Debris Removal Additional Limit of Insurance*	CP 04 15 06 95
-	*Cause of Loss Special Form*	CP 10 30 06 95
-	*Radioactive Contamination*	CP 10 37 07 88
-	*Pier and Wharf Additional Covered Causes of Loss*	CP 10 70 07 88
-	*Loss Payable Provisions*	CP 12 18 06 95
-	*Commercial Property Coverage Part Renewal*	CP 12 40 10 85
-	*Joint or Disputed Loss Agreement*	CP 12 70 09 96
-	*Business Income From Dependent Properties Broad Form*	CP 15 08 06 95
-	*Utility Services—Direct Damage*	CP 04 17 06 96
-	*Utility Services—Time Element*	CP 15 45 06 95
-	*Flood Insurance—National Flood Insurance Form*	

The following internal peril or Boiler & Machinery provision needs to also be included either by endorsement or separate policy to cover testing and exposures:

-	*Boiler & Machinery Coverage Form*	BM 00 25 06 95
-	*Object Definition #6 Comprehensive*	BM 00 31 06 95
-	*Business Interruption—Valued Coverage*	BM 15 25 06 95
-	*Consequential Damage*	BM 15 28 06 95
-	*Utility Interruption*	BM 15 35 02 91
-	*Mortgage Holders*	BM 99 33 09 88
-	*In Use or Connected Ready for Use*	BM 99 36 11 85
-	*Coverage Limitation Changes*	BM 99 38 09 88
-	*Explosion Coverage Extension*	BM 99 40 06 95
-	*Joint or Disputed Loss Agreement*	BM 09 43 09 96

The limit of insurance should equal the total project cost including soft costs, site work and actual construction. Business Interruption limits and sub-limits required by endorsements are to be determined.

Errors & omissions liability insurance providing coverage against claims for bodily injury, property damage and cleanup costs arising from errors, omissions or negligence in risk assessment, remedial action plan design, lab testing or other aspects of professional work resulting in a release of contamination, with limits not less than <<*insert currency amount*>> per occurrence and <<*insert currency amount*>> in the aggregate per annum.

Professional liability insurance with a minimum limit of <<*insert currency amount*>> per occurrence and <<*insert currency amount*>> in the aggregate.

EXHIBIT 21.2.1(a)

FORM OF LENDER CONSENT AND AGREEMENT[325]

THIS CONSENT AND AGREEMENT is dated as of [_____] and is by and among [_____] (the "Contractor"), [_____] (the "Sponsor") and [_____], acting as [collateral] agent (the "Agent") for the Financing Parties (as herein defined).

WHEREAS, the Sponsor was organized to develop, construct, own, operate and maintain a <<*insert type*>> facility to be located in [_____] (the "Project");

WHEREAS, in connection with the financing of the Project, the Sponsor has entered into certain agreements (as amended, modified and/or supplemented from time to time, the "Financing Agreements") with certain financial institutions providing financing[, risk guarantees] and/or otherwise extending credit to the Sponsor in connection with the Project;

WHEREAS, in connection with the Project, the Sponsor and the Contractor have entered into the Fixed Price, Lump Sum Turnkey Engineering, Procurement and Construction Contract dated as of [_____] (as amended, modified and/or supplemented or replaced from time to time, the "Contract");

WHEREAS, in order to secure the Sponsor's obligations under the Financing Agreements, the Sponsor has agreed to irrevocably transfer and assign for security purposes, and grant a security interest in, all of the Sponsor's right, title, and interest in the Contract (pursuant to certain security agreements (the "Security Agreements") for the benefit of the secured Parties referred to in such Security Agreements (collectively, [including the Agent,] the "Financing Parties")); and

WHEREAS, it is a condition precedent to the making of loans and the extensions of credit under the Financing Agreements that the Contractor shall have executed and delivered this Consent and Agreement.

NOW THEREFORE, in consideration of the foregoing and as an inducement for the Sponsor's agreeing to enter into the Contract and for other good and valuable consideration, the receipt and adequacy of which are hereby acknowledged, the parties hereby agree as follows:

1. Capitalized terms used in this Consent and Agreement without definition shall have the respective meanings ascribed thereto in the Contract. In this Consent and Agreement, unless the context expressly indicates otherwise, terms defined in the singular shall have the same meanings when used in the plural and vice versa; references to any law (whether statutory, administrative or otherwise) shall include all provisions amending, replacing, succeeding or supplementing such law; references to agreements shall include all amendments, modifications and supplements thereto and replacements thereof to the extent permitted by this Consent and Agreement and the Financing Agreements; and references to persons or entities shall include such person's or entity's successors and assigns to the extent permitted by this Consent and Agreement and the Financing Agreements.

2. The Contractor hereby acknowledges and agrees for the benefit of the Financing Parties that:

(a) Upon receipt by the Contractor of written notice from the Agent that an event of default under the Financing Agreements has occurred and is continuing, the Agent and/or any Financing Party designated by the Agent shall be entitled, but shall not be obligated, to exercise any and all rights of the Sponsor under the Contract, and the Contractor shall comply in all respects with such exercise. Without limiting the generality of the foregoing, the Agent and/or any Financing Party designated by the Agent shall have the full right and power to enforce directly against the Contractor all obligations of the Contractor under the Contract and otherwise to exercise all remedies thereunder and to make all demands and give all notices and make all requests required or permitted to be made by the Sponsor under the Contract. [Without limiting any other rights or powers granted to the Agent hereunder or under any of the Financing Agreements, the Sponsor shall, under a power of attorney in substantially the form attached hereto as Exhibit B, constitute and appoint the Agent, or any other person or entity empowered by the Agent, acting for and on behalf of the Sponsor and any Financing Party and each successor or assign of the Agent and any Secured Party, the Sponsor's true and lawful attorney in fact (the "<u>Attorney</u>"), with full power (in its name or otherwise), valid until all obligations of the Sponsor under the Financing Agreements are discharged, to enforce any and all rights of the Sponsor under the Contract in accordance with its terms, including, without limitation:

   (i) to ask, require, demand, receive and give acquittance for any and all moneys and claims for moneys due and to become due under or arising out of the Contract;

  (ii) to exercise rights and elect remedies thereunder, to endorse any checks or other instruments or orders in connection therewith; and

 (iii) to file any claims or take any action or institute any proceedings in connection therewith which the Attorney may deem to be necessary or advisable;

<u>provided</u>, <u>however</u>, that the Attorney shall not take any such action or exercise any such rights contemplated by this Article 2(a) except upon the occurrence and during the continuation of an event of default under the Financing Agreements. The Agent shall provide notice to the Sponsor promptly after taking any such action, <u>provided</u>, <u>that</u> the failure of the Agent to provide such notice shall not affect (i) the rights or remedies of the Agent hereunder or under such power of attorney, nor (ii) the effectiveness of any such action. Such power of attorney shall be a power coupled with an interest and shall be irrevocable.][326]

3. Notwithstanding anything contained in the Contract to the contrary, the Contractor shall not, without the prior written consent of the Agent, take any action to:

(a) (i) cancel or terminate, or suspend performance under, the Contract unless such is expressly provided therein and the Contractor first complies with Article 3 hereof, or (ii) consent to or accept any cancellation, termination or suspension of the Contract unless such consent or acceptance is expressly required thereunder and the Contractor first complies with Article 3 hereof;

(b) exercise any of its rights set forth in the Contract to cancel or terminate, or suspend its performance under, the Contract [{due to the Sponsor's failure to perform any of its obligations under the Contract}], unless the Contractor shall have delivered to the Agent at least <<*insert number*>> days' prior written notice of its intent to exercise such right, specifying the nature of the default or event giving rise to such right (and, in the case of a payment default, specifying the amount thereof) and permitting the Agent (and/or any of the other Financing Party) or any of their designee(s) or assignee(s) to cure such default or circumstance within such aforementioned period by performing or causing to be performed (including, without limitation, causing the Sponsor to perform) the obligation in default or eliminating such circumstance [(<u>provided</u> <u>that</u> the Contractor may temporarily suspend performance, but not cancel or terminate the Contract, following the date <<*insert number*>> days after it shall have delivered to the Agent the aforesaid notice in respect of a payment default)];

(c) amend, supplement or otherwise modify or waive any provision of the Contract including, without limitation, change orders thereunder; or

(d) sell, assign, delegate or otherwise dispose of (by operation of law or otherwise) any part of its interest in the Contract.

In furtherance of the foregoing clause (b), the Contractor agrees that, notwithstanding anything contained in the Contract to the contrary, if (i) physical possession of the Project or any portion thereof is necessary to cure any default (other than a payment default) or circumstance giving rise to the Contractor's request to suspend its performance under or terminate the Contract and (ii) the Agent (and/or any Financing Party) or any of their designee(s) or assignee(s) commences proceedings or other actions necessary to obtain or secure possession of the Project within the above-prescribed period, then the Contractor will not take any action to terminate or suspend its performance under the Contract on account thereof so long as the Agent (and/or a Financing Party) or their designee(s) or assignee(s) diligently pursues such proceedings or other actions necessary to obtain or secure such possession of the Project and all payment defaults of the Sponsor under the Contract have been cured within such above-prescribed period. The Agent shall provide notice to the Contractor promptly after the commencement of any such proceeding or the taking of any such other action, <u>provided</u> <u>that</u> the failure of the Agent to provide such notice shall not affect the effectiveness of any such proceeding or action. If the Agent or its designee(s) or assignee(s) is precluded from securing physical possession of the Project or from exercising remedies or curing any default by virtue of applicable law or otherwise, the foregoing time periods shall be extended by the period of such preclusion. No curing of or attempt to cure, by the Agent or any other Financing Party, any of the Sponsor's defaults under the Contract shall be construed as an assumption by the Agent or any such Financing Party of any covenants, agreements or obligations of the Sponsor under the Contract.

4. With respect to the Contract, the Contractor shall deliver to the Agent, concurrently with the delivery thereof to the Sponsor, a copy of each notice from the Contractor to the Sponsor of any default, termination, arbitration or force majeure event under the Contract.

5. The Contractor irrevocably consents to the transfer and assignment by way of security of the Sponsor's right, title and interest in, and obligations under, the Contract. The Contractor agrees that upon delivery of a written notice from the Agent regarding the exercise of the Financing Parties' remedies under the Security Agreements after the occurrence and during the continuance of an event of default under the Financing Agreements, the following will apply:

(a) In the event that the Agent or any of its designee(s) or assignee(s) elects to succeed to the Sponsor's interest under the Contract, the Agent or such designee(s) or assignee(s) may elect by written notice delivered to the Contractor to assume the Sponsor's rights and obligations under the Contract, including any payment obligations under the Contract theretofore accrued but excluding any other obligations or liabilities that may have accrued prior to such foreclosure or assignment (any right in respect of such excluded obligation being hereby expressly waived by the Contractor). Except as otherwise set forth in the preceding sentence, none of the Agent nor any other Financing Party shall be liable for the performance or observance of any of the obligations or duties of the Sponsor under the Contract, and the assignment by way of security of the Contract by the Sponsor to the Financing Parties pursuant to the Security Agreements shall not give rise to any duties or obligations whatsoever on the part of any of the Financing Parties owing to the Contractor.

(b) Until such time as the Agent (or any Financing Party as provided herein) seeks to exercise remedies under the Security Agreements in respect of the Contract after the occurrence and during the continuance of an event of default under the Financing Agreements, and gives written notice as provided herein, the Contractor shall, except as otherwise provided in this Consent and Agreement, continue to deal directly with the Sponsor with respect to its obligations to the Sponsor under the Contract. Notwithstanding anything else herein, the assignment of the Contract pursuant to this Article 5 shall not relieve the Sponsor of any obligations arising thereunder.

(c) Upon the exercise (as contemplated above) by the Agent or any of its designee(s) or assignee(s) of any of the remedies under the Security Agreements in respect of the Contract after the occurrence and during the continuance of an event of default under the Financing Agreements, the Agent or any of its designee(s) or assignee(s) may assign its rights and interests and the rights and interests of the Sponsor under the Contract to any Person, if such Person shall assume liability for all of the obligations of the Sponsor, including any payment obligations, under the Contract theretofore accrued but excluding any other obligations or liabilities that may have accrued prior to such foreclosure or assignment (any right in respect of such excluded obligation being hereby expressly waived by the Contractor). Upon such assignment and assumption in respect of the Contract, the Agent and all other Financing Parties shall be relieved of all obligations under the Contract.

6. In the event that:

(a) the Contract is rejected by a trustee, liquidator, debtor in possession or similar person or entity in any bankruptcy, insolvency or similar proceeding involving the Sponsor;

(b) the Contract is terminated as a result of any bankruptcy, insolvency or similar proceeding involving the Sponsor; or

(c) the assignment by way of security of the Contract hereunder is ineffective or challenged for any reason whatsoever and, if within <<*insert number*>> days after such rejection, termination, ineffectiveness or challenge, the Agent or any of its designee(s) or assignee(s) shall so request and shall certify in writing to the Contractor that it intends to perform the obligations of the Sponsor as and to the extent required under the Contract (as if the Contract had not been rejected or terminated, but otherwise only to the extent such obligations would be undertaken had such person or entity succeeded to the Sponsor thereunder pursuant to Article 5 above), the Contractor will execute and deliver to the Agent or any such designee(s) or assignee(s), a new Contract for the balance of the remaining term under the original Contract before giving effect to such rejection or termination, and such new Contract shall contain the same conditions, agreements, terms, provisions and limitations as the original Contract (except for any requirements which have been fulfilled by the Sponsor and the Contractor prior to such rejection, termination, ineffectiveness or challenge or which are not required to be undertaken by such person or entity as aforesaid). References in this Consent and Agreement to the "Contract" shall be deemed also to refer to the new Contract in replacement of the Contract. The Contractor hereby indemnifies the Agent from and against any damages that may be suffered by the Financing Parties for breach of the Contractor's obligation hereunder.

7. In the event that the Agent or its designee(s), or any purchaser, transferee, grantee or assignee of the interests of the Agent or its designee(s) in the Project, assume or are liable under the Contract (as contemplated above or otherwise), liability in respect of any and all obligations of any such person or entity under the Contract shall be limited solely to such person's or entity's Ownership interest in the Project, and the Contract shall expressly exclude any liability (other than as regards payment obligations contemplated to be assumed above) for acts or omissions of the Sponsor or any other predecessor of such person's or entity's interest in the Project (and no officer, director, employee, shareholder, affiliate or agent thereof shall have any liability with respect thereto).

8. All references in this Consent and Agreement to the "Agent" shall be deemed to refer to the Agent and/or any attorney in fact or assignee thereof acting on behalf of the Financing Parties (regardless of whether so expressly provided), and all actions permitted to be taken by the Agent under this Consent and Agreement may be taken by any such attorney in fact or assignee.

9. In connection with any cure of the Sponsor's default(s) under the Contract or any assumption of the Sponsor's liabilities thereunder, only those obligations and liabilities arising expressly under the Contract shall be required to be cured or assumed, as the case may be, and there shall be no obligation to cure or assume any non-contractual liability that may have arisen.

10. The Contractor shall duly and timely perform all of its obligations and responsibilities under the Contract.

11. As soon as available, but in any event within <<*insert number*>> days after the close of each of its fiscal years, the Contractor shall furnish to the Agent (in sufficient

number of copies for distribution to each of the Financing Parties) its balance sheet as of the end of such fiscal year with the related statements of income, retained earnings and cash flows for such fiscal year, in each case setting forth comparative figures for the preceding fiscal year and certified by independent certified public accountants of recognized international standing, which certification shall state that such statements fairly and truly present its financial position and results of operation as of the close of such fiscal year and that its results of operations and its cash flows for the period then ended are prepared in accordance with generally accepted accounting principles consistently applied.

12. Promptly upon the request of the Sponsor or the Agent, the Contractor agrees that it shall provide good standing certificates (to the extent applicable to it and available in its jurisdiction of formation), incumbency certificates, resolutions and such other corporate documents necessary to evidence approval of execution, due authorization, delivery and performance by the Contractor of the Contract and this Consent and Agreement, in each case certified by an authorized officer of the Contractor.

13. All payments to be made by the Contractor to the Sponsor under the Contract (if any) from and after the date hereof shall be made in lawful money of the currency specified in the Contract, directly to the Agent, for deposit into the [NAME OF ACCOUNT] (Account No. [_____]), at [_____] or to such other Person and/or at such other address as the Agent may from time to time specify in writing to the Contractor. All payments by the Contractor shall be accompanied by a notice from the Contractor stating that such payments are made under the Contract and identifying the relevant provision thereof under which such payment was made. The Contractor shall not, without the prior written consent of the Agent, make any payments to or for the benefit of the Sponsor other than as contemplated pursuant to the first sentence of this Article 13.

14. The Contractor hereby represents and warrants to the Agent and each of the other Financing Parties that:

(a) The Contractor is duly organized under the laws of the jurisdiction of its formation and is duly qualified to do business and is in good standing in all jurisdictions where necessary in light of the business it conducts and the property it owns and the business that it intends to conduct and the property that it intends to own in light of the transactions contemplated by the Contract and this Consent and Agreement.

(b) The Contractor has the full power, authority and legal right to execute, deliver and perform its obligations hereunder and under the Contract. The execution, delivery and performance by the Contractor of this Consent and Agreement and the Contract and the consummation of the transactions contemplated hereby and thereby have been duly authorized by all necessary corporate, shareholder and governmental action. This Consent and Agreement and the Contract have been duly executed and delivered by the Contractor and constitute the legal, valid and binding obligations of the Contractor, enforceable against the Contractor in accordance with their respective terms. The Contractor has not assigned, transferred or hypothecated the Contract or this Consent and Agreement or any interest herein or therein.

(c) The execution, delivery and performance by the Contractor of this Consent and Agreement and the Contract do not and will not (i) require any consent or approval of the board of directors (or similar body) of the Contractor or any shareholder of the Contractor or of any other Person which has not been obtained and each such consent or approval that has been obtained has not been modified and is in full force and effect, (ii) result in, or require the creation or imposition of any lien, security interest, charge or encumbrance upon or with respect to any of the assets or properties now owned or hereafter acquired by the Contractor, (iii) violate any provision of any law, rule, regulation, order, writ, judgment, decree, determination or award having applicability to the Contractor or any provision of the certificate of incorporation or by laws or other constituting documents of the Contractor or (iv) conflict with, result in a breach of or constitute a default under any provision of the certificate of incorporation, by laws or other constituting documents or any resolution of the board of directors (or similar body) of the Contractor or any indenture or loan or credit agreement or any other agreement, lease or instrument to which it is a party or by which it or its properties and assets are bound or affected. The Contractor is not in violation of any such law, rule, regulation, order, writ, judgment, decree, determination or award referred to in clause (iii) above or its certificate of incorporation or by laws or in breach of or default under any provision of its certificate of incorporation or by laws or any agreement, lease or instrument referred to in clause (iv) above.

(d) Each government approval required for the execution, delivery or performance of this Consent and Agreement and the Contract by the Contractor has been validly issued and duly obtained, taken or made, is not subject to any condition, does not impose restrictions or requirements inconsistent with the terms hereof or thereof, as the case may be, is in full force and effect and is not subject to appeal. The Contractor has no reason to believe that any government approval that has been issued will be revoked, modified, suspended or not renewed on substantially the same terms.

(e) As of the date of this Consent and Agreement the Contract is in full force and effect and has not been amended, supplemented or modified, except as amended by this Consent and Agreement.

(f) There is no action, suit or proceeding at law or in equity by or before any government authority, arbitral tribunal or other body now pending or, to the best knowledge of the Contractor, threatened against or affecting the Contractor or any of its properties, rights or assets which (i) if adversely determined, individually or in the aggregate, could reasonably be expected to have a material adverse effect on its ability to perform its obligations under this Consent and Agreement or the Contract or (ii) questions the validity, binding effect or enforceability of this Consent and Agreement or the Contract or any action taken or to be taken pursuant hereto or thereto or any of the transactions contemplated hereby or thereby.

(g) Neither the Contractor nor, to the best knowledge of the Contractor, the Sponsor, is in default of any of their respective obligations under the Contract. The Contractor and, to the best knowledge of the Contractor, the

Sponsor, have complied with all conditions precedent to the respective obligations of such party to perform under the Contract. No event or condition exists which would either immediately or with the passage of any applicable grace period or giving of notice, or both, enable the Contractor, or, to the best knowledge of the Contractor, the Sponsor, to terminate or suspend the Contractor's obligations under the Contract.

(h) The Contractor is subject to civil and commercial law with respect to its obligations hereunder and under the Contract, and the execution, delivery and performance of this Consent and Agreement and the Contract by it constitute private and commercial acts rather than public or governmental acts. Neither it nor any of its property has any immunity from jurisdiction of any court or from set-off or any legal process whether through service or notice, attachment prior to judgment, attachment in aid of execution, execution or otherwise.

15. Each of the representations and warranties made by the Contractor in the Contract is:

(a) incorporated herein by reference as fully and to the same extent as if set forth herein in its entirety;

(b) true and correct as of the date of this Consent and Agreement with the same force and effect as if made on and as of such date (or, if stated to have been made solely as of an earlier date, was true and correct as of such earlier date); and

(c) made for the express benefit of the Financing Parties.

16.

(a) The Contractor hereby acknowledges the right of the Agent (or any of its designee(s) or assignee(s)) from time to time) to exercise all rights pursuant to this Consent and Agreement on behalf of the Financing Parties. The Contractor hereby acknowledges notification pursuant to Article 9–103 of the UCC of the security interest of the <<Agent/Collateral Trustee>> in the Contract under the Security Agreements. Contractor hereby agrees to deliver to the Agent[and each Financing Party] a legal opinion, dated as of the date hereof and in form and substance acceptable to the Agent, addressing the matters so requested by [it] [them].

(b) No failure or delay on the part of the Contractor, the Sponsor, the Agent, any Financing Party or any agent or designee of any of the foregoing to exercise, and no course of dealing with respect to, any right, power or privilege hereunder shall operate as a waiver thereof, and no single or partial exercise of any right, power or privilege hereunder shall preclude any other or further exercise thereof or the exercise of any other right, power or privilege.

(c) The remedies of the Agent and each of its respective designee(s) and/or assignee(s) provided herein are cumulative and not exclusive of any remedies provided by law. In addition, the Agent may exercise its rights in respect of the Contract in such order as the Agent may deem expedient.

(d) All notices, requests and other communications provided for herein (including, without limitation, any modifications of, or waivers or consents under, this Consent and Agreement) shall be given or made in writing delivered to the intended recipient at the "Address for Notices" specified below its name on the signature pages hereof or, as to any party hereto, at such other address as shall be designated by such party in a written notice to each other party hereto. Except as otherwise provided in this Consent and Agreement, all such communications shall be deemed to have been duly given when transmitted by telecopier (and confirmed by a confirmation report) or personally delivered or, in the case of a mailed notice, upon receipt, in each case given or addressed as aforesaid.

(e) This Consent and Agreement may be amended, waived or modified only by an instrument in writing signed by [the Sponsor] Contractor and the Agent[; provided that the written consent of the Sponsor shall also be required if such amendment, waiver or modification affects the Sponsor's rights or obligations under the Contract]. Any waiver shall be effective only for the specified purpose for which it is given.

(f) This Consent and Agreement may be executed in any number of counterparts, all of which when taken together shall constitute one and the same instrument, and any of the parties hereto may execute this Consent and Agreement by signing any such counterpart. This Consent and Agreement shall become effective at such time as the Agent shall have received counterparts hereof signed by all of the intended parties hereto.

(g) If any provision hereof is invalid and unenforceable in any jurisdiction, then, to the fullest extent permitted by law, (i) the other provisions hereof shall remain in full force and effect in such jurisdiction, (ii) the invalidity or unenforceability of such provision hereof in such jurisdiction shall not affect the validity or enforceability of such provision in any other jurisdiction and (iii) to the extent practicable, invalid or unenforceable provisions shall be replaced by valid and enforceable solutions having the same economic effect on the parties as was intended by the invalid or unenforceable provisions.

(h) Headings appearing herein are used solely for convenience and are not intended to affect the interpretation of any provision of this Consent and Agreement.

(i) The agreements of the parties hereto are solely for the benefit of the Contractor, the Sponsor, the Agent and the other Financing Parties, and shall be binding upon and inure to the benefit of the respective successors and assigns of each of the foregoing parties. No Person (other than the foregoing parties, including their respective successors and assigns) shall have any rights hereunder.

(j) This Consent and Agreement shall be governed by, and construed in accordance with, the law of [LAW OF THE STATE OF NEW YORK[327]].

(k) [Each of the parties hereto hereby submits to the nonexclusive jurisdiction of the United States District Court for the Southern District of New York and of any New York State Court sitting in New York City for the purposes of all legal proceedings relating to the execution, validity or enforcement of this Consent and Agreement. Each of the parties hereto hereby irrevocably waives, to the fullest extent permitted by law, any objection which it may now or hereafter have to the laying of the venue of any such proceeding brought in such a court and any claim that any such proceeding brought in such a court has been brought in an inconvenient forum. With respect to any such proceedings in the courts of the United States District Court for the Southern District of New York or any New York State Court sitting in New York City, (i) the Contractor appoints [_____], whose address is presently [_____], to receive for and on its behalf service of process in such jurisdiction in any such enforcement proceedings, (ii) the Contractor agrees to maintain in New York City a duly appointed process agent, notified to the Agent for purposes of the foregoing clause (i) and (iii) the Contractor agrees that failure by such process agent to give notice of any process to it shall not impair the validity of such service or of any judgment based thereon].[328]

(l) **[EACH OF THE PARTIES HERETO HEREBY IRREVOCABLY WAIVES, TO THE FULLEST EXTENT PERMITTED BY LAW, ANY AND ALL RIGHT TO TRIAL BY JURY IN ANY LEGAL PROCEEDING ARISING OUT OF OR RELATING TO THIS CONSENT AND AGREEMENT.]**

(m) Nothing in this Consent and Agreement shall affect the right of the Agent or any other Person to serve legal process in any other manner permitted by applicable Law or affect the right of the Agent or any other person or entity to commence legal proceedings or otherwise sue the Contractor in any other jurisdiction.

(n) To the extent that the Contractor may (whether in [_____], in any other jurisdiction in which its property is located [or in the State of New York or the United States of America], each a "<u>Relevant Jurisdiction</u>") be entitled to claim for itself or its property sovereign immunity or any other type of immunity (howsoever described) from suit or judgment and to the extent that in any such Relevant Jurisdiction there may be attributed such immunity (whether or not claimed), the Contractor hereby irrevocably and expressly agrees not to claim, and waives to the fullest extent permitted by the laws of the Relevant Jurisdiction, such immunity and hereby irrevocably and expressly agrees that it and its property in any such Relevant Jurisdiction are and shall be subject to execution on a judgment and attachment (whether provisional or final, through service or notice, attachment before judgment, attachment in aid of execution or otherwise) on account of the obligations incurred by it under this Consent and Agreement. Without limiting the generality of the foregoing, the Contractor agrees that the waivers set out in this Article 16(n) shall have the fullest scope permitted under the Foreign Sovereign Immunities Act of 1976 of the United States and are intended to be irrevocable for purposes of such Act.

(o) This Consent and Agreement is made and executed in the English language. Any translation hereof made shall be inadmissible for any purpose.

(p) [Sponsor acknowledges and agrees that the Contractor is authorized to act in accordance with the Agent's exercise of Sponsor's rights in accordance with the terms and conditions set forth in this Consent and Agreement, and that the Contractor shall bear no liability to Sponsor in connection therewith]. Each signatory accepting this Consent and Agreement shall be deemed a party hereto for all purposes hereof.

(q) In the event of conflict or inconsistency between the Contract and this Consent and Agreement, the terms and conditions of this Consent and Agreement shall control.

(r) This Consent and Agreement supersedes all prior agreements, written or oral, among the parties with respect to the subject matter of this Consent and Agreement.

(s) Each party agrees that from time to time on or after the date hereof, it shall in good faith and at the reasonable request of the other party, execute and deliver or cause to be executed and delivered such instruments and documents, and take such further action, as may be reasonably required in order more effectively to consummate the terms of this Consent and Agreement.

[(t) Joint and Several Provision—All representations and warranties of the Contractor contained herein shall be deemed to have been made jointly and severally by [the Offshore Contractor and Onshore Contractor], and the [Offshore Contractor and the Onshore Contractor] shall be jointly and severally liable for all agreements, undertakings, liabilities and obligations of [the Onshore Contractor (in the case of the Offshore Contractor) or the Offshore Contractor (in the case of the Onshore Contractor)] and the Contractor hereunder.][329]

IN WITNESS WHEREOF, the undersigned by its officer duly authorized has caused this Consent and Agreement to be duly executed and delivered as of the date first written above.

[CONTRACTOR][CONSORTIUM MEMBER]
By:_____
Name:
Title:
Address for Notices:
Telecopier No.:
Telephone No.:
Attention:

[SPONSOR]
By:_____
Name:
Title:

By:_____
Name:
Title:
Address for Notices:
Telecopier No.:
Telephone No.:
Attention:

Accepted:

[_____],

as Agent

By:_____
Name:
Title:
Address for Notices:
Telecopier No.:
Telephone No.:
Attention:

ANNEX I

[EXHIBIT 21.2.1(b)

FORM OF ASSIGNMENT AND ASSUMPTION AGREEMENT]

This Assignment and Assumption Agreement (this "Agreement"), dated effective as of [_____], (the "Effective Date"), is entered into by and among [_____] (the "Counterparty"), [_____], a [Delaware] corporation ("Sponsor") and [_____], a [Delaware] corporation ("Contractor").

WITNESSETH:

WHEREAS, the Sponsor and Contractor entered into an Engineering, Procurement and Construction Contract (the "Contract").

WHEREAS, Sponsor entered into the _____ Agreement (as defined in the Contract) (the "Assigned Contract");

WHEREAS, Sponsor now desires to assign the Assigned Contract to Contractor and Contractor desires to receive an assignment of the Assigned Contract.

NOW, THEREFORE, in consideration of the mutual covenants and agreements contained herein and other valuable consideration, the receipt and sufficiency of which are hereby acknowledged, Sponsor, Seller and Contractor, intending legally to be bound, agree as follows:

1. **Assignment and Assumption**. Effective as of the date all parties (including the Counterparty) have executed this Agreement and unless and until Contractor becomes insolvent or Sponsor terminates the Contract in accordance therewith and pays the Contractor all amounts due thereunder in which case this assignment shall cease to be of any force and effect and become null and void as if it had never been entered into, Sponsor does hereby assign, transfer and convey to Contractor, all of its right, title and interest in, to and under the Assigned Contract. Contractor hereby accepts such assignment, transfer and conveyance and assumes all of the obligations and liabilities of Sponsor under the Assigned Contract arising after the date hereof and agrees to perform and observe all of the terms, covenants, agreements, obligations and conditions therein contained

in Sponsor's part to be kept, performed and observed and to be bound by all of the terms and conditions of the Assigned Contract after the date hereof, including, but not limited to, the payment of all payments under the Assigned Contract due after the date hereof.

2. **Representations and Warranties**. Each of Sponsor and Contractor represents and warrants to the other that (i) it has full power and legal right and authority to execute and deliver this Agreement and to perform the provisions of this Agreement; (ii) the execution, delivery and performance of this Agreement have been duly authorized by all action, corporate, partnership, limited liability company or otherwise, and do not violate any provision of its charter, by laws, limited liability company agreement or other organizational documents or any contractual obligations or requirement of law binding on it; and (iii) this Agreement constitutes its legal, valid and binding obligation, enforceable against it in accordance with its terms.

3. **Consent**. The Counterparty consents to the assignment by Sponsor and assumption by Contractor of the Assigned Contract.

4. **Release**. The Counterparty and Contractor each hereby release Sponsor from all obligations and liabilities under the Assigned Contract.

5. **Binding Effect**. This Agreement shall be binding upon and inure to the benefit of the parties hereto and their respective successors and permitted assigns.

6. **Governing Law**. This Agreement shall be governing by and construed and enforced in accordance with the laws of the State of New York.

7. **Interpretation**. The headings of the Articles contained in this Agreement are solely for convenience of reference and shall not affect the meaning or interpretation of this Agreement.

8. **Counterparts**. This Agreement may be executed in two or more counterparts, each of which shall be deemed an original, but all of which together shall constitute one and the same instrument.

9. **Severability**. If any provision of this Agreement shall be declared by any court of competent jurisdiction illegal, void or unenforceable, the other provisions shall not be affected, but shall remain in full force and effect.

10. **No Third-Party Beneficiaries**. Nothing in this Agreement shall be deemed to stipulate any benefit for third-parties not signatories hereto.

IN WITNESS WHEREOF, the parties hereto have executed this Agreement as of the date and year first above written.

[_____]

By:_____

Name:_____

Title:_____

[ADDRESS]
Telecopy: [_____]
Telephone: [_____]

Attention:_____

[_____]

[_____]

By:_____

_____

Name:_____

_____

Title:_____

[Counterparty]

By:_____

_____

Name:_____

_____

Title:_____

_____

EXHIBIT 21.2.1(c)

FORM OF CONTRACTOR "IN-HOUSE" COUNSEL LEGAL OPINION

EXHIBIT 21.2.1(e)

FORM OF CONTRACTOR "OUTSIDE" COUNSEL LEGAL OPINION

To the Persons listed on Annex I

Re:_____ (the "Project")

Ladies and Gentlemen:

We have acted as counsel to _____ (the "Contractor") and _____ (the "Parent Guarantor" and, together with the Contractor, the "Companies"), in connection with the Project to be constructed by _____ (the "Sponsor"). This opinion is being provided in connection with the transactions (the "Transactions") contemplated by the terms of the Fixed Price, Lump Sum, Turnkey Engineering, Procurement, and Construction Contract (the "Contract") dated as of <<   >>, 2___, by and among the Sponsor and the Consent (as defined below).

In connection with this opinion, we have examined originals or copies, certified or otherwise identified to us, of the following:

(a) the Contract;

(b) the Consent and Agreement, dated as of <<*insert date*>> (each, a "Consent");

(c) the [Certificate of Formation and Limited Liability Company Agreement] of the Contractor and the Parent Guarantor; and

(d) the resolutions of the Contractor and the Parent Guarantor authorizing the execution and delivery of the Contract and the relevant Consent duly adopted by <<the _____>>.

The documents referred to in items (a), (b), (c) and (d) above are hereinafter collectively referred to as the "<u>Governing Documents</u>", the documents referred to in items (c) and (d) above are hereinafter collectively referred to as the "<u>Organizational Documents</u>" and the Contract and the Consent are hereinafter collectively referred to as the "<u>Documents</u>." In addition, we have examined and are familiar with originals or copies, certified or otherwise identified to us, of such other documents as we have deemed necessary or appropriate as a basis for the opinions set forth below.

In our examination we have assumed, without independent verification, the genuineness of all signatures, the authenticity of all documents submitted to us as originals, the conformity to original documents of all documents submitted to us as certified or photostatic copies, and the authenticity of the originals of such copies. In rendering the opinions expressed below, we have further assumed, without any independent investigation or verification of any kind, that each document we have examined is the valid and binding obligation of each party thereto other than the Companies. We have relied, with your consent and without independent investigation, upon certificates of officers and representatives of the Companies and of government officials as to matters of fact not independently established by us.

Based upon the foregoing and subject to the limitations, qualifications, exceptions and assumptions set forth herein, we are of the opinion that:

(a) the Contractor is [a limited liability company] duly formed, validly existing and in good standing under the laws of << >> and is duly qualified to transact business in << >>. The Parent Guarantor is [a limited partnership] duly formed, validly existing and in good standing under the laws of the << >> and is duly qualified to transaction business in << >>.

(b) the Contractor has full [limited liability company] power and authority, and the Parent Guarantor has full [limited partnership] power and authority, to enter into and deliver, and perform its respective obligations under, each Document.

(c) the Contractor has taken all necessary [limited liability company] action, and the Parent Guarantor has taken all necessary [limited partnership] action, to authorize the execution, delivery and performance by it of each Document and each of the Documents has been duly authorized, executed and delivered by each Company in accordance with the Organizational Documents.

(d) Each Document constitutes the valid and binding obligation of the Companies, enforceable against each Company in accordance with its terms[, except as enforcement thereof may be limited by bankruptcy, insolvency, reorganization, moratorium or other similar laws affecting enforcement of creditors' rights generally and by general principal of equity (regardless of whether enforcement is sought in a proceeding in equity or at law)].

(e) The execution and delivery by the Companies of the Documents do not and will not contravene the Organizational Documents or any applicable provision of any law, regulation, ruling, order or decree known to us of any governmental authority, to which or by which the Companies or any of their respective property or assets is subject or bound. The execution and delivery by the Companies of the Documents do not and will not[, to our knowledge,] conflict with, result in any breach of, or constitute a default under, or result in the creation or imposition of (or the obligation to create or impose) any lien or encumbrance upon any of the property or assets of the Companies pursuant to any provision of any securities issued by the Companies, or any indenture, mortgage, deed of trust, contract, undertaking, document, instrument or other agreement to which either Company is a party or by which it or any of its property or assets is bound.

(f) No consent, order, authorization, waiver, approval or any other action by, or registration, declaration or filing with, any governmental authority (collectively, the "Approvals"), is required to be obtained by the Companies in connection with the execution or delivery of the Documents or if required or desirable, have been obtained or effected (as appropriate) and are in full force and effect.

(g) There is no action, suit or proceeding pending or threatened against or affecting the Companies before any court, governmental or regulatory authority or arbitrator, which would affect the enforceability of the Documents or the Companies' ability to perform under the Documents.

(h) No taxes, duties, charges or levies are payable in [_____] or [_____] in connection with the execution and delivery of the [_____] Documents.

The foregoing opinions are subject to the following:

A. We express no opinion as to matters governed by any laws other than the laws of the <<State>> of the Contractor's incorporation [and the [federal] law of the [United States of America]].

B. The qualification of any opinion or statement herein by the use of the words "to our knowledge," "known to us" or words of similar import, means that during the course of representation as described in this opinion, no information has come to the attention of the attorneys in this firm involved in the transactions described which would give such attorneys current actual knowledge of the existence of the facts so qualified. Except as set forth herein, we have not undertaken any investigation to determine the existence of such facts, and no inference as to our knowledge thereof shall be drawn from the fact of our representation of any party or otherwise.

This opinion speaks as of its date and we undertake no, and hereby disclaim any, duty to advise as to changes of law or fact coming to our attention after the date hereof. Assignees of, or participants in, the interests of the Persons listed on Annex I may rely on this opinion as if it were addressed to them.

Very truly yours,

Additional Riders Containing Alternative Provisions for Certain Circumstances

1. If Pre Notice to Proceed Work is called for, the following riders may be necessary.

Add in **ARTICLE 1:**
"**Guaranteed Facility Provisional Acceptance Date**" shall mean:

(a) if the number of Interim Days is fewer than <<*insert number*>>, the date which is <<*insert number*>> Days after the Pre NTP Notice Date;

(b) if the number of Interim Days is more than <<*insert number*>> but fewer than <<*insert number*>> Days, the date which is <<*insert number*>> Days after the Work Commencement Date; or
(c) if the number of Interim Days exceeds <<*insert number*>> Days, the date which is <<*insert number*>> Days after the Work Commencement Date.

"**Guaranteed [Facility] [Unit One] Commissioning Completion Date**" shall mean:

(a) if the number of Interim Days is fewer than <<*insert number*>>, the date which is <<*insert number*>> Days after the Pre NTP Notice Date;
(b) if the number of Interim Days is more than <<*insert number*>> but fewer than <<*insert number*>> Days, the date which is <<*insert number*>> Days after the Work Commencement Date; or
(c) if the number of Interim Days exceeds <<*insert number*>> Days, the date which is <<*insert number*>> Days after the Work Commencement Date.

"**Guaranteed Unit Two Commissioning Completion Date**" shall mean:

(a) if the number of Interim Days is fewer than <<*insert number*>>, the date which is <<*insert number*>> Days after the Pre NTP Notice Date;
(b) if the number of Interim Days is more than <<*insert number*>> but fewer than <<*insert number*>> Days, the date which is <<*insert number*>> Days after the Work Commencement Date; or
(c) if the number of Interim Days exceeds <<*insert number*>> Days, the date which is <<*insert number*>> Days after the Work Commencement Date.

"**Interim Days**" is the number of Days that have elapsed between the Pre NTP Notice Date and the Work Commencement Date.
"**Pre NTP Notice**" is the written notice given by the Sponsor to the Contractor to begin the Pre NTP Work.
"**Pre NTP Notice Date**" is the date upon which the Pre NTP Notice is given by the Sponsor to the Contractor.
"**Pre NTP Work**" is that portion of the Work described in Exhibit <<*insert number*>> to be done by the Contractor after the Pre NTP Notice Date.
In **ARTICLE 2: INTENT AND COMMENCEMENT OF THE WORK:** Add a new 2.3: "2.3 Pre NTP Work. On the Pre NTP Notice Date, the Contractor will commence performing all Pre NTP Work and shall be compensated therefor pursuant to Article <<*insert number*>>."
In **ARTICLE 4:**
ADD A NEW ARTICLE 4.7
"Deadline Adjustments. Without derogating from any of the Sponsor's rights set forth in Article 3.6, the Sponsor may, if the number of Interim Days exceeds <<*insert number*>> Days, issue a Change Order subtracting <<*insert number*>> Days from each of the Guaranteed Facility Provisional Acceptance Date, the Guaranteed Unit One Commissioning Completion Date and the Guaranteed Unit Two Commissioning Completion Date if the Sponsor agrees to pay the Contractor's costs of keeping

its team mobilized during the period after <<*insert number*>> Days from the Pre NTP Notice Date. Contractor may not request an Time Adjustment or an Price Adjustment based upon the reason that the Interim Days are fewer than <<*insert number*>>."

In **ARTICLE 5:**

Change: "Contract Date" to "Pre NTP Notice Date" in Article 5.1.

Add "perform Pre NTP Work" in Article 5.1(a)(ii).

In **ARTICLE 7:**

Add to 7.1: "(g) not give Pre NTP Notice unless it expects [to give Notice to Proceed][330] [Financial Closing to occur] within [_____] ([___]) Days thereof [and, after the Pre NTP Notice Date, keep the Contractor informed with respect to progress regarding achievement of Financial Closing];"

In **ARTICLE 10:**

Add "and Pre NTP Work" in parenthetical in first sentence to 10.1.

Add "other than Pre NTP Work" in the second sentence of 10.15.

Add a new 10.1.7:

"10.1.7 Pre NTP Payment. [Only in the case of the Pre NTP Work, the Sponsor agrees to pay to the Contractor, if the Work Commencement Date does not occur by the <<*insert number*>> Day after the Pre NTP Notice Date, <<*insert number*>> percent of all Contractor's actual [unaffiliated third party] costs therefor (as submitted by Contractor in accordance with Article 10.3.1), up to a maximum of <<*insert currency and amount*>> on the <<*insert number*>> Day after the Pre NTP Notice Date provided that any such amount in respect thereof that the Sponsor has paid shall reduce the amount of the first Milestone Payment made hereunder by an equal amount]. or ["If the Work Commencement Date has not occurred by [_____], 2___, the Sponsor shall reimburse the Contractor, upon written demand therefor, <all> <___ percent of> the actual third party costs incurred by the Contractor in connection with the Site Assessment up to a maximum of <<*insert currency and amount*>>]."[331]

2. Additional Riders for Article 3.

Choice A

["**Overseas Engineering and Services Contract Price**" means the fixed lump sum of <<*insert currency and amount*>> payable to the Contractor for engineering and design services occurring outside the country in which the <Site><Route> is located with respect to the Work.]

["**Overseas Lump Sum Prices**" means the sum of (i) the Overseas Engineering and Services Contract Price, (ii) the Overseas Procurement Contract Price, and (iii) the Overseas Transportation Contract Price.]

["**Overseas Procurement Contract Price**" means the fixed lump sum of <<*insert currency and amount*>> payable to the Contractor for procurement services such as the sale of materials, supplies, equipment and machinery occurring outside the country in which the <Site><Route> is located with respect to Work.]

["**Overseas Transportation Contract Price**" means the fixed lump sum of <<*insert currency and amount*>> payable to the Contractor for transportation services occurring outside the country in which the <Site><Route> is located with respect to Work.]

"The Notice to Proceed":

(a) shall not be issued before:
  (i) the Day upon which the Sponsor achieves Financial Closing;
  (ii) the Sponsor has satisfied the Contractor that the insurance required to be maintained by the Sponsor under this Contract is in effect in the form contemplated in this Contract [and Exhibit 15]; and
  (iii) the Contractor shall have carried out [a comprehensive ground conditions survey] [the Site Assessment] [that does not disclose any extraordinary below ground physical conditions which an experienced contractor could not reasonably have been expected to foresee on the basis of the information or data available to it at the Contract Date, which extraordinary physical conditions would cause the Contractor to incur significant additional cost and expense and require significant additional time to perform its obligations under the Contract]; and
(b) shall be a full release of the Contractor to commence the Work [within <<*insert number*>> Days thereof].

Choice B

"The Contractor shall commence the Work no later than <<*insert number*>> Days after the Work Commencement Date so long as all of the following conditions are satisfied:

(a) all Sponsor Permits legally required for commencement of the Work on the Site have been obtained; and
(b) the Sponsor shall have paid to the Contractor the first Milestone Payment listed in Exhibit L and provided to the Contractor written certification [from the agent for the Financing Parties] that the Sponsor has satisfied all the conditions precedent to the effectiveness of their loan agreements except the condition for payment to the Contractor of the first Milestone Payment hereunder (if such payment is a condition)."

Choice C

"The Sponsor shall not give Notice to Proceed until:

(a) <<*insert number*>> Days after the Contract Date; and
(b) the Sponsor shall have provided to the Contractor reasonable evidence that the Sponsor can pay the Contract Price in accordance herewith."

3. Additional Riders for Article 10

Choice A

["**Domestic Contract Price**" means the fixed lump sum of <<*insert currency amount*>> payable to the Contractor for engineering, procurement, civil works, construction and erection, installation and testing rendered in the country of the <Site> <Route>with respect to the Work].

"**The Overseas Procurement Contract Price**": The Overseas Procurement Contract Price shall be the full and sole compensation of the Contractor for all its procurement services and obligations hereunder, but may be adjusted by additions or deletions expressly authorized under this Contract. Unless otherwise stated herein the

(a) the Overseas Procurement Contract Price shall not be adjusted for changes in costs or any other matters; and

(b) all initial spare parts specified in this Contract are included in the Works and the cost of such parts is included in the Overseas Procurement Contract Price, and if the Contractor uses any spare parts, the Contractor, at its expense, shall replenish such spare part within <<*insert number*>> Days or as soon as practicable if such spare part is not readily available within such <<*insert number*>> Day period.

"**The Overseas Transportation Contract Price**": The Overseas Transportation Contract Price shall be the full and sole compensation of the Contractor for all its air or maritime transportation services and obligations hereunder, but may be adjusted by additions or deletions expressly authorized under this Contract. Unless otherwise stated herein the Overseas Transportation Contract Price shall not be adjusted for changes in costs or any other matters.

"**The Domestic Contract Price**" : The Domestic Contract Price shall be the full and sole compensation of the Contractor for all its engineering, procurement, civil works, construction, erection and installation services and obligations hereunder rendered in the Country, but may be adjusted by additions or deletions expressly authorized under this Contract. Unless otherwise stated herein the Domestic Contract Price shall not be adjusted for changes in costs or any other matters.

"**[Adjustment of Lump Sum Prices**": The Contractor may adjust the Overseas Lump Sum Prices and the Domestic Contract price, respectively, to reflect actual services rendered and materials supplied provided in no case shall the sum of the Overseas Lump Sum Prices plus the Domestic Contract Price exceed <<*insert currency amount*>> and the Contractor shall complete the Works and achieve the Actual Project Acceptance Date irrespective of their cost of achieving such.]

An adjustment for piling [may be][will be made] in the amount of <<*insert currency and amount*>> multiplied by the result (whether positive or negative) of <<*insert aggregate lengths of all piling expected*>> minus the aggregate length of all piles driven on the Site irrespective of their driven depth which adjustment shall be paid or refunded promptly upon completion of piling work;

As full consideration to the Contractor for the full and complete performance of the Work (including the Preliminary Services [and Pre NTP Work]) and all costs incurred in connection therewith, the Sponsor shall pay, and the Contractor shall accept, the sum of <<*insert currency and amount*>> (the "Base Price"), as such sum may be adjusted in the remainder of this Article 10 or pursuant to any Price Adjustment incorporated in a Change Order (the "Contract Price"). Until the Pre NTP Notice Date, the Base Price shall escalate (if at all) on the following basis:

(a) if the Pre NTP Notice Date occurs before <<*insert date*>>, the Base Price shall be multiplied by the First Period Escalator;

(b) if the Pre NTP Notice Date occurs after <<*insert date*>> but before <<*insert date*>>, the Base Price shall be multiplied by the Second Period Escalator; or

(c) if the Pre NTP Notice Date occurs after <<*insert date*>> but before <<*insert date*>>, the Base Price shall be multiplied by the Third Period Escalator.

Should any adjustment be made to the Base Price pursuant to the preceding sentence, the result shall be referred to as the "Pre NTP Adjusted Base Price." If the Pre NTP Notice Date and the Work Commencement Date do not occur on the same Day, the Base Price or the Pre NTP Adjusted Base Price, as the case may be, shall be multiplied by the Interim Escalator. [Should the Pre NTP Notice Date occur after <<*insert number*>>, the Parties shall attempt in good faith to agree upon a reasonable increase to the Contract Price using the formulas set forth above as a basis for discussions.]

"**Adjusted CPI**" means the percentage change in the <<Consumer Price Index for All Urban Consumers (seasonally adjusted) for all items less food and energy as listed in "Table A Percentage changes in CPI for Urban Consumers (CPIU) Seasonally adjusted" under Expenditure Category "All Items less food and energy" in the column "Unadjusted 12 mos. ended ____" as most recently announced prior to the Pre NTP Notice Date in the Consumer Price Index Summary issued monthly by the Bureau of Labor Statistics of the U.S. Department of Labor which can currently be found on the world wide web at http://www.bls.gov/news.release/cpi.nws.htm>>.

"**Elapsed Days**" is the number of Days that have elapsed between <<*insert date*>> and the Pre NTP Notice Date.

"**First Period Escalator**" shall be:

(a) Adjusted CPI multiplied by the Elapsed Days divided by 365; plus

(b) [100] percent.

"**Interim Days**" is the number of Days that have elapsed between the Pre NTP Notice Date and the Work Commencement Date.

"**Interim Escalator**" shall be:

(a) Adjusted CPI multiplied by the Interim Days divided by 365; plus

(b) 100 percent[, (unless the Interim Days exceed 365, in which case 365 above shall be changed to [730])].

"**Pre NTP Adjusted Base Price**" is defined in [Article 8.1].

"**Second Period Escalator**" shall be:

(a) Adjusted CPI multiplied by the Elapsed Days divided by 365; plus

(b) 100 percent.

"**Third-Period Escalator**" shall be:

(a) Adjusted CPI multiplied by the Elapsed Days divided by 365; plus

(b) 100 percent.

Choice B

"As full consideration to the Contractor for the full and complete performance of the Work (including the Preliminary Services) and all costs incurred in connection therewith the Sponsor shall pay, and the Contractor shall accept, the sum of <<*insert currency and amount*>> (the "<u>First Currency Base Price</u>") and <<*insert currency and amount*>> (the "<u>Second Currency Base Price</u>"), as such sum may be adjusted in this Article 8.1.1 or pursuant to any Price adjustment incorporated in a Change Order or pursuant to a directed change under Article 3 (the "<u>Contract Price</u>"). If the Sponsor has directed a change pursuant to Article 3, the undisputed cost therefor shall not be added to the Contract Price but shall be paid in <<*insert currency*>> to the Contractor on a percentage completion basis therefor until any Adjustment related thereto is resolved and the Sponsor shall pay any such amounts due pursuant to invoices rendered no more often than monthly. Provided that the Sponsor is permitted to convert <<*insert currency*>> into <<*insert currency*>> and there is sufficient availability of <<*insert currency*>>, the Contractor shall have the right to elect (by indicating such in the relevant Payment Application) that the Sponsor pay up to <<*insert number*>> percent of any portion of the First Currency Base Price which is due in <<*insert currency*>>. If the Contractor makes such an election, the relevant portion of the Final Price invoiced in the relevant Payment Application will be converted from <<*insert currency*>> into <<*insert currency*>> at the Exchange Rate on the date of payment by the Sponsor. If the Work Commencement Date occurs after <<*insert number*>>, the First Currency Base Price shall be multiplied by the <<*insert currency*>> Escalator (the "<u>Revised <<*insert currency*>> Base Price</u>"). Whenever any portion of the Second Currency Base Price is due hereunder, such portion thereof shall be adjusted by multiplying it by the <<*insert currency*>> Escalator as of the date corresponding to such Milestone Payment on the Milestone Payment Schedule (the "<u>Provisional Second Currency Price</u>"); provided that once all Indicators have been announced for a date which is after the date of calculation for a Second Currency Escalator, the Revised <<*insert currency*>> Escalator shall be calculated and such portion of the Second Currency Base Price paid in connection with such <<*insert currency*>> Escalator shall be multiplied by the Revised Second Currency Escalator and if the result thereof is different from the Provisional Second Currency Price, then either the amount by which such product exceeds the Provisional Second Currency Price shall be paid by the Sponsor to the Contractor pursuant to the next invoice in respect of a Payment Application or the amount by which such product is less than the Provisional Second Currency Price shall be paid by the Contractor to the Sponsor within <<*insert number*>> Days.

"<u><<*insert currency*>></u>**Elapsed Days**" shall be the number of Days, if any, that have elapsed between <<*insert date*>> and the Work Commencement Date.

"**Dollar Escalator**" means _____ multiplied by the <<*insert currency*>> Elapsed Days divided by [365].

"**Final Dollar Price**" shall be the <<*insert currency*>> Base Price if such is in effect on the Work Commencement Date or the Revised <<*insert currency*>> Price if such is in effect on the Work Commencement.

"<u><<*insert currency*>></u> **CPI**" shall be _____.

"<u><<*insert currency*>></u> **PPI**" shall be _____.

"<u><<*insert currency*>></u> **CPI**" shall be _____.

"<u><<*insert currency*>></u> **PPI**" shall be _____.

"**Indicators**" shall mean all of <<*insert currency*>> PPI, <<*insert currency*>> CPI, <<*insert currency*>> CPI, <<*insert currency*>> PPI, <<*insert currency*>> CPI and <<*insert currency*>> PPI.

"**<<*insert currency*>> CPI**" shall be the <<_____>> Consumer Price Index for the [year] announced by the Main Statistical Office of the Government of Poland.

"**<<*insert currency*>> PPI**" shall be the _____ Producer Price Index for the [year] announced by the Main Statistical Office of the Government of _____.

"**Revised <<*insert currency*>> Based Price**" is defined in Article 8.1.1.

"**Revised <<*insert currency*>> Escalator**" relating to any <<*insert currency*>> Escalator already calculated shall be:

0.6RCPI + 0.4RPPI

where:

RCPI	=	RCPIP (RCPIF + RCPIG)
RCPIP	=	the RCPIP Index on such date
RCPIF	=	the RCPIF Index on such date
RCPIG	=	the RCPIG Index on such date
RCPIP Index	=	the <<*insert currency*>> CPI on the date of payment of the Provisional <<*insert currency*>> Price relating to the <<*insert currency*>> Escalator in question as determined by linear interpolation based on two points, the first of which is the <<*insert currency*>> CPI most recently announced prior to the date of invoicing of the Provisional <<*insert currency*>> Price referred to above and the second of which is the <<*insert currency*>> CPI most recently announced after the date of invoicing of the Provisional <<*insert currency*>> Price referred to above.
RCPIF Index	=	the <<*insert currency*>> CPI on the date of payment of the Provisional <<*insert currency*>> Price relating to the <<*insert currency*>> Escalator in question as determined by linear interpolation based on two points, the first of which is the <<*insert currency*>> CPI most recently announced prior to the date of invoicing of the Provisional <<*insert currency*>> Price referred to above and the second of which is the <<*insert currency*>> CPI most recently announced after the date of invoicing of the Provisional <<*insert currency*>> Price referred to above.
RCPIG Index	=	the <<*insert currency*>> CPI on the date of payment of the Provisional <<*insert currency*>> Price relating to the <<*insert currency*>> Escalator in question as determined by linear interpolation based on two points, the first of which is the <<*insert currency*>> CPI most recently announced prior to the date of invoicing of the provisional <<*insert currency*>> Price referred to above and the second of which is the <<*insert currency*>> CPI most recently announced after the date of invoicing of the provisional <<*insert currency*>> Price referred to above.
RPPIG Index	=	the <<*insert currency*>> PPI on the date of payment of the Provisional <<*insert currency*>> Price relating to the <<*insert currency*>> Escalator in question as determined by linear interpolation based on two points, the first of which is the <<*insert currency*>> PPI most recently announced prior to the date of invoicing of the Provisional <<*insert currency*>> Price referred to above and the second of which is the <<*insert currency*>> PPI most recently announced after the date of invoicing of the provisional <<*insert currency*>> Price referred to above.

| RPPIF Index | = | the <<*insert currency*>> PPI on the date of payment of the Provisional <<*insert currency*>> Price relating to the <<*insert currency*>> Escalator in question as determined by linear interpolation based on two points, the first of which is the <<*insert currency*>> PPI most recently announced prior to the date of invoicing of the Provisional <<*insert currency*>> Price referred to above and the second of which is the <<*insert currency*>> PPI most recently announced after the date of invoicing of the Provisional <<*insert currency*>> Price referred to above. |
| RPPIP Index | = | the <<*insert currency*>> PPI on the date of payment of the Provisional <<*insert currency*>> Price relating to the <<*insert currency*>> Escalator in question as determined by linear interpolation based on two points, the first of which is the <<*insert currency*>> PPI most recently announced prior to the date of invoicing of the Provisional <<*insert currency*>> Price referred to above and the second of which is the <<*insert currency*>> PPI most recently announced after the date of invoicing of the Provisional <<*insert currency*>> Price referred to above. |

"<<*insert currency*>> **Base Price**" is defined in Article 8.1.1.
"<<*insert currency*>> **Escalator**" on any date shall be:
0.6CPI + 0.4PPI

where:

CPI	=	CPlp (CPIF + CPIG)
CPIP	=	the CPlp Index on such date
CPIF	=	the CPlf Index on such date
CPIG	=	the CPIG Index on such date
CPIP Index	=	<<*insert currency*>> CPI on the date most recently announced prior to the date of the <<*insert currency*>> Escalator being calculated divided by <<*insert currency*>> CPI on the Contract Date.
CPIF Index	=	<<*insert currency*>> CPI on the date most recently announced prior to the date of the <<*insert currency*>> Escalator being calculated divided by <<*insert currency*>> CPI on the Contract Date.
CPIG Index	=	<<*insert currency*>> CPI on the date most recently announced prior to the date of the <<*insert currency*>> Escalator being calculated divided by <<*insert currency*>> CPI on the Contract Date.
PPI	=	PPlp (PPIF + PPIG)
PPIP	=	the PPlp Index on such date
PPIF	=	PPIF Index on such date
PPIG	=	PPIG Index on such date
PPIP Index	=	<<*insert currency*>> PPI on the date most recently announced prior to the date of the <<*insert currency*>> Escalator being calculated divided by <<*insert currency*>> PPI on the Contract Date.
PPIF Index	=	<<*insert currency*>> PPI on the date most recently announced prior to the date of the <<*insert currency*>> Escalator being calculated divided by <<*insert currency*>> PPI on the Contract Date.
PPIG Index	=	<<*insert currency*>> PPI on the date most recently announced prior to the date of the <<*insert currency*>> Escalator being calculated divided by <<*insert currency*>> PPI on the Contract Date.

Choice C

"8.1.1 Provided that the Sponsor is permitted to convert <<*insert currency*>> into [currency] and there is sufficient availability of [currency], the Contractor shall have the right to elect (by indicating such in the relevant Payment Application) that the Sponsor pay up to [__] percent of any portion of the Contract Price which is due in <<*insert currency*>> be paid in [currency]. If the Contractor makes such an election, the relevant portion of the Final <<*insert currency*>> Price invoiced in the relevant Payment Application will be converted from <<*insert currency*>> into [currency] at the Exchange Rate on the date of payment by the Sponsor.

8.1.2 [At any time up to and including the Work Commencement Date, upon written notice to the Contractor (a "[currency] Notice"), the Sponsor may elect to pay up to __ percent of the Final <<*insert currency*>> Price in [currency] and the percentage of the Final <<*insert currency*>> Price so elected to be paid in [currency] shall be referred to as the "Elected [currency] Percentage."]

8.1.3 If a [currency] Notice has been given, whenever any portion of the Contract Price is due hereunder in accordance with the Milestone Payment Schedule, in order to:

(a) determine the [currency] component thereof, the Final <<*insert currency*>> Price shall be multiplied by the Elected [currency] Percentage which product shall be multiplied by the percentage of the Final <<*insert currency*>> Price related to such milestone payment (set forth on Exhibit L) which product shall be multiplied by the Election Day Exchange Rate which product shall be multiplied by the [currency] Factor; and

(b) determine the Dollar component thereof, the Final <<*insert currency*>> Price shall be multiplied by the <<*insert currency*>> Percentage which product shall be multiplied by the percentage of the Final Dollar Price related to such Milestone Payment (set forth on Exhibit L).

"[currency] Factor" for any Milestone Payment in question shall be (rate) x where:

x = the number of Days elapsed between the date of the [currency] Notice and the Work Commencement Date plus the number of Days listed next to the relevant milestone to which the Milestone Payment relates, the sum of which addition shall then be divided by 365.

"[currency] Notice" is defined in Article 8.1.

"**CPI**" shall be the [country] Consumer Price Index announced by _____.

["**<<*insert currency*>> Elapsed Days**" shall be the number of Days, if any, that have elapsed between <<*insert date*>>, and the Work Commencement Date].

["**Dollar Escalator**" shall be:

(a) [3] percent multiplied by the Dollar Elapsed Days divided by 365; plus
(b) [100] percent.

"<u>*<<insert date>>*</u>**Percentage**" shall be 100 percent minus the Elected [currency] Percentage.

"**Elected[currency] Percentage**" is defined in Article 8.1.

"**Election Day Exchange Rate**" shall be the Exchange Rate in effect on the date the [currency] Notice is sent to the Contractor.

"**Exchange Rate**" for any Day shall be the number of [currency] that can be purchased for one *<<insert currency>>* according to the rate announced therefor by the National Bank of [country]."

Choice D

[Should the Thermal Customer in connection with the Thermal Interconnection Facilities require Equipment beyond that included or reasonably implied therefor in Technical Specifications (such as, for example additional pumps, pipes or valves), the Sponsor shall, if the Contractor, on an open book basis, can demonstrate that it will incur additional labor and Equipment costs, add to the Final *<<insert currency>>* Price one hundred percent of the cost thereof up to a maximum of *<<insert currency and amount>>* and fifty percent of the cost thereafter up to a maximum of *<<insert currency and amount>>*, and all costs beyond *<<insert currency and amount>>* shall be borne and paid entirely by the Contractor and under no circumstances shall the Contractor be entitled to request an Adjustment therefor].

Notes

1 Tax planning in many jurisdictions may deem it propitious that this contract will be split into an on-shore and an off-shore agreement in which case a "coordination" agreement should be employed to "unite" the separate contracts in order to preserve the intent of seeking an EPC contractor as the sole party responsible for all the Work. The coordination agreement will, among other things, provide that neither contractor may claim an excuse from performance under its contract as a result of the other contractor's actions or inaction. At a minimum, the coordination agreement should be signed by a parent/entity of both contractors and ideally, if tax considerations permit, by both contractors as well.

2 "Between" is grammatically proper if there are only two parties to the contract.

3 "Among" is grammatically proper if there are more than two parties to the contract.

4 See Volume I, Chapter 2, "Choosing a Legal Entity."

5 See Volume I, Chapter 3, "The EPC Contractor's Approach."

6 For expansion projects only.

7 Including this phrase is slightly to the Sponsor's disadvantage because the Contract is unlikely to contain all the requirements for the Facility and some requirements may have to be surmised from industry custom. See Volume I, Chapter 6, "Intent."

8 See Volume I, Chapter 19, "Eligible Banks" and "Bank Guaranties." In addition to, or as an alternative for a list, object criteria could be proposed as a selection method if candidates cannot be identified to the Parties' satisfaction. The Sponsor should evaluate whether or not it wants to impose a continuing ratings requirement on the Contractor that the Acceptable Issuer, Acceptable Bondsman, and/or Parent Guarantor maintains the rating each held on the date that the Contract is signed because if Financing Parties are involved, and the Required Rating is not maintained, this default under this Contract would likely also have been made an event of default under the Sponsor's loan agreements that could cause Financing Parties to stop lending. This issue would likely not arise if this "continuing" ratings requirement were not imposed.

 9 See Volume I, Chapter 19, "Performance Bonds." In addition to, or as an alternative for a list, object criteria could be proposed as a selection method if candidates cannot be identified to the Parties' satisfaction.

10 Five different levels of details are commonly used for project management schedules. In general, a Level I schedule is the least detailed of the levels and shows only the main activities necessary to complete a project. A Level II schedule shows the scope of work to achieve the schedule's milestone events. A Level III schedule shows the deliverables to achieve the scope of work. These three levels are the summary levels. A Level IV schedule shows the tasks needed to complete the deliverables, and at Level V the tasks will be fully resourced. A Level I schedule is typically one or two pages, a Level II schedule about 50 activities, and a Level III schedule about 250 pages. A Level V detail for a large project can be many thousands of activities.

11 See Volume I, Chapter 11, "Developing the Baseline Project Schedule."

12 See Volume I, Chapter 4, "Formal Bid Solicitations."

13 For electric transmission lines, pipelines, railroads, highways, etc.

14 See Volume I, Chapter 12, "Change Orders."

15 As drafted, this definition includes industry boards and similar bodies and most EPC contractors will resist such an all-encompassing definition.

16 See Volume I, Chapter 3, "Consortia."

17 This concept can be used instead of "Site" but is typically only used when the Contractor is responsible for expanding an Existing Plant or the Facility is only one part of a much larger project.

18 See Volume I, Chapter 11, "Developing the Baseline Project Schedule."

19 See Volume I, Chapter 11, "Schedule Updates and Progress Reports."

20 Add any other utility providers.

21 See Volume I, Chapter 5, "The Environmental Impact Study."

22 See Volume I, Chapter 5, "The Environmental Impact Study."

23 See Volume I, Chapter 5, Table 5.2.

24 For expansion projects only.

25 For expansion projects only.

26 For expansion projects only.

27 For expansion projects only.

28 For expansion projects only.

29 See Volume I, Chapter 4, "Financing."

30 See Volume I, Chapter 12, "Force Majeure."

31 Not typical.

32 Not typical.

33 See Volume I, Chapter 4, "Functional Analysis" and Volume I, Chapter 9, "The Functional Specification."

34 Public utilities typically are required to observe very high standards.

35 See Volume I, Chapter 6, "Prudent Practices."

36 See Volume I, Chapter 8, "Hazardous Materials."

37 Add other facilities if other products like desalinated water, etc.

38 See Volume I, Chapter 18, "Hidden or Latent Defects."

39 See Volume I, Chapter 18, "Hidden or Latent Defects."

40 See Volume I, Chapter 5, "Limited Notice to Proceed."

41 See Volume I, Chapter 5, "Notice to Proceed."

42 See Volume I, Chapter 19, "Parent Guarantors."

43 See Volume I, Chapter 19, "Parent Guarantors."

44 See Volume I, Chapter 11, "Schedule Updates and Progress Reports."

45 See Volume I, Chapter 8, "Site or Route Survey."

46 The <Site> <Route> Assessment can also be described as the survey that an experienced contractor would undertake to confirm that the Work can be performed for the Contract Price in accordance with the Baseline Project Schedule. See Volume I, Chapter 8, "Site or Route Survey."

47 The Sponsor Information usually contains the relevant technical information and exhibits from the project agreements but not the provisions containing the tariff that the Sponsor will earn under the Offtake Agreement or the all-in project cost (i.e., EPC price plus development and financing costs) so that the Contractor cannot figure out the Sponsor's profit margins.

48 See Volume I, Chapter 7, "Subcontractors."

49 The Warranty Notification Period can be subdivided into different periods to cover different portions of the work (such as foundations or custom software). It is common for warranties on sophisticated items such as combustion turbines to run from arrival at site and not commencement of commercial operation because vendors do not want to take the risk that their equipment is not installed promptly. Sometimes different units or trains, etc. will have different warranty notification periods if they are placed into service at different times.

50 Contractors typically request "sunset" provisions. It is also common to have separate warranty periods for separate items like software.

51 Move into alphabetical order depending upon name of currency.

52 This paragraph can be modified at the Contractor's request to include Prudent Utility Practices or utility standards or independent power producer standards because Contractors may think this language is too open-ended. Utility standards are generally better for the Sponsor than independent power producer standards because utilities often opt redundancy as a result of high statutorily mandated reliability concerns. Independent power producers may be more concerned about lowering costs (because they cannot pass them along to the rate payers as a public utility may be able to do).

53 It is crucial to understand that generally the Functional Specifications are prepared by the Sponsor and the Sponsor's Engineer, and the Functional Specifications are not a laundry list of parts but functional in nature and specify the general requirements of the Facility such as the output required, its capability to burn certain types of fuel, the load points at which it can operate (i.e., full load, 75 percent load), the type of technology, and the general quality of materials and standards. It is simply impossible to list all the items that are required to build the Facility at the time this Contract is executed because the Facility has not been designed yet. The Functional Specifications do not and should not contain design documents, just the Facility's design requirements. It is the Contractor's job to design the Facility according to what the Sponsor has specified. In other words, it is like a buyer telling a dealer I want a car with 200 horsepower that gets 35 miles per gallon during city driving, will pass the air omissions tests, have antilock brakes, power windows, etc. but the buyer doesn't design the car, order the components from the vendors thereof or assemble and deliver it. If the Sponsor has actually provided any preliminary design drawings (as is sometimes the case), the Contractor should expressly represent and warrant that it will not rely on or use such documents in any way but perform its own design engineering, otherwise the Contractor will have an excellent basis to escape the Performance Guarantees in this Contract. The Functional Specifications are usually not (and in fact cannot be) detailed enough to specify all details of the Facility, which is why it is important to leave in the Contract the language about the Contractor supplying everything implied in the Contract. A good example is usually arguments in cases where the Contractor designs a system with one pump or three valves and the Sponsor reviews the design and says we want two pumps or 10 valves and the Functional Specifications do not address that level of detail. For this reason, the Contractor should never be entitled to request an Adjustment for a dispute over the Functional Specifications, otherwise the Contractor could fail to employ due care in design and then request an Adjustment if the Sponsor objects to a design. It is the Contractor's job to review the Functional Specifications before signing the Contract and then object or forever hold its peace after its signature.

54 This provision should only be included if the Sponsor will be vigilant in reviewing the Work while it is being performed otherwise the Sponsor may be held to have waived its rights to insist that time is, in fact, of the essence.

55 See Volume I, Chapter 6, "Additional Terms."

56 See Volume I, Chapter 6, "Philosophy of the EPC Contract."

57 See Volume I, Chapter 5, "Notice to Proceed."

58 The three Days are here to allow the Sponsor to give Notice to Proceed, which will be required by the Financing Parties to achieve Financial Closing if the Sponsor is financing the Project, but allow the Contractor to be sure that Financial Closing has occurred and Financing Parties' funding is available before commencing the Work.

59 Sometimes the Contractor may not be able to obtain an executed Parent Guarantee from its parent until this Contract is executed.

60 See Volume I, Chapter 9, "Sourcing of Equipment."

61 See Volume I, Chapter 7, "Subcontractors" and Chapter 13, "Subcontractor Difficulties."

62 The Contractor will usually request a threshold, which is a good compromise; the Sponsor should not be too concerned with Subcontractors unless there is special equipment or technology required for the Facility.

63 Use only if the Contractor requests. Could include at the Contractor's request. If you are representing the Financing Parties, remind them to check creditworthiness of major Vendors.

64 This Article 4.2.2, except for clause (b), could be deleted if the Contractor insists because it is not really practical for the Sponsor to police this clause as there may be hundreds or thousands of Subcontractors.

65 See Volume I, Chapter 7, "Third-Party Beneficiaries."

66 This warranty assignment to the Sponsor is typically not carried out in practice unless there is special equipment or technology to be included in the Facility (such as in the case of a solar power plant).

67 This clause is not typical unless there is a Subcontract that is central to the Contractor's successful performance.

68 New York court cases have held that the inclusion of this clause can obligate the Contractor to obtain the prerequisites and requisites to carry out the Work (such as licenses, etc.).

69 Only use if Environmental Report exists.

70 See Volume I, Chapter 8, "Site or Route Survey."

71 Bear in mind that the Contractor usually must make deposits on major equipment with its purchase awards therefor.

72 To Sponsor's disadvantage but reasonable to include. This addresses the problem that not all Interconnection Requirements will be listed in the Functional Specifications or even codified by local utilities. Here the Contractor may be worried about the requirements of local utilities, which often (outside the United States) are not written; for example, requirements of water, sewage and telephone companies specifying depth of pipes, etc., or grid connections requiring multiple tie-ins. Adding this provision will entitle the Contractor to make an Adjustment Claim in the case that these requirements change under the Change in Law regime herein.

73 See previous footnote.

74 What goes here depends upon whether or not spare parts are included in the Contract Price and whether the Sponsor is financing its project to purchase spare parts under the EPC Contract because the Sponsor will be borrowing some percentage of their purchase price instead of paying for them in full if they are purchased as an operating cost, because "long term" lenders usually finance only initial capital costs. This financing of spare parts that the Sponsor desires to have on hand when operations first begin should improve the project's internal rate of return.

75 Usually the outline of this training program is put into the Functional Specifications but sometimes it can be a separate exhibit.

76 See Volume I, Chapter 9, "The EPC Contractor's Training Obligations during Commissioning."

77 Sometimes "nominal" liquidated damages are attached to non-delivery of these design documents; however if these damages do not bear a reasonable relation to the actual damages estimated to be suffered, this provision may not be enforceable and theoretically could invalidate the entire contract.

78 List other contractors like environmental consultants, etc.

79 One could add start up fuel, back up fuel, water, limestone or Consumables (such as lubrication oil in the case of a reciprocating engine performance test on a ship, etc.) The idea here is that the Sponsor pays for all this fuel and may or may not be receiving electricity revenues for the "infirm" power generated during the testing period but the Contractor has no incentive not to "burn up" fuel if the Contractor is not paying for it.

80 Use if there is a grid system operator involved.

81 This gives the Sponsor time to notify the Financing Parties' Engineer, whom the Financing Parties will want present.

82 Use only if Offtake Agreement requires.

83 See previous footnote.

84 See Volume I, Chapter 9, "The EPC Contractor's Obligations during the Design Phase."

85 See Volume I, Chapter 9, "Supervision."

86 Design work and component assembly often will take place away from the Site.

87 See Volume I, Chapter 9, "Security and Surveillance."

88 The Contractor should be running a Site that has proper safety training and procedures and should not allow anyone who is not familiar with these on the Site (even Sponsor personnel) and the Contractor should be required to remove anyone not properly indoctrinated from the Site. Generally, insurance will cover injury to all personnel (whoever employs them) on the Site so long as the Contractor maintains proper procedures.

89 See Volume I, Chapter 9, "Quality Control."

90 Sometimes this program is put in a separate exhibit.

91 Only applicable for coal-fired power plants.

92 See Volume I, Chapter 11, "Monitoring the Progress of the Work."

93 A cost loaded schedule is only appropriate where payment is made based on progress of the Work (i.e. percentage complete of individual Work activities); where (as in the standard form of this agreement) payment is based on a Milestone Payment Schedule, references in this article to cost loading of schedule activities are unnecessary.

94 The Contractor is probably less likely to be accountable for any further delays if the Sponsor has approved the Contractor's recovery plan.

95 It may not be desirable for the Sponsor to approve the Remedial Plan because the time that the Sponsor takes to review the Remedial Plan may subject the Sponsor to a claim of Owner delay or even operate to excuse the Contractor of certain obligations if the Remedial Plan is not successful. See previous footnote.

96 Note that a Change in Law endorsement to the construction all risk insurance can be purchased. The definition of Law could be drafted so that there will be no Adjustment for a Change in Law outside the country where the Site is located (or a Change in Law outside the state if the Facility is in the United States [but could be drafted to permit a change in federal law) because the Sponsor does not (usually) tell the Contractor from whom to purchase its Equipment or in what country the Equipment should be manufactured.

97 To Sponsor's disadvantage, but reasonable to include.

98 This Article is necessary (i) if ECA funding will be availed of because financing eligibility is based upon where Equipment is designed, manufactured, and assembled before shipping and (ii) for assessing the premiums under marine cargo insurance (because insurers must know the value of the largest water shipment and how much Equipment is being shipped by water) and construction all risk insurance which usually covers ground transportation.

99 See Volume I, Chapter 9, "Sourcing of Equipment."

100 This Article can be deleted if the Sponsor is not interested in shipping, but insurers often require prior notice and inspection of marine cargo and certification of the classification by a classification society of the ship's condition before shipping to see that cargo is properly packed and the vessel is properly maintained and equipped or coverage on the shipment may be voided.

101 See Volume I, Chapter 9, "Shipping and Delivery Schedules."

102 See Volume I, Chapter 9, "Union Labor."

103 See Volume I, Chapter 9, "Coordination with Other Contractors."

104 Generally, if the Sponsor is financing its project, it is to the advantage of the Sponsor to purchase spare parts under the EPC Contract because the Sponsor is borrowing to

pay the Contract Price and thus also financing the spare parts, which normally are not financed by vendors on attractive terms. See Volume I, Chapter 21.

105 Not typical, except occasionally in cases of relatively smaller contractors or very large projects given the contractor's undertakings in the contract as compared to its size.

106 The Contractor may want to clarify what is meant by "competitor."

107 See Volume I, Chapter 11, "Rights of Inspection."

108 Make sure that the Functional Specifications do not include provisions for review that are inconsistent with the rest of this Contract. Often, Contractors propose to attach to the Contract a list of the only documents that may be reviewed by the Sponsor, but the Sponsor may want to resist this approach because, with thousands of documents, lists could eschew important documents and then the Sponsor might have the opportunity to review documents.

109 From a practical point of view, the most grave errors can occur during the design phase of a Project, so the Sponsor should have rights to critique or even reject design documents, but bear in mind that if the Sponsor requires design changes despite the Contractor's objections, there is a serious risk that the Contractor could be excused from the Performance Guarantees because the Contractor may argue that the change prevented their achievement—from a legal point of view, it is not desirable for the Sponsor to have rejection rights, but practically, it is important to have these rights. Errors that occur in the erection phase can be simply fixed mechanically by disassembly, but, if, for example, the coal handling system is improperly sized in the design stage, the Project will never operate properly once it is built and the only solution is to redesign, demolish and reconstruct that system, all of which could take years.

110 See Volume I, Chapter 11, "Project Accounting and Audit."

111 See Volume I, Chapter 10, "The Sponsor's Obligations Under The EPC Contract."

112 See Volume I, Chapter 14 ["'No Damages for Delay' Provisions"].

113 Coordinate this with the Sponsor's obligations to give the Fuel Supplier and Fuel Transporter notice of when it wants deliveries to begin.

114 Only for coal-fired power plants.

115 See Volume I, Chapter 14, "The Owner's Failure to Discharge Its Responsibilities."

116 Add this only if the Contractor requests but it is better to leave it out to give the Sponsor more freedom to administer the Contracts.

117 Not typical.

118 This section may be combined with Article 5.28, "Coordination with Other Contractors" above.

119 This section may be combined with Article 5.28, "Coordination With Other Contractors" above.

120 See Volume I, Chapter 15, "The Owner's Right To Suspend or Terminate the Work."

121 See Volume I, Chapter 15, "The Owner's Right to Suspend or Terminate Work."

122 See Volume I, Chapter 13, "Adequate Assurance."

123 Very unusual.

124 See Volume I, Chapter 12, "Directed Changes."

125 Contractors like to cap these changes to make sure that the Sponsor can afford them. This approach protects lenders too because it prevents a Sponsor from directing a change in the Work (which it will not be able to do under a properly drafted loan agreement anyway). Another compromise would be to require the Sponsor to provide adequate evidence to the Contractor that it has the creditworthiness to afford the payment for the Change Order.

126 Not typical and not necessary if the Contractor makes its relationship with the Sponsor exclusive for the Facility.

127 See Volume I, Chapter 16, "Testing and Completion of the Work."

128 This definition assumes the Unit or Facility has been commissioned by the Contractor and it is now ready to undergo the Performance Tests. (For most facilities, this is more than mechanical completion because the Unit or Facility has already been "hot" tested and/or synchronized with the Grid.)

129 Add other products if any (such as desalinated water).

130 If this clause is not included, it implies that either there is to be no Punch List or that all work on the Punch List has been completed.

131 Add only if the Sponsor Information contains testing requirements, etc.

132 Inclusion of this clause is not typical.

133 Many Contractors will not take the risk that a third party (such as the Offtaker) can decide whether or not the Contractor has performed under the contract, but it is preferable not to relieve the Contractor from paying Delay Damages until this condition has been met. Contractors may accept this type of condition if it is clearly objective in terms of testing requirements, but not if it is open to the subjective discretion of the Offtaker.

134 Inclusion of this clause is not typical.

135 Inclusion of this clause is not typical.

136 See Volume I, Chapter 16, "Final Acceptance."

137 See Volume I, Chapter 16, "The Punch List." Occasionally, the Punch List is divided into two or more categories corresponding to disputed and undisputed items and if undisputed items are valued at less than a certain amount the Contractor can achieve Provisional Acceptance despite the dispute.

138 This provision is not typical but it is sometimes included in projects in very remote areas that will have no value whatsoever if their performance is seriously impaired.

139 See Chapter 16, "Operating Revenues during the Early Operations Period."

140 If there is to be a bonus for early completion, it could be inserted in this Article. (See Volume I, Chapter 17, "Early Completion Bonuses").

141 See Volume I, Chapter 16, "Early Operation."

142 Note that the builder's all risk insurance may no longer offer coverage once the Facility is operated commercially. See Volume I, Chapter 23 "Builder's Insurance."

143 Preferably, Liquidated Damages are payable in a so-called "hard" currency or indexed to a "hard" currency because they may not come due for years after the Contract Date, and inflation, devaluation or exchange rates can all have a serious impact on the value of these liquidated payments.

144 It may be necessary under the circumstances to create a concept of a "Unit Two" Work Commencement Date, which may be tied to the Contract Date and/or the Work Commencement Date for Unit One.

145 See previous footnote.

146 Since liquidated damages are not usually tied to Commissioning Completion unless the Sponsor has to pay liquidated damages under the Offtake Agreement for missing this milestone (which is rare because the milestone is usually substantial completion under the Offtake Agreement) or the Sponsor wants to operate the Unit or the Facility early, the Contractor will always request this type of relief. It is OK to grant it here but not at Provisional Acceptance because at that point the Sponsor needs to be able to fund its debt service reserve accounts and cover its principal loan repayments and the liquidated damages must cover this and the Sponsor's loss of equity return in addition to liquidated damages payable to the Offtaker under the Offtake Agreement.

147 Note that this provision must be coordinated with the Contract Article 8.10 above covering the Punch List.

148 Not very typical. If the Contractor desires to provide for an escrow arrangement, the Contractor would be best served by providing that the escrow agreement signing by the Sponsor is a condition to the Sponsor's ability to give Notice to Proceed.

149 Unless the Sponsor expects that the EPC contractor will not hold its price quotation firm or there is limited notice to proceed engineering work to be done, there is little reason for the Sponsor to execute this Contract until the day before the Sponsor determines to proceed with its project or Financial Closing in the case that the Sponsor is financing its project, because once the Sponsor locks in the Contract Price, the only way to improve the deal is to lower financing costs.

150 Please see riders for different price options for indexation and foreign currencies.

151 See Volume I, Chapter 21, "Payment for the Work."

152 This factor will be greater than "1" and represent an administrative "markup" for the Contractor on the Assigned Agreements.

153 This is usually included if the Contract has been negotiated on an "open book" basis, in which the Contractor shows the Sponsor all its costs and then they agree on the Contract Price. It is also helpful to include an exhibit like this so that if the Contractor runs into trouble or for some valid reason requests a Price Adjustment, the Sponsor will have some basis to assess how much additional money the Contractor is should receive.

154 See Volume I, Chapter 21, "Payment for the Work."

155 See Volume I, Chapter 21, "Payments for Unresolved Change Orders."

156 Use only if the Contractor is a consortium.

157 Use only if Pre NTP Work or Site Assessment to be paid for by the Sponsor if Notice to Proceed is not given by a certain date.

158 Note that the definition of Contract Price includes Change Orders so Price Adjustments are picked up here and do not need to be referred to separately.

159 Use only if payment is to be made for Pre NTP Work or the Site Assessment or preliminary design.

160 See Volume I, Chapter 21, "Spare Parts."

161 Sometimes, for some parts on the list, a "takeout" price may be furnished. This provides the Sponsor with the opportunity to remove that part from the list of spare parts for the Facility, and, accordingly, the Contractor will reduce the Contract Price by that amount.

162 See Volume I, Chapter 21, "Milestone Payments."

163 Note: if invoices come at different times during the month, the Sponsor will have to give more than one borrowing request under its loan agreement, which may not be permitted under the loan agreement.

164 See Volume I, Chapter 21, "Retainage."

165 If the Sponsor has financed its project, it may not be allowed to borrow or make payments more than once per month, so this period may be too short.

166 See Volume I, Chapter 21, "Release of Retainage."

167 See Volume I, Chapter 21, "Payments Withheld."

168 Not typical in lump sum projects except for any part of the Work that was not part of the lump sum price such as so-called "provisional sums."

169 See Volume I, Chapter 21, "The Right to Verify Progress."

170 See Volume I, Chapter 21, "Disputed Payments."

171 See Volume I, Chapter 21, "Payment Currency."

172 This addresses the fact that in some countries local companies can only pay each other legally in their own currency. If the Sponsor is domiciled in the local jurisdiction and so is the Contractor or a Consortium Member, the Consortium Member or the Contractor may not be permitted to pay Liquidated Damages to the Sponsor in another currency, so either the liquidated damages must be indexed to a "hard" currency or such payment must be the responsibility of a Consortium Member that can pay in the desired currency. See footnote above.

173 This provision is not typical and see Volume I, Chapter 21, "Suspension of Milestone Payments."

174 See Volume I, Chapter 21, "Special Payment Arrangements for Export Credit Agencies."

175 See Volume I, Chapter 21, "Special Payment Arrangements for Export Credit Agencies."

176 Not typical in very large projects.

177 Often, outside the United States, bank guaranties are given instead of a letter of credit. Generally, an on-demand bank guaranty is not preferable to a letter of credit. A Contractor might suggest that an on-demand guaranty is the same as a letter of credit, and also will cost less than a letter of credit. An on-demand guaranty will cost less because it is not as good as a letter of credit. A guaranty assumes that an underlying obligation exists. A letter of credit can be drawn if the proper document is presented, and there are no defenses as there might be with a guarantee, which is subject to the law of suretyship in most common law jurisdictions. A bond from a bonding company is even less desirable—the Sponsor may simply end up with the need to bring a lawsuit against the bondsman in addition to the Contractor.Bear in mind that as the result of Change Orders the Contract Price may change, and therefore fixing the numeric amount of the Performance LC in the Contract text may be less desirable than stating its amount as a percentage of the Contract Price. See Volume I, Chapter 19, "Performance Bonds."

178 See Volume I, Chapter 19, "Parent Guarantees."
179 Use a guarantee from each parent if the Contractor is a consortium or joint venture, but make sure each guarantee is joint and several for the entire Contract Price and performance of the Contract.
180 Sponsor guaranties are not so typical in financed projects because lenders usually are making a large percentage of the funds available if the conditions contained in their loan agreement to each borrowing are met.
181 See Volume I, Chapter 12, "Unforeseen Circumstances, Adjustments and Change Orders."
182 In most projects, the Contractor is more often than not the Asserting Party.
183 Under Article 14.2, the Contractor has to rebuild the Facility if it is destroyed because it has the risk of loss, so it is necessary to make clear that it cannot submit an Adjustment Claim for these costs.
184 The concept here is that, if the Contractor needs more time to build the Facility, nothing can change that. However, if a natural disaster requires the Contractor to spend more money, it is a commercial decision who bears this risk. This issue goes both ways in the current market. Another compromise would be entitling the Contractor to a Price Adjustment for events occurring in the country where the Site or Route is located, but not in the foreign jurisdictions where the Contractor chose to manufacture equipment. Note that an extension to the construction all risk insurance can be purchased for force majeure events and the Contractor can purchase this coverage.
185 See Volume I, Chapter 8, "Site or Route Survey."
186 The Contractor generally needs about 60 to 90 days to conduct the Site Assessment, and then time to prepare an Adjustment for time and price if it finds a sub-surface problem. Most Contractors do not want the Sponsor to give notice to proceed until the Adjustment is granted. This is basically O.K. because if the Sponsor doesn't like the result, it does not have to give Notice to Proceed and could exercise its optional termination rights under Article 13. In fact, Financing Parties usually won't permit the Sponsor to give Notice to Proceed anyway because a severe Site or Route problem could raise the project cost beyond the Sponsor's earning capability, and if construction has already started, the only way for the Financing Parties to get their investment back would be to complete the project at a yet unknown cost which could be far beyond the contingency in the pro forma upon which they based their decision to extend credit to the Sponsor. It is always best to have the Contractor carry out the Site Assessment because the Contractor won't be able to "blame" unforeseen conditions on the Sponsor.
187 Sometimes the Contractor is given *force majeure* relief for earthquakes before permanent structures are complete (since they may not yet be reinforced properly).
188 See Volume I, Chapter 8, "Hazardous Materials."
189 Generally, the Contractor is the best-placed party to carry out cleanup because the Sponsor typically does not have the resources or the knowledge to do so and will not want to risk causing delays (which might result from a new contractor arriving on the Site to dispose of the material). It is usually preferable to have the Contractor deal with this type of problem. Most contractors will accept this burden (or, at least, a threshold value below which the Contractor will handle the problem). If the Contractor is still not agreeable, it may be agreeable to be responsible for "simple" materials like oil spills and asbestos, but not radioactive isotopes or heavy metals, etc. Also note that the definition of "Hazardous Materials" includes Antiquities whose discovery can cause significant delays in some cases.
190 See Volume I, Chapter 22, "Representations and Warranties."
191 If either Party is a governmental agency, the other Party should determine whether or not the governmental agency has any type of legal immunity and whether the governmental agency has the power to waive such immunity. Sometimes, in addition to a formal written waiver, the governmental agency might give a warranty that its breach, in connection with the Contract, is not protected by any immunity.
192 The Contractor cannot represent to more than a visual investigation if it has not yet performed a soil survey the Site or Route to conduct the Site or Route Assessment, which is most often the case.

193 If the Site or Route Assessment has been performed, this Article should be modified to include any obstacles in the underground conditions mentioned therein, etc., and the Sponsor should review the environmental report and Site or Route Assessment to modify this Article appropriately and those reports should be attached as exhibits to this Contract.

194 See Volume I, Chapter 18, "The EPC Contractor's Warranties of the Work."

195 See Volume I, Chapter 18, "Remedies for Breach of General Warranty."

196 Not typical and probably not necessary because a warranty covering the Work can likely be purchased from the person rendering the Work and the cost thereof will be charged back to the Contractor.

197 This is generally not a favorable clause for the Sponsor because it may be hard to gauge the value of a service that was not properly performed or, worse, the replacement cost of an item may be inconsequential compared to the damage or disruption it has made or caused.

198 See Volume I, Chapter 18, "Subcontractor Warranties."

199 The Sponsor may want to consider buying an extended warranty directly from certain Vendors in certain cases, so appropriate provisions would need to be drafted into this clause.

200 See Volume I, Chapter 18, "Scheduling of Warranty Work."

201 See Volume I, Chapter 18, "Limitation of Warranties."

202 See Volume I, Chapter 18, "Warranty Exclusions."

203 The manuals should not call for excessive preventative maintenance.

204 Not typical.

205 This Article should always be reviewed by the Sponsor's insurance adviser.

206 See Volume I, Chapter 12, "Risk of Loss."

207 The construction all risk insurance will cover damage as a result of negligent acts of the Sponsor, and the Contractor will therefore be protected, but the Sponsor could offer to pay any deductible. The Contractor is responsible for security and safety on Site, and it should remove any Person that is a menace, no matter who employs such Person.

208 See Volume I, Chapter 12, "Risk of Loss."

209 Delete if the Contractor is purchasing construction all risk and marine cargo insurance.

210 Many contractors will insist on this, which shifts the risk from them to the Sponsor for events like war that are are uninsurable and for which the Contractor would not have insurance proceeds to cover the loss.

211 See Volume I, Chapter 22, "Title and Other Legal Matters."

212 Sometimes this is appropriate for taxation or importation regulations.

213 It is possible to use an exhibit instead of, or in addition to, this Article—see Exhibit 14.6. This Article should always be reviewed by the Sponsor's insurance advisor.

214 See Volume I, Chapter 23, "Indemnification and Insurance."

215 See Volume I, Chapter 23, "Third-Party Liability Insurance."

216 This "gap" allows the Sponsor not to put insurance in place until it is needed and, therefore, save money.

217 See Volume I, Chapter 23, "Builder's Insurance."

218 This allows the Sponsor time to put the insurance in place.

219 See Volume I, Chapter 23, "Marine Cargo Insurance."

220 Policy coverage should always be, at least, value of largest single shipment.

221 See Volume I, Chapter 23, "Automobile Insurance."

222 See Volume I, Chapter 23, "Advance Loss of Profits (Delayed StartUp)."

223 See Volume I, Chapter 23, "Environmental Cleanup Insurance."

224 See Volume I, Chapter 23, "Aircraft Insurance."

225 This is crucial because the Sponsor's Advance Loss of Profits Insurance or Delayed Start Insurance will not respond unless the Construction All Risk Insurance is the primary policy.

226 See Volume I, Chapter 23, "Deductibles."

227 See Volume I, Chapter 23, "Cooperation and Claims."

228 This is crucial because if a material fact such as unproven technology is not disclosed to the insurer, it may escape liability under the policy by the terms of the policy or common law.

229 This should generally not be given to the Contractor because it may raise the Sponsor's rates.

230 See Volume I, Chapter 23, "Indemnification and Insurance."

231 The Contractor may want to limit the indemnity to the Sponsor instead of all Sponsor Persons, which is generally acceptable because any Sponsor Person is free to sue the Sponsor for its loss and then the Sponsor can in turn seek recovery from the Contractor.

232 The Contractor will try to delete these words, which is generally acceptable because all this will be covered by insurance anyway.

233 In New York state, indemnities usually must be specific about what they cover in order to be enforceable.

234 This provision protects Sponsor Persons that have losses in a very inflationary currency because they might pay out for the loss long before they are reimbursed by the Contractor.

235 See Volume I, Chapter 13, "The Concept of Default."

236 See Volume I, Chapter 13, "The Concept of Default."

237 Not typical.

238 The Contractor will usually insist that this date be no earlier than the date upon which the Delay Damages would be exhausted if the Contractor has run into delays.

239 This provision may not be desirable if there are Financing Parties because their loan agreements often provide that a default under the EPC contract is also an automatic default under the loan agreements if the Sponsor decides to "ignore" or waive the Sponsor's rights against the Contractor for the Contractor's invoking this protection, so the Sponsor should consider carefully what defaults are included in the EPC contract.

240 Not typical.

241 See note 5.

242 Not typical.

243 Not typical.

244 See Volume I, Chapter 13, "The Owner's Remedies" and "The Owner's Options."

245 The Contractor may insist some Articles can't be terminated, such as Article [___] on dispute resolution.

246 The Contractor may want to modify the foregoing provisions so that they become the exclusive remedy of the Sponsor. The Sponsor may not want to accommodate this request (although in U.K. law–governed EPC contracts, it is common to assent to this request) because the remedies listed are the most typical remedies but are not usually exhaustive of those available under the law, and if there are other remedies under law the Sponsor may want to preserve them. (Of course, under Article 18 the Contractor's liability is nonetheless limited except as provided therein.)

247 If the Sponsor is financing the construction of its Facility, the Financing Parties will want to see as few default events for the Sponsor as possible. (This is obviously true for all agreements in a project financing.) The fewer defaults there are for the Sponsor to trigger, the less risk lenders have of losing the EPC contract which is their ultimate security for the Facility's construction and proper performance. The Sponsor's only significant obligations under an EPC contract are generally to pay money and purchase insurance. Even if the Sponsor is insolvent, the lenders may continue to pay the Contractor Milestone Payments because, if the project remains half built, the lenders will *never* recover their loans. Also, the Sponsor often will trigger its own defaults under any Offtake Agreement that it has if it does not deliver service or product on time. Therefore, lenders may still pay the Contractor promptly so no construction delays occur as a result of non-payment. Keep in mind that, no matter what a construction contract says, on a practical level, as soon as a contractor or equipment vendor is not paid, it usually stops working or shipping and this will likely cause delays.

248 See Volume I, Chapter 14, "The Owner's Failure to Discharge Its Responsibilities."

249 The Sponsor may want to protect itself in the case it must cease construction as a result of litigation against it or its Permits from authorities, non-governmental groups, or citizens.

250 See Volume I, Chapter 20, "Limitations on Overall Liability and Contract Expiration."

251 This Article does not contain any cap on Facility Buy Down Amounts or Unit Buy Down Amounts because these are paid *only* at the Contractor's election under this Contract in Article 7 if the Contractor has met the minimum performance levels and therefore it can excuse itself from the guaranteed levels. This is different from most EPC contracts under which the Sponsor can assess Buy Down Amounts, and therefore the Contractor needs to have a cap on these amounts. However, in all cases the Contractor *never* has the benefit of a cap for performance (as opposed to delay) damages if it has not met the minimum levels because in that case, the facility is in such bad shape it usually cannot operate well enough to even repay the Financing Parties. This Contract does not have an overall Liquidated Damages cap because the Buy Down Amounts are paid only at the Contractor's option. If you change the Contract so that the Sponsor can assess Buy Down Amounts, it is fair to cap Delay Damages and Buy Down Amounts at 30 percent of the Contract Price in the aggregate. Typically, Delay Damages and Buy Down Amounts are capped at 30 percent of the Contract Price, which means that if Delay Damages equal to 20 percent of the Contract Price were assessed by the Sponsor, the Contractor would never be liable for Buy Down Amounts in excess of 10 percent of the Contract Price. Sometimes a Contractor will request subcaps. For example, a 20 percent subcap on Delay Damages would mean that the Sponsor could never assess Delay Damages exceeding 20 percent of the Contract Price—even if there were no Buy Down Amount to be assessed because the Facility has met all the guaranteed points. Also keep in mind that as the result of Change Orders (or the Sponsor's electing optional work if the Contract contemplates such), the Contract Price may change so any caps should always be stated as percentages of the Contract Price and not as a fixed amount.

252 See Volume I, Chapter 20.

253 Bear in mind that unlike during the original contractual negotiation of the Contract in which the Sponsor has asked contractors to bid a Contract Price for the Work and not examined how much the Contractor expected to earn as profit on the Work, now that the Contractor has executed the Contract and asks for a Price Adjustment, the Contractor could try to "re-trade" the "deal" that it signed. Therefore, a good Adjustment mechanism requires a Contractor to list *all* its new costs so the Sponsor can review them. This is why Adjustments are done on a costs plus profit–like general contracting and not a lump sum basis. This is also why a very burdensome and detailed Form of Adjustment Claim is attached to this Contract. If the Contractor wants more money, it must identify every item that it will be using and who will be putting them together for how long.

254 See Volume I, Chapter 12, "Equitable Adjustments."

255 This is highly undesirable for the Contractor, especially if the delay causes a more pro-tracted delay than the first delay or a delay pushes the Contractor into a season in which it cannot perform Work or into a Permit "blackout" period during that period.

256 Another issue to consider here is whether a Change Order affecting the Contract Price should increase the stated amount under the Performance LC. This Contract does not currently deal with this rubric. A fair compromise here may be that the Performance LC need not be increased if the Change Order is (a) less than a threshold amount when aggregated with all other Change Order amounts, or (b) the result of a Sponsor action or failure to perform hereunder, or (c) a Sponsor directed change.

257 See Volume I, Chapter 24, "Dispute Resolution and Governing Law."

258 The Contract and the Parent Guarantee should be governed by and at the same proceed-ing and under the same law too.

259 See Volume I, Chapter 24, "Mutual Discussions."

260 If ultimate parents are not the Parties, an undertaking letter from them that they agree to comply with this provision may be desirable.

261 See Volume I, Chapter 24, "Arbitration." Often, EPC contracts contain a provision that disputes of a "technical nature" be resolved by a technical expert. This is not advisable because there are no rules, procedures, etc. established and what is "technical" can become subject to an arbitration first if the parties do not agree and prolong the process further.

262 For arbitration information in the United States, go to www.adr.org.

263 This scheme is usually only used in very large or complex projects. See Volume I, Chapter 24, "Dispute Resolution Boards."

264 The decision to employ a one- or three-member Dispute Review Board will depend on the cost, complexity and duration of the Facility, as well as the Parties' commitment to the process and sensitivities to the costs associated with the process.

265 See Volume I, Chapter 24, "Pendency of Disputes."

266 Generally not necessary for U.S. deals.

267 This is not typical.

268 Delete if English law.

269 Make sure there are no other agreements like extended maintenance between the parties that should not be canceled, etc.

270 Usually Letters of Intent are signed by different parties because the Sponsor and the Contractor were not formed yet.

271 See Volume I, Chapter 7, "Third-Party Beneficiaries."

272 See Volume I, Chapter 13, "Remedies at Law vs. Remedies in Equity" and "Exclusivity of Remedies."

273 Make sure to carve out arbitration if it is used.

274 See Volume I, Chapter 24, "Litigation."

275 Only use if no Party is domiciled in New York.

276 Rule 9 of the Rules of Practice for the Commercial Division of New York state courts allows parties to agree to have a Commercial Division lawsuit heard on an expedited basis. The rule permits parties to agree in a contract that any lawsuit arising out of or related to the contract will be heard under the Commercial Division's "accelerated adjudication" process. The accelerated adjudication rule sharply limits discovery and requires that a case be ready for trial within nine months. The rule also deems both parties to have waived important rights (including the right to any interlocutory appeal and the right to a jury trial). If the parties agree to accelerated adjudication, all pre-trial proceedings, including all discovery, pre-trial motions and mandatory mediation, must be completed and the parties ready for trial within nine months from the date of filing of a Request of Judicial Intervention. There is no requirement, however, that the court actually hold the trial within a certain time frame. Unless the parties agree otherwise, discovery is limited to seven interrogatories, five requests to admit and seven depositions per side. Document requests must be "restricted in terms of time frame, subject matter and persons or entities to which the requests pertain," and electronic discovery must be "narrowly tailored to include only those individuals whose electronic documents may reasonably be expected to contain evidence that is material to the dispute." Additionally, "where the costs and burdens of e-discovery are disproportionate to the nature of the dispute or the amount in controversy, or to the relevance of the materials requested, the court will either deny such requests or order disclosure on condition that the requesting party advance the reasonable cost of production to the other side, subject to the allocation of costs in the final judgment." Parties that agree to have disputes heard as accelerated actions are "deemed ... to have irrevocably waived":

- any objections based on lack of personal jurisdiction or forum non conveniens;
- the right to a jury trial;
- the right to recover punitive damages; and
- the right to interlocutory appeal.

This means that a defendant cannot appeal from the denial of a motion to dismiss and neither party can appeal from a temporary restraining order, preliminary injunction, or other interim relief.

277 Delete if arbitration.

278 Refer to notice provision.

279 See Volume I, Chapter 24, "Language."

280 Coordinate this with Article 19.23.

281 Use only if Sponsor Information is supplied or the agreements concerning the Project are attached as an exhibit to this Contract in order to reflect the likelihood that these

agreements often are amended over time and the Contractor cannot agree to be bound by changes in these agreements which would change the Contractor's scope unless the Contractor has a right to an Adjustment for a change in a project agreement. However, it is better for the Sponsor to direct the Contractor to make a change under Article 3.5 for any amendment to the project agreements and then let the Contractor then make an Adjustment Claim rather than have the Contractor decide what changes to make and then request an Adjustment.

282 English law does not have a concept a contract is construed against its draftsman.

283 Use only if Project Manager under Article 7.3.

284 See Volume I, Chapter 25, "Assignment of the EPC Contractor."

285 It is not advantageous for the Contractor to permit this.

286 See Volume I, Chapter 4.

287 This time period is important because these items will be a condition precedent to the first drawing under the loan agreement of the Financing Parties so the Contractor should not be able to delay Financial Closing (which the Contractor may not particularly worry about because the Contract Price may already be escalating).

288 See Volume I, Chapter 22, "Legal Opinions."

289 Although confidentiality provisions are often written for the benefit of both Parties, the Sponsor is purchasing the Work and unless there are technological secrets embodied in the Work (such as a gas liquefaction process), the Sponsor may want to avoid placing confidentiality provisions on itself. The public disclosure of the Contract Price may, however, be more sensitive for the Contractor. See Volume I, Chapter 25, "Publicity, Confidentiality and Proprietary Information."

290 See Volume I, Chapter 25, "Publicity, Confidentiality and Proprietary Information."

291 See Volume I, Chapter 25, "Indiscreet Conduct."

292 Add other desired legislation such as the U.K. Bribery Act.

293 Although not a legal requirement, counterparts are probably best avoided and it is recommended that every single page of the Contract and its exhibits be hand-initialed by all signatories in order to avoid unnecessary disputes as to which version of the contract is the final execution version.

294 See Volume I, Chapter 12, "Equitable Adjustments."

295 It may be to Sponsor's benefit to have Financing Parties' agent approve, in which case add signature line for agent.

296 See Volume I, Chapter 18, "The EPC Contractor's Warranties of the Work."

297 It is not advisable to use any city that is not a relatively stable and established banking jurisdiction.

298 See Volume I, Chapter 7.

299 This exhibit will contain either a copy of the site assessment if it has already been performed or the scope if it is to be performed. It is best to let the Contractor decide how to actually do the assessment in terms of what subcontractors to hire and what equipment and methods to use; otherwise, the Contractor will claim that the Sponsor is to blame if the Contractor finds a problem after the cutoff date for an Adjustment Claim under Article 19.

300 Note that the title company may have its own form of Indemnity Agreement. If that is the case, then Contractor should sign the form of indemnity provided by the title company. In any other case, use this form.

301 As an alternative, cancellation costs can be separately broken down in particular work packages.

302 Probably best to leave this in.

303 Only use if the Contractor requests and gives an Availability Guarantee.

304 Only used if the Contractor gives an Availability Guarantee.

305 The idea here is to make sure that the manuals do not call for an excessive number of cleanings, inspections or lubricant applications or replacements.

306 One could change this test to one for the Facility instead of each Unit if the Contractor really objects.

307 Contractors usually prefer to measure Facility availability instead of Unit availability, but by Unit is more precise for the Sponsor because it evaluates the performance of each unit individually to detect problems.

308 Offset damages between Units only if the Contractor requests.

309 Typically used in ECA financing or "open book" pricing where Contractor provides all component prices.

310 The Exhibit should always be carefully read and revised to reflect the deal payment structure, particularly if ECAs are involved because they usually require the borrower (Sponsor) to pay the equipment vendor (not the Contractor) whose equipment they were financing directly and not have the Sponsor pay the Contractor who pays the vendor. This way they are sure the proceeds of their loans go to the right indigenous party whose operations their country wants them to finance.

311 May be required if an ECA or Vendor is involved (if Sponsor purchased a turbine before signing the Contract) because ECAs usually can only make loans if the proceeds go directly to the equipment vendor of their home country and vendors do not want their money flowing through the EPC contractor because they do not want to be exposed to more credit risk than they have to by needing to consider the EPC contractor's credit standing.

312 To be used only if [insert currency] Notice given.

313 See Volume I, Chapter 19, "Standby Letters of Credit."

314 For New York law.

315 See Volume I, Chapter 19, "Parent Guarantors."

316 This may need to be adapted if the Contractor has Consortium Members.

317 It is preferable for Sponsor not to include this unless the Sponsor and the Parent Guarantor have entered into other agreements between themselves apart from this Parent Guarantee.

318 It is preferable for Sponsor not to include this unless the Sponsor and the Parent Guarantor have entered into other agreements between themselves apart from this Parent Guarantee.

319 Only necessary if the guarantor is not a New York entity.

320 Use arbitration if that is what governs EPC Contract and joinder and consolidation provisions in the EPC Contract to make sure the Parent Guarantor is joined in any Dispute under the Contract.

321 It is best for Sponsor not to include this because the EPC Contract itself usually will not have a defined term. Many Parent Guarantors will typically limit the duration of and cap their liability under the Parent Guaranty.

322 This provision is not typical unless the EPC contractor is relatively small or the project is very large.

323 English law form.

324 See Volume I, Chapter 23.

325 See Volume I, Chapter 4, "Financing."

326 Not typical.

327 This Consent and Agreement is prepared under New York law. Financing Parties will generally require this Consent and Agreement to be governed by New York law [or English law, for which this form would have to be modified] or the law that governs their loan agreement if different.

328 Always conform this section to Contract provisions for law.

329 Use if the Contractor has Consortium Members.

330 This is better for the Sponsor in case it wants to start construction before it obtains financing or if it decides not to obtain financing. But note it is always better to use a sliding date tied to a moving end point and not a calendar date.

331 This is better for the Sponsor in case it wants to start construction before it obtains financing or if it decides not to obtain financing. But note it is always better to use a sliding date and not a calendar date.

Supplemental Provisions for Certain Types of Projects

Power Plant Specific Provisions[1,2,3]

1. <u>A Minimum Net Electrical Output Test</u> that comprises the sum of the power output generated by the combustion turbines and steam turbines, measured at the high side of the transformers and corrected to guarantee conditions. Testing will be performed without operation of the evaporative coolers for both power plants and without supplemental duct burners in operation for [_____], but with full operation of supplemental duct burners for [_____], which is consistent with the operational basis of the guarantees. When the auxiliary loads required by equipment that must be in operation at the test conditions exceed the auxiliary loads necessary at the guarantee conditions, the difference in kW will be added to the Net Electrical Output. Acceptance criteria for the Minimum Net Electrical Output test is <<95>> percent of Guaranteed Net Electrical Output.
2. <u>A Maximum Net Plant Heat Rate Test</u> that compares the ratio of the plant heat input, to the Net Electrical Output. Heat input is calculated as the sum of all the fuel flows entering the test boundaries, multiplied by the heating value of the fuel stream. Minimum acceptance criteria for the Maximum Net Plant Heat Rate Test is <<105>> percent of the Guarantee Net Plant Heat Rate. Testing will be performed without operation of the evaporative coolers and with supplemental duct burner operations described for the net electrical output test.
3. <u>An Emissions Guarantee Test</u> will be performed to demonstrate that during the thermal performance tests, environmental and noise emissions levels can be achieved within the guaranteed limits. The Emissions Test (emission stack test) is to be performed on each CTG prior to or concurrent with the net output and heat rate tests. The Emission Test is to be performed with the HRSG duct burners in service for [_____] (steam cycle phase). The acceptance criteria shall be that the stack emissions are within the requirements of and do not exceed emission guarantees in the respective EPC Contracts. Emission testing methodology is further evaluated in the Environmental Consultant's Report.
4. <u>Demonstration Performance Tests (Required for Substantial Completion)</u>. As a prerequisite to achieving Facility Provisional Acceptance, the [_____] Contractor will conduct two Demonstration Performance Tests, demonstrating that the Facility can operate continuously at 50 and 75 percent of base load. These tests are to be conducted in accordance with Offtake Agreement requirements.

"**Unit 60% Load Gas Btu Consumption Guarantee**" means [___].
"**Unit Heat Rate Buy Down Cap**" shall be the sum of:

(a) the difference between the Unit Heat Rate achieved during the Performance Tests when determining if the guarantee set forth in Article 7.2.4 has been met minus _____ kJ/kWh multiplied by $_____/kJ/kWh; plus

(b) the difference between the Unit Heat Rate achieved during the Performance Tests when determining if the guarantee set forth in Article 7.2.5 has been met minus _____ kJ/kWh multiplied by $_____/kJ/kWh; plus

(c) the difference between the Unit Heat Rate achieved during the Performance Tests when determining if the guarantee set forth in Article 7.2.6 has been met minus _____ kJ/kWh multiplied by $_____/kJ/kWh;

(c) the Facility Net Electrical Output guaranteed in Article <<*insert number*>>, the Contractor need:

  (i) not pay any Facility Buy Down Amount in respect thereof if the Contractor has already paid to the Sponsor Unit Buy Down Amounts in excess of such Facility Buy Down Amount for the Units' failure to achieve the Unit Net Electrical Output guaranteed for each of them under Article <<*insert number*>>; or

  (ii) only pay a Facility Buy Down Amount in respect thereof in the amount by which the Facility Buy Down Amount therefor exceeds the sum of all Unit Buy Down Amounts paid to the Sponsor for each of Unit One and Unit Two failing to achieve the Unit Net Electrical Output guaranteed for each of them under Article <<*insert number*>>;

(d) the Facility Net Electrical Output guaranteed in Article <<*insert number*>>, the Contractor need:

  (i) not pay any Facility Buy Down Amount in respect thereof if the Contractor has already paid to the Sponsor Unit Buy Down Amounts in excess of such Facility Buy Down Amount for the Units' failure to achieve the Unit Net Electrical Output guaranteed for each of them under Article <<*insert number*>>; or

  (ii) only pay a Facility Buy-Down Amount in respect thereof in the amount by which the Facility Buy-Down Amount therefor exceeds the sum of all Unit Buy-Down Amounts paid to the Sponsor for each of Unit One and Unit Two failing to achieve the Unit Net Electrical Output guaranteed for each of them under Article <<*insert number*>>;

(e) the Facility Heat Rate guaranteed in Article 7.2.1, the Contractor need:

  (i) not pay any Facility Buy Down Amount in respect thereof (ignoring the provisions for reduction thereof in Article <<*insert number*>> if the Contractor has already paid to the Sponsor Unit Buy Down Amounts in excess of such Facility Buy Down Amount for failure of the Units to achieve the Unit Heat Rate guaranteed for each of them under Article <<*insert number*>>; or

  (ii) only pay a Facility Buy Down Amount in respect thereof in the amount by which the Facility Buy Down Amount therefor exceeds the sum of all Unit Buy Down Amounts paid to the Sponsor for each of Unit One and Unit Two failing to achieve the Unit Heat Rate guaranteed for each of them under Article <<*insert number*>>;

(f) the Facility Heat Rate guaranteed in Article <<*insert number*>>, the Contractor need:
   (i) not pay any Facility Buy Down Amount in respect thereof (ignoring the provisions for reduction thereof in Article <<*insert number*>> if the Contractor has already paid to the Sponsor Unit Buy Down Amounts in excess of such Facility Buy Down Amount for failure of the Units to achieve the Unit Heat Rate guaranteed for each of them under Article <<*insert number*>>; or
   (ii) only pay a Facility Buy Down Amount in respect thereof in the amount by which the Facility Buy Down Amount therefor exceeds the sum of all Unit Buy Down Amounts paid to the Sponsor for each of Unit One and Unit Two failing to achieve the Unit Heat Rate guaranteed for each of them under Article <<*insert number*>>; or
(g) the Facility Heat Rate guaranteed in Article <<*insert number*>>, the Contractor need:
   (i) not pay any Facility Buy Down Amount in respect thereof (ignoring the provisions for reduction thereof in Article <<*insert number*>> if the Contractor has already paid to the Sponsor Unit Buy Down Amounts in excess of such Facility Buy Down Amount for failure of the Units to achieve the Unit Heat Rate guaranteed for each of them under Article <<*insert number*>>; or
   (ii) only pay a Facility Buy Down Amount in respect thereof in the amount by which the Facility Buy Down Amount therefor exceeds the sum of all Unit Buy Down Amounts paid to the Sponsor for each of Unit One and Unit Two failing to achieve the Unit Heat Rate guaranteed for each of them under Article <<*insert number*>>.

"Each of the Performance Guarantees and the Facility Minimum Performance Levels [and Unit Minimum Performance Levels for each Unit] is subject to the Facility [and each Unit] being operated under the conditions set forth in Exhibits 1G, 1H and 1K.
   ["**Facility Heat Rate Buy Down Cap**" shall be the sum of:

(a) the difference between the Facility Heat Rate achieved during the Performance Tests when determining if the guarantee set forth in Article <<*insert number*>> has been met minus <<*insert number*>> kJ/kWh multiplied by <<*insert currency and amount*>> /kJ/kWh; plus
(b) the difference between the Facility Heat Rate achieved during the Performance Tests when determining if the guarantee set forth in Article <<*insert number*>> has been met minus <<*insert number*>>kJ/kWh multiplied by <<*insert currency and amount*>>/kJ/kWh; plus
(c) the difference between the Facility Heat Rate achieved during the Performance Tests when determining if the guarantee set forth in Article <<*insert number*>> has been met minus <<*insert number*>> kJ/kWh multiplied by <<*insert currency and amount*>>/kJ/kWh].

   ["**Unit Heat Rate**" shall mean: <<*insert amount*>>].
   ["**Unit Heat Rate Buy Down Cap**" shall be the sum of:

(a) the difference between the Unit Heat Rate achieved during the Performance Tests when determining if the guarantee set forth in Article <<*insert number*>>

has been met minus <<*insert amount*>> kJ/kWh multiplied by <<*insert currency and amount*>>/kJ/kWh; plus

(b) the difference between the Unit Heat Rate achieved during the Performance Tests when determining if the guarantee set forth in Article <<*insert number*>> has been met minus <<*insert amount*>> kJ/kWh multiplied by <<*insert currency and amount*>>/kJ/kWh; plus

(c) the difference between the Unit Heat Rate achieved during the Performance Tests when determining if the guarantee set forth in Article <<*insert number*>> has been met minus <<*insert amount*>> kJ/kWh multiplied by <<*insert currency and amount*>>/kJ/kWh].

["**Unit Net Electrical Output**" for a Unit shall mean the electrical output of such Unit when operated on Fuel, exported to the Power Purchaser at <<*insert voltage*>>kV, measured in kilowatts at the high voltage side of the Unit's generator transformer when the other Unit is not operating].[4]

Gas-Fired Power Plants

["**CTG**" shall mean the dual fuel, natural gas and oil-fired combustion turbine generators and associated equipment as described in the <<Vendor Purchase Agreement>>.][5]

["**EOH**" shall have the meaning ascribed thereto in Article 7.6.]

[**Equivalent Operating Hours during Tests on Completion**. The number of equivalent operating hours ("**EOH**") for each gas turbine shall not exceed <<*insert number*>> EOH prior to Facility Provisional Acceptance. In the event Contractor exceeds such amount of EOH, Contractor shall pay to Sponsor [liquidated] damages as follows [_____].]

["**Gas Turbine Purchase Agreement**" means <<*refer to title of Agreement*>>.][6]

["**Gas Turbine Spare Parts**" means all parts of the gas turbine shaft excluding its generator. It includes items such as the gas turbine blades, vanes, installation material for blade and vanes, minor inspection consumables, control spares, bricks, brick holders, burners, inner casing, mixing chambers. It excludes the parts for the gas turbine auxiliaries (skids, inlet filter, etc.).][7]

"**Lower Heating Value**" shall mean that number of Btus produced by the complete combustion at a constant pressure of fourteen and six hundred ninety six thousandths pounds per square inch absolute (14.696 psia) of one cubic foot (1 cu. ft.) gas at sixty nine degrees Fahrenheit (69° F) with excess air at the same pressure and temperature as the said gas when the products of combustion are cooled to 60 degrees Fahrenheit (60° F), the water formed by such combustion remains in a gaseous state, and the products of such combustion contain the same total mass of water vapor as the said gas and air before combustion.

["**Turbine**" shall mean a <<*insert vendor model*>> turbine generator system, complete with all components including <<*insert equipment list*>>.]

["**Turbine Provisional Acceptance**" shall have the meaning ascribed thereto in [Article 8<<__>>].]

["**Turbine Provisional Acceptance Date**" shall mean the date on which <<*insert vendor*>> achieves Turbine Provisional Acceptance for a Turbine.]

Combined Cycle

"**Back Pressure**" means the pressure measured at the exhaust duct of each simple cycle combustion turbine of the Existing Plant, with the Existing Plant and the Combined Cycle Conversion operating in combined cycle mode at Base Load, expressed in millimeters of column of water (H_2O).

"**Back Pressure Verification Measurements**" means verification measurements (conducted in accordance with Exhibit [___]) for each simple cycle combustion turbine of the Existing Plant, while the Integrated Facility is operating at Base Load, to verify the Back Pressure.

"**Baseline Existing Plant Levels**" means the levels determined in the baseline testing conducted by [_____] and the Contractor of the Existing Plant to establish its and each combustion turbine's base load electrical output, exhaust mass flow, heat rate and noise and emissions levels, information to be attached as Exhibit [___].

"**By-pass Stacks Replacement**" means all the activities (e.g., engineering, procurement, demolition, removal, proper disposal, construction and commissioning) related to the replacement of the existing stacks of the two simple cycle generating units currently in operation in ([_____] and [_____]) with by-pass stacks including hydraulic diverter damper and guillotine. [_____] and [_____] are described in Exhibit [___].

["**Combined Cycle Conversion**" means machinery and apparatus intended to form or forming part of the Permanent Works wherever located necessary and desirable to add and Integrate a steam turbine to operate with <<one, two or all three>> of the combustion turbine units of the Existing Plant all as further described herein (including the Functional Specifications)].

"**Combined Cycle Conversion Contract**" is defined in the Preamble hereof and includes the Schedules, the Functional Specifications and the Exhibits and such documents as may be expressly incorporated herein.

"**Combined Cycle Performance Guarantee Level(s)**" means the following levels for the Integrated Facility, which shall be measured during the Combined Cycle Net Electric Output Test while the Works are operating in compliance with all applicable Laws and corrected to the reference conditions set forth in Exhibit [____]: (a) for Combined Cycle Net Electric Output at least <<*insert number*>>; and (b) [list other levels]. The foregoing values are based on a total net electric output of all gas turbines running a combined cycle of <<*insert number*>>, <<*insert number*>>; and <<*insert number*>>. If at Performance Tests the total net electric output of all gas turbines running a combined cycle varies, in plus or minus, then the Combined Cycle Net Electric Output indicated above will be adjusted proportionally.

"**Combined Cycle Taking-Over Levels**" means the following levels for the Integrated Facility demonstrated simultaneously during the Combined Cycle Net Electric Output Test and Back Pressure Verification Measurements, while the Integrated Facility is operating in compliance with all Applicable Laws of [_____] and Applicable Permits and corrected as set forth in Exhibit [___]: (a) no less than 98 percent of the Combined Cycle Performance Guarantee Levels for Net Electric Output, and (b) no more than <<*insert number*>> millimeters H_2O of Back Pressure.

["**CTG STG Purchase Agreements**" shall mean, collectively, the agreements between the Sponsor and <<*insert name*>> providing for the purchase of the CTGs and the STG.]

"**Combined Cycle Conversion**" means machinery and apparatus intended to form or forming part of the Work wherever located necessary and desirable to add and Integrate a steam turbine to operate with <<one, two or all three>> of the combustion turbine units of the Existing Plant all as further described herein (including the Functional Specifications).

"**Combined Cycle Performance Guarantee Level(s)**" means any or all of the following levels for the Integrated Facility, all of which shall be measured simultaneously during Combined Cycle Heat Rate Test and Combined Cycle Net Electric Output Test (as applicable) while the Works are operating in compliance with all applicable Laws and corrected to the reference conditions set forth in Exhibit 1Q:

(a) for Combined Cycle Net Electric Output at least <<*insert number*>> kW; and[8]
(b) for Combined Cycle Heat Rate a decrease below those levels set out in the Baseline Existing Plant Levels of at least <<*insert number*>> kJoule/kW.

"**Interface**" The Contractor guarantees that the Existing Plant Shutdown Hours will not exceed the following timeframes:

(a) for [_____] gas turbine generating unit, [__] hours.
(b) for [_____] gas turbine generating unit, [__] hours.
(c) for [_____] gas turbine generating unit, [__] hours.

By the first (1st) Day of the <<*fourth (4th)*>> calendar month after the Combined Cycle Notice to Commence, the Contractor shall notify the Sponsor in writing of the schedule of dates in which the above mentioned Existing Plant Shutdown Hours will be executed, for each generating unit, required for Contractor to perform the Work.

Contractor agrees that only one combustion turbine generating unit may be taken out of service at a time. The Contractor must give the Sponsor at least <<*inset number*>> Days advance written notice of each actual date for which the Contractor is requesting an Existing Plant Outage so the Sponsor and the Contractor can agree upon a schedule therefor.

The Contractor shall be allowed to schedule up to three (3) Existing Plant Outages per combustion turbine generating unit and which must be commenced at or after <<*insert number*>> p.m. <<*insert day of the week*>> and at or before <<*insert number*>> <<*insert day of the week*>>. If the Contractor requires Existing Plant Shutdown Hours in addition to those allowed in the preceding sentence, then the Contractor shall pay to [_____] an amount equal to <<*insert amount and currency*>> per hour for each Existing Plant Shutdown Hour in excess of the applicable allowed Existing Plant Shutdown Hours specified above.

The Contractor guarantees that the Existing Plant Shutdown Hours will not exceed the following timeframes:

		By-pass Stacks Replacement	Integration and Testing
(a)	for [_____] gas turbine generating unit, [__] hours.	<<*insert number*>> hours	<<*insert number*>> hours
(b)	for [_____] gas turbine generating unit, [__] hours.	<<*insert number*>> hours	<<*insert number*>> hours
(c)	for [_____] gas turbine generating unit, [__] hours.	Not applicable	<<*insert number*>> hours

(a) By-pass Stacks ReplacementRegarding the Existing Plant Shutdown Hours to implement the By-pass Stacks Replacement, the Parties agree that the related Existing Plant Outage for each combustion turbine generating unit [{_____}] must occur between [{_____}] and [{_____}]. The Existing Plant Shutdown Hours mentioned in the table above to implement the By-pass Stacks Replacement shall be utilized by the Contractor in only one Existing Plant Outage for each combustion turbine generating unit.If the Contractor requires Existing Plant Shutdown Hours in addition to those allowed in the table above, within the time-frame established in the preceding paragraph, then the Contractor shall pay to the Sponsor an amount equal to <<*insert amount and currency*>> per hour for each Existing Plant Shutdown Hour in excess. If the Existing Plant Shutdown Hours in addition to those allowed in the table above occur after [_____], then the Contractor shall pay to the Sponsor an amount equal to <<*insert amount and currency*>> per hour for each Existing Plant Shutdown Hour in excess.

(b) [the Thermal Interconnection Facilities have been completed and approved by the Sponsor in writing for compliance with the Functional Specifications [and the Sponsor Information;]

(c) Integration and Testing
The Contractor shall be allowed to schedule up to ten (10) Existing Plant Outages per combustion turbine generating unit for Integration and any other Existing Plant Outage required to conduct tests or to perform any other activity by the Contractor which is required to complete the Works. If the Contractor requires Existing Plant Shutdown Hours in addition to those allowed in the table above for Integration, then the Contractor shall pay to the Sponsor an amount equal to <<*insert amount and currency*>> per hour for each Existing Plant Shutdown Hour in excess if such excess occurs between [{_____}] and [{_____}].

If the Existing Plant Shutdown Hour in excess occurs between the first day of [{_____}] and the last calendar day of [{_____}], then the Contractor shall pay to the Sponsor an amount equal to <<*insert amount and currency*>> per hour for each Existing Plant Shutdown Hour in excess.

If the Contractor exceeds the number of scheduled Existing Plant Outages described in the first paragraph of this section (b), per combustion turbine generating unit of the Existing Plant, then the Contractor shall pay to the Sponsor an amount equal to <<*insert amount and currency*>> per hour per each additional Existing Plant Outage incurred by the Contractor. The Existing Plant Shutdown Hours in excess, if any, will be liquidated separately in addition to the described Existing Plant Outage compensation.

Cogeneration

"**Facility Buy Down Amounts**" are set forth in Article 7.2 and with respect to any particular Performance Guarantee shall be calculated in accordance with Article 7.2.

"**Facility Heat Rate**" shall mean:

(a) [3.7 multiplied by (the heat added to the steam and water of both Unit boilers (kW)) multiplied by 100 divided by the boiler efficiency measured as a percentage based on the net calorific value of Fuel according to DIN 1942; all divided by

(b) the sum of the electrical power (MW_e) measured at the high-voltage side of the generator transformer and delivered by both Units to the Site Electrical Interconnection Point and the thermal power (MW_{th}) from both Units at the Thermal Interconnection Point].[9]

"**Facility Minimum Performance Levels**" shall be met when the Facility is capable of achieving all of the following:

(a) [simultaneously] generating a Facility Net Electrical Output of _____ MW_e [and a Facility Net Thermal Output of ____ MW_{th}] at a Facility Heat Rate of no more than _____ kJ/kWh;

(b) [simultaneously] generating a Facility Net Electrical Output of _____ MW_e [and a Facility Net Thermal Output ____ MW_{th}] at a Facility Heat Rate of no more than _____ kJ/kWh;

(c) generating a Facility Net Electrical Output _____ MW_e at a Facility Heat Rate of no more than _____ kJ/kWh when operating in full condensing mode; and

(d) achieving each of the guaranteed levels set forth in Article 7.2.7.

"**Facility Net Electrical Output**" shall mean the electrical output of the Facility when operating on Fuel (net of Facility electrical consumption) at [110] kV, measured in MW_e at the Site Electrical Interconnection Point.[10]

["**Facility Net Thermal Output**" shall mean the thermal energy output of the Facility measured in MW_{th} at the Thermal Interconnection Point when the Facility is operating on Fuel.][11]

["**Facility Net Thermal Output**" shall mean the thermal energy output of the Facility measured in Btu at the Thermal Interconnection Point when the Facility is operating on Fuel.][12]

"**New Facility Minimum Performance Level**" shall be met when the New Facility has achieved[, when all chillers are operating at full load,] all of the following:

(a) simultaneously achieving both an Adjusted Facility Net Electrical Output of at least <<insert number percent >> of the New Facility Full Load Gas Electrical Output Guarantee and an Adjusted Facility Net Thermal Output of at least <<*insert number*>> of the Facility Full Load Gas Thermal Output Guarantee when operating on Gas during the Performance Tests at an Adjusted Facility Btu Consumption not in excess of <<*insert number*>> of the New Facility Full Load Gas Btu Consumption Guarantee;

(b) simultaneously achieving both an Adjusted Unit Net Electrical Output of at least <<*insert number*>> of the Unit One Full Load Oil Electrical Output Guarantee and an Adjusted Unit Thermal Output of at least <<*insert number*>> of the Unit Full Load Oil Thermal Output Reference Guarantee when operating on Oil during the Performance Tests at an Adjusted Facility Btu Consumption of no more than <<*insert number*>> of the Unit Full Load Oil Btu Consumption Guarantee;

(c) an Adjusted New Facility Btu Consumption of no more than <<*insert number*>> of the New Facility <<sixty percent (60)>> Load Gas Btu Consumption Guarantee during the Performance Tests at <<sixty percent (60)>> CTG and full STG load;

(d) each Unit has achieved the Unit Minimum Performance Level; and

(e) the New Facility and each Unit can meet the Emissions Guarantee during operations.

"**Unit Minimum Performance Level**" shall be met when a Unit has achieved all of the following:

(a) simultaneously achieving both an Adjusted Unit Net Electrical Output of at least ninety-seven percent (97%) of the Unit Full Load Gas Electrical Output Guarantee and an Adjusted Unit Thermal Output of at least ninety-seven percent (97%) of the Unit Full Load Gas Thermal Output Guarantee when operating on Gas during the Performance Tests at an Adjusted Unit Fuel Consumption of no more than one hundred and three percent (103%) of the Unit Full Load Gas Btu Consumption Guarantee;

(b) simultaneously achieving an Adjusted Unit Net Electrical Output of at least ninety-seven percent (97%) of the Unit Full Load Oil Electrical Output Guarantee when operating on Oil during the Performance Tests at an Adjusted Btu Consumption of no more than one hundred and three percent (103%) of the Unit Full Load Oil Btu Consumption Guarantee; and

(c) achieving the Emissions Guarantee.

"**Unit Minimum Performance Levels**" shall be met when the Unit in question is capable of achieving all of the following:

(a) [simultaneously] generating a Unit Net Electrical Output of ___ MW_e [and a Unit Net Thermal Output of ___MW_{th}] at a Unit Heat Rate of no more than _____ kJ/kWh;

(b) [simultaneously] generating a Unit Net Electrical Output of ___ MW_e [and a Unit Net Thermal Output of ___ MW_{th}] at a Unit Heat Rate of no more than _____ kJ/kWh;

(c) generating a Unit Net Electrical Output of ___ MW_e at a Unit Heat Rate of no more than _____ kJ/kWh when operating in full condensing mode; and

(d) achieving each of the guaranteed levels set forth in Article 7.2.7.[13]

The Contractor guarantees that the Facility[, when operating in full condensing mode,] shall be capable of generating a Facility Net Electrical Output of _____ MW_e at a Facility Heat Rate[14] of no more than _____ kJ/kWh. If the Facility has achieved each of the Facility Minimum Performance Levels but failed to achieve the performance guaranteed in the previous sentence, the Contractor must, in full satisfaction of its obligations under the previous sentence, make a payment to the Sponsor by the last Day of the Optional Facility Guarantee Cure Period calculated pursuant to the next two sentences. The Facility Buy Down Amount to be paid by the Contractor for the Facility failing to achieve the Facility Heat Rate of <<*insert number*>> kJ/kWh shall be equal to the product of:

(a) the difference (if positive) between the Facility Heat Rate achieved during the Performance Tests when determining if the guarantee set forth in this Article 7.2.1 has been met minus <<*insert number*>> kJ/kWh; multiplied by
(b) <<insert currency and amount>>/kJ/kWh.

The Facility Buy Down Amount to be paid by the Contractor for the Facility failing to achieve a Facility Net Electrical Output of <<*insert amount*>> MW$_e$ shall be equal to the product of:

(a) the difference (if positive) between <<*insert amount*>> MW$_e$ minus the Facility Net Electrical Output achieved during the Performance Tests when determining if the guarantee set forth in this Article 7.2.1 has been met; multiplied by
(b) <<insert currency and amount>>/kW.

The Contractor guarantees that the Facility shall be [simultaneously] capable of generating a Facility Net Electrical Output of <<*insert number*>> MW$_e$ [and a Facility Net Thermal Output of <<*insert number*>> MW$_{th}$] at Facility Heat Rate of no more than <<*insert number*>> kJ/kWh. If the Facility has achieved each of the Facility Minimum Performance Levels but failed to achieve the performance guaranteed in the previous sentence, the Contractor must, in full satisfaction of its obligations under the previous sentence, make a payment to the Sponsor by the last Day of the Optional Facility Guarantee Cure Period calculated pursuant to the next two sentences. The Facility Buy Down Amount to be paid by the Contractor for the Facility failing to achieve a Facility Heat Rate of <<*insert number*>> kJ/kWh shall be equal to the product of:

(c) the difference (if positive) between the Facility Heat Rate achieved during the Performance Tests when determining if the guarantee set forth in this Article 7.2.2 has been met minus <<*insert number*>> kJ/kWh; multiplied by
(d) <<insert currency amount>>/kJ/kWh.

The Facility Buy Down Amount to be paid by the Contractor for the Facility failing to achieve a Facility Net Electrical Output of <<*insert number*>> MW$_e$ shall be equal to the product of:

(a) the difference (if positive) between <<*insert number*>> MW$_e$ minus the Facility Net Electrical Output achieved during the Performance Tests when determining if the guarantee set forth in this Article 7.2.2 has been met; multiplied by
(b) <<insert currency and amount>>/kW.[15]

The Contractor guarantees that the Facility shall be capable of [simultaneously] generating a Facility Net Electrical Output of <<*insert number*>> MW$_e$ [and a Facility Thermal Output of <<*insert number*>> MW$_{th}$] at a Facility Heat Rate of no more than <<*insert number*>> kJ/kWh. If the Facility has achieved each of the Facility Minimum Performance Levels but failed to achieve the performance guaranteed in the previous sentence, the Contractor must, in full satisfaction of its obligations under the previous sentence, make a payment to the Sponsor the last Day of the Optional Facility Guarantee Cure Period calculated pursuant to the next two sentences. The Facility Buy Down Amount to be paid by the Contractor for the Facility failing to achieve a Facility Heat Rate of <<*insert number*>> kJ/kWh shall be equal to the product of:

(c) the difference (if positive) between the Facility Heat Rate achieved during the Performance Tests when determining if the guarantee set forth in this Article 7.2.3 has been met minus <<*insert number*>> kJ/kWh; multiplied by

(d) <<insert currency and amount>>/kJ/kWh.

The Facility Buy Down Amount to be paid by the Contractor for the Facility failing to achieve a Facility Net Electrical Output of <<*insert number*>> MW$_e$ shall be equal to the product of:

(a) the difference (if positive) between <<*insert number*>> MW$_e$ minus the Facility Net Electrical Output achieved during the Performance Tests when determining if the guarantee set forth in this Article 7.2.3 has been met; multiplied by

(b) <<insert currency and amount>>/kW.

The Contractor guarantees that [when operated in full condensing mode] each Unit shall be capable of generating a Unit Net Electrical Output of <<*insert number*>> MW$_e$ at a Unit Heat Rate of no more than <<*insert number*>> kJ/kWh. If a Unit has achieved each of the Unit Minimum Performance Levels but failed to achieve the performance guaranteed in the previous sentence, the Contractor must, in full satisfaction of its obligations under the previous sentence for that particular Unit, make a payment to the Sponsor the last Day of the Optional Facility Guarantee Cure Period calculated pursuant to the next two sentences. The Unit Buy Down Amount to be paid by the Contractor for a Unit failing to achieve a Unit Heat Rate of <<*insert number*>> kJ/kWh shall be equal to the product of:

(a) the difference (if positive) between the Unit Heat Rate achieved during the Performance Tests when determining if the guarantee set forth in this Article 7.2.4 has been met minus <<*insert number*>> kJ/kWh; multiplied by

(b) <<insert currency and amount>>/kJ/kWh.

The Unit Buy Down Amount to be paid by the Contractor for a Unit failing to achieve a Unit Net Electrical Output of <<*insert number*>> MW$_e$ shall be equal to the product of:

(a) the difference (if positive) between <<*insert number*>> MW$_e$ minus the Unit Net Electrical Output achieved during the Performance Tests when determining if the guarantee set forth in this Article <<*insert number*>> has been met; multiplied by

(b) <<insert currency and amount>>/kW.

The Contractor guarantees that each Unit shall be [simultaneously] capable of generating a Unit Net Electrical Output of <<*insert number*>> MW_e [and a Unit Net Thermal Output of <<*insert number*>> MW_{th}] at a Unit Heat Rate of no more than <<*insert number*>> kJ/kWh. If a Unit has achieved each of the Unit Minimum Performance Levels but failed to achieve the performance guaranteed in the previous sentence, the Contractor must, in full satisfaction of its obligations under the previous sentence with respect to that particular Unit, make a payment to the Sponsor the last Day of the Optional Facility Guarantee Cure Period calculated pursuant to the next two sentences. The Unit Buy Down Amount to be paid by the Contractor for the Unit failing to achieve a Unit Heat Rate of <<*insert number*>> kJ/kWh shall be equal to the product of:

(a) the difference (if positive) between the Unit Heat Rate achieved during the Performance Tests when determining if the guarantee set forth in this Article 7.2.5 has been met minus <<*insert number*>> kJ/kWh; multiplied by
(b) <<insert currency and amount>>/kJ/kWh.

The Unit Buy Down Amount to be paid by the Contractor for a Unit failing to achieve a Unit Net Electrical Output of <<*insert number*>> MW_e shall be equal to the product of:

(a) the difference (if positive) between <<*insert number*>> MW_e minus the Unit Net Electrical Output achieved during the Performance Tests when determining if the guarantee set forth in this Article 7.2.5 has been met; multiplied by
(b) <<insert currency and amount>>/kW.

The Contractor guarantees that each Unit shall be capable of [simultaneously] generating a Unit Net Electrical Output of <<*insert number*>> MW_e [and a Unit Net Thermal Output of <<*insert number*>> MW_{th}] at a Unit Heat Rate of no more than <<*insert number*>> kJ/kWh. If the Unit has achieved each of the Unit Minimum Performance Levels but failed to achieve the performance guaranteed in the previous sentence, the Contractor must, in full satisfaction of its obligations under the previous sentence with respect to that particular Unit, make a payment to the Sponsor the last Day of the Optional Facility Guarantee Cure Period calculated pursuant to the next two sentences. The Unit Buy Down Amount to be paid by the Contractor for a Unit failing to achieve a Unit Heat Rate of <<*insert number*>> kJ/kWh shall be equal to the product of:

(a) the difference (if positive) between the Unit Heat Rate achieved during the Performance Tests when determining if the guarantee set forth in this Article 7.2.6 has been met minus <<*insert number*>> kJ/kWh; multiplied by
(b) <<insert currency and amount>>/kJ/kWh.

The Unit Buy Down Amount to be paid by the Contractor for a Unit failing to achieve a Unit Net Electrical Output of <<*insert number*>> MW_e shall be equal to the product of:

(a) the difference (if positive) between <<*insert number*>> MW$_e$ minus the Unit Net Electrical Output achieved during the Performance Tests when determining if the guarantee set forth in this Article 7.2.6 has been met; multiplied by
(b) <<insert currency and amount>>/kW.

The Contractor guarantees that, for each Unit, [at any load point above <<*insert number*>> percent], the Facility will be capable of simultaneously achieving all of the following: [16]

(a) the emissions from the Facility shall not exceed the emissions delineated in Exhibits 1G, 1H and 1K;
(b) all emissions rates from the Facility shall not exceed the emission standards of the relevant Governmental Authorities and shall fully comply with all applicable Laws [including the Environmental Standards];
(c) Facility noise levels shall be as set forth in Exhibits 1G, 1H and 1K;
(d) the purity of steam provided by the boiler of each Unit shall be as set forth in Exhibits 1G, 1H and 1K;
(e) wastewater quality other than sanitary waste shall not be worse than that set forth in Exhibits 1G, 1H and 1K; and
(f) the Facility shall be designed such that clean rain water runoff will meet all limitations set forth in applicable Laws for effluent emissions.

[The Contractor guarantees that [at <<*insert number*>> percent TMCR], each Unit shall be capable of achieving the hourly SO$_2$ emission guarantee of <<*insert number*>> mg/Nm^3 set forth in Exhibits 1G, 1H and 1K without consuming more than <<*insert number*>> kilograms of limestone in its furnace during the same hour, provided, that, if the Facility has achieved each of the Facility Minimum Performance Levels but a Unit has failed to achieve the performance guaranteed in this sentence, the Contractor must, in full satisfaction of its obligations under this sentence with respect to such Unit, make a payment of liquidated damages to the Sponsor the last Day of the Optional Facility Guarantee Cure Period calculated pursuant to the next sentence. The Unit Buy Down Amount to be paid by the Contractor for a Unit failing to meet the performance guaranteed in the previous sentence shall be <<*insert currency amount*>> multiplied by the amount by which the Limestone Ratio exceeds 1[; provided, however, that the amount by which a Unit's Limestone Ratio is less than 1 may be subtracted from the amount by which the Limestone Ratio of the other Unit exceeds 1 to reduce the amount of liquidated damages payable as a result of such deficiency].][17]
[The Contractor guarantees that each Unit will pass the [720-hour] reliability test described in the Functional Specifications.]
[Remedies for Failure to Achieve Performance Guarantees. If the Contractor has achieved Facility Provisional Acceptance by the date which is <<*insert number*>> Days after the Guaranteed Facility Provisional Acceptance Date and by written notice to the Sponsor within <<*insert number*>> Days of the Facility Provisional Acceptance Date elects to conduct repairs and/or additional Performance Tests, the Contractor shall have a period from the Facility Provisional Acceptance Date until <<*insert number*>>

Days [after the end of the First Maintenance Period for Unit Two] to try to achieve the Performance Guarantees (the "Optional Facility Performance Guarantee Cure Period").] [If, at the end of the Optional Facility Performance Guarantee Cure Period the Contractor shall have demonstrated through the relevant Performance Tests that the [Facility Net Thermal Output,] the Facility Net Electrical Output, and/or the Facility Heat Rate have improved from those levels which formed the basis of the Facility Buy Down Amount for such Performance Guarantee, then the Sponsor shall, within <<insert number>> Days after the end of the Optional Facility Performance Guarantee Cure Period, refund to the Contractor the Facility Buy Down Amounts in respect of all Performance Guarantees (if any) that have improved, but only to the extent of such improvement and only after deducting any actual losses suffered by the Sponsor as a result of not having the benefit of such improvement during the Facility Optional Performance Guarantee Cure Period including increased Fuel costs and lost profits as a result of failure to produce electrical output at the guaranteed levels and any actual losses (including lost profits) suffered by the Sponsor as a result of loss of use of the Facility or any part thereof during the Contractor's activities.]

[If the Contractor has achieved Unit Provisional Acceptance for Unit One by the date which is <<insert number>> Days after the Guaranteed Unit One Provisional Acceptance Date and by written notice to the Sponsor within <<insert number>> Days of the Unit Provisional Acceptance Date for Unit One elects to conduct repairs and additional Performance Tests, the Contractor shall have a period from the Unit Provisional Acceptance Date for Unit One until <<insert number>> Days [after the end of the First Maintenance Period for Unit One] to try to achieve the Performance Guarantees (the "Optional Unit One Performance Guarantee Cure Period"). If, at the end of the Optional Unit One Performance Guarantee Cure Period the Contractor shall have demonstrated through the relevant Performance Tests that [the Unit Net Thermal Output, Limestone Ratio,] the Unit Net Electrical Output [,] and/or the Unit Heat Rate have improved from those levels which formed the basis of the Unit Buy Down Amount for such Unit Performance Guarantee, then the Sponsor shall, within <<insert number>> Days after the end of the Optional Unit One Performance Guarantee Cure Period, refund to the Contractor the Unit Buy Down Amounts in respect of all Unit Performance Guarantees (if any) that have improved, but only to the extent of such improvement and only after deducting any actual losses suffered by the Sponsor as a result of not having the benefit of such improvement during the Optional Unit One Performance Guarantee Cure Period including increased Fuel costs and lost profits as a result of failure to produce electrical output at the guaranteed levels and any actual losses (including lost profits) suffered by the Sponsor as a result of loss of use of the Facility or any part thereof during the Contractor's activities.]

[If the Contractor has achieved Unit Provisional Acceptance for Unit Two by the date which is <<insert number>> Days after the Guaranteed Unit Two Provisional Acceptance Date and by written notice to the Sponsor within <<insert number>> Days of the Unit Provisional Acceptance Date for Unit Two elects to conduct repairs and additional Performance Tests, the Contractor shall have a period from the Unit Provisional Acceptance Date of Unit Two until <<insert number>> Days [after the end of the First Maintenance Period for Unit Two] to try to achieve the Unit Performance Guarantees (the "Optional Unit Two Performance Guarantee Cure Period"). If, at the end of the Optional Unit Two Performance Guarantee Cure Period the Contractor shall have demonstrated through the relevant Performance Tests that the Unit Net Thermal Output, the Limestone Ratio, the Unit Net Electrical Output, and/or the Unit Heat Rate has improved from those levels which formed the basis of the Unit Buy Down Amount for such Unit Performance Guarantee, then the Sponsor shall, within <<*insert number*>> Days after the end of the Optional Unit Two Performance Guarantee Cure Period, refund to the Contractor the Unit Buy Down Amounts in respect of all Unit Performance Guarantees (if any) that have improved, but only to the extent of such improvement and only after deducting any actual losses suffered by the Sponsor as a result of not having the benefit of such improvement during the Optional Unit Two Performance Guarantee Cure Period including increased Fuel costs and lost profits as a result of failure to produce electrical output at the guaranteed levels and any actual losses (including lost profits) suffered by the Sponsor as a result of loss of use of the Facility or any part thereof during the Contractor's activities.][18]

[The Contractor shall have the right to reduce:

(a) the cumulative Facility Buy Down Amounts it would otherwise be required to pay to satisfy its obligations with respect to the guaranteed Facility Heat Rate under Articles <<*insert number*>> by the amount by which such cumulative Facility Buy Down Amounts exceed the Facility Heat Rate Buy Down Cap;

(b) the cumulative Unit Buy Down Amounts for Unit One that it would otherwise be required to pay to satisfy its obligations with respect to the Unit Heat Rate for Unit One guaranteed under Articles <<*insert number*>> by the amount by which such cumulative Unit Buy Down Amounts for Unit One exceed the Unit Heat Rate Buy Down Cap for Unit One; and

(c) the cumulative Unit Buy Down Amounts for Unit Two that it would otherwise be required to pay to satisfy its obligations with respect to the Unit Heat Rate for Unit Two guaranteed under Articles <<*insert number*>> by the amount by which such cumulative Unit Buy Down Amounts for Unit Two exceed the Unit Heat Rate Buy Down Cap for Unit Two.][19,20]

If the Contractor elects to pay Facility Buy Down Amounts, then, in the case of:

(a) the Facility Net Electrical Output guaranteed in Article <<*insert number*>>, the Contractor need:
   (i) not pay any Facility Buy Down Amount in respect thereof if the Contractor has already paid to the Sponsor Unit Buy Down Amounts in excess of such Facility Buy Down Amount for the Units' failure to achieve the Unit Net Electrical Output guaranteed for each of them under Article <<*insert number*>>; or
   (ii) only pay Facility Buy Down Amounts in respect thereof in the amount by which the Facility Buy Down Amount therefor exceeds the sum of all Unit Buy Down Amounts paid to the Sponsor for each of Unit One and Unit Two failing to achieve the Unit Net Electrical Output guaranteed for each of them under Article <<*insert number*>>;

["**Facility Heat Rate**" shall mean:

(a) [(i) 3.7 multiplied by (ii) the heat added to the steam and water of both Unit boilers (kW) (iii) multiplied by 100 (iv) divided by the boiler efficiency measured as a percentage based on the net calorific value of Fuel according to DIN 1942; and which result shall be divided by

(b) the sum of the electrical power (MW_e) measured at the high voltage side of the generator transformer and delivered by both Units to the Site Electrical Interconnection Point plus the thermal power (MW_{th}) from both Units at the Thermal Interconnection Point.][21]

(c) [Any reference to any heat rate herein, unless expressly stated otherwise, refers to [lower] [higher] heating value].][22]

STEAM PURITY

The purity of steam provided by each boiler shall not be worse than that shown below when each boiler is operated at any load up to and including 100% BMCR with the boiler water quality within the limits shown in the Functional Specifications.

Operating and Ambient Conditions

SiO_2	max.	mg/kg	0.02
Na + K	max.	mg/kg	0.01
Fe	max.	mg/kg	0.02
Cu	max.	mg/kg	0.003

Concentration
2500
1000
800

During Performance Tests, the coal, limestone, sand, raw water properties shall be as specified above, and main steam pressure and temperature shall all be as specified in Article 5.0 of Exhibit Y and the generator power factor, grid voltage and frequency shall be as close as reasonably practical to the design specifications set forth therefor in Exhibit Y but tests shall be valid so long as they are within the ranges specified in the Functional Specifications. In addition, the Facility shall operate within the variations permitted by the test codes. All test measurements used to determine the guaranteed performance shall be average of values recorded during the relevant measuring period.

The Performance Guarantees for each Unit and the Facility are conditional upon the following:

(i) Net Electrical Outputs.
    Burning coal with properties in the ranges shown in Table 5.1 of Exhibit S at the ambient, district heating and electrical conditions shown in item 10.1.(vii) of Exhibit S.

(ii)   Net Heat Rates.

Burning the Design and Performance Base Coal shown in Table 5.1 of Exhibit S at the ambient, district heating and electrical conditions shown in item 10.1.(vii) of Exhibit S.

(iii) Emissions.

Burning coal with properties in the ranges shown in Table 5.1 of Exhibit S and feeding the boiler with any limestone with properties in the ranges shown in Table 5.2 of Exhibit S, the hourly average concentrations of SO_2, NO_x, particulate and CO in the stack gases shall be based on 6% O_2 in dry flue gas.

(iv)   Limestone Consumption.

Burning the Design and Performance Base Coal shown in Table and feeding the boiler with the Parent Guarantee limestone shown in Table 5.2 of Exhibit S, with each boiler of the Unit operating at 100% TMCR.

(v)   Waste Water Quality.

Raw water within the raw water analysis stated in Table 5.4 of Exhibit S.

(vi)   Steam Purity.

Based on boiler water quality as below:

pH_{25}			8.5 - 9.4
SiO_2 content	<	mg/kg	0.35
Acid capacity	<	mmol/kg	0.05
Na and K content	<	mg/kg	8
Conductivity (25°C after neutralization)	<	mS/m	4
PO_4 content		mg/kg	2 to 6
Consumption of $KMnO_4$	<	mg/kg	5

(vii) Ambient and District Heating System Conditions.

Ambient and district heating system conditions as shown below:

Thermal Load for Facility with 2 Units Operating	MW_{th}	0	110	360
Thermal Load for Facility with 1 Unit Operating	MW_{th}	0	55	180
DH Supply Temperature	°C	---	82	131
DH Return Temperature	°C	---	60	60
Dry Bulb Temperature	°C	15	13	4.6
Relative humidity	%	70	70	70
Atmospheric pressure	mbar	1013	1013	1013

(viii) Boiler Test Conditions.

The boiler shall be cleaned using the boiler cleaning devices. In addition, the Contractor and the Project Company shall agree that the boiler is in a suitable condition for testing.

Test code	DIN 1942
Reference temperature	15°C
Average air temperature of outlet at combustion fans	40°C
Envelope as DIN 1942	Normal
Heat credits	None for auxiliary power consumption
Boiler blowdown and root blowing	Zero
Radiation loss	As DIN 1942, Figure 6

Heat in auxiliary steam shall be calculated as product of steam and difference of inlet and outlet steam enthalpies at boiler system boundary.

(ix) Net Electrical Output.Net Electrical Output being measured at the generator transformer high voltage terminals after deduction of the station auxiliary power load at the conditions of normal operating grid voltage and with the generator operating at 0.8 power factor or as close as possible thereto as grid conditions permit.

The continuous auxiliary power consumers are the following:
primary and secondary air fans
induced draft fan
high pressure blower
fuel feeding equipment after day silos
bottom ash handling equipment
electrostatic precipitator (without heaters and rappers)
limestone handling equipment after day silos
cooling towers
cooling water pumps
service cooling water pumps
district heating pumps (not in condensing mode)
water treatment
raw water pumps
air compressors
waste water treatment
heating and ventilation
feed water pumps
condensate pumps
district heating condensate pumps (not in condensing mode)
district heating make - up pump (not in condensing mode)
district heating feedwater pump (not in condensing mode)
preheater condensate pump
lighting
other auxiliaries required for normal continuous operations in accordance with the Contractors Operating Information.

In addition to the continuous auxiliary power consumption calculations include the following non-continuously operated equipment. The power consumption of following consumers are calculated as an average over 24 hours:

Belt feeders from the coal receiving hopper of rail trucks unloading station. Operation time is between hours 10.00–18.00.

Belt conveyors to the stacker / reclaimer. Operation time is between hours 10.00–18.00.

Stacker / reclaimer. Operation time is between hours 06.00–10.00 and 18.00–22.00 for reclaiming and 10.00–18.00 for stacking.

Belt conveyors from Stacker / reclaimer to crushing station. Operation time is between hours 06.00–10.00 and 18.00–22.00.

Crushers. Operation time is between hours 06.00–10.00 and 18.00–22.00.

Belt conveyors from the crushing station to boiler house silos. Operation time is between hours 06.00–10.00 and 18.00–22.00.

Limestone pneumatic transportation from the storage silo to boiler house silo. Operation time is 14 hours/47 hours. This consumption is calculated as an average over 47 hours.

1. DEFINITIONS

"**Adjusted Facility Btu Consumption**" for any period shall mean the number of Btus consumed by the New Facility during the period in question.

"**Adjusted Facility Net Electrical Output**" shall mean the electrical output of the New Facility when operating on Fuel, as the case may be, measured in kW at the Site Electrical Interconnection Point as adjusted to the Reference Conditions using the Correction Curves.

"**Adjusted Facility Btu Consumption**" for any period shall mean the number of Btus consumed by the New Facility during the period in question.

"**Adjusted Facility Net Electrical Output**" shall mean the electrical output of the New Facility when operating on Gas or Oil, as the case may be, measured in kW at the Site Electrical Interconnection Point as adjusted to the Reference Conditions using the Correction Curves.

"**Adjusted Facility Net Thermal Output**" for a Unit shall mean the net thermal output of such Unit, measured in pounds per hour at the Site Thermal Interconnection Point when the other Unit is not operating as adjusted to Reference Conditions using the Correction Curves.

"**Adjusted Unit Btu Consumption**" for a Unit for any period shall the number of Btus consumed by such Unit during the period in question when the other Unit is not in service.

"**Adjusted Unit Net Electrical Output**" shall mean the electrical output of a Unit when operating on Gas or Oil as the case may be, measured in kW at the Site Electrical Interconnection Point as adjusted to the Reference Conditions using the Correction Curves when the other Unit is not in service.

"**Adjusted Unit Net Thermal Output**" for a Unit shall mean the net thermal output of such Unit, measured in pounds per hour at the Site Thermal Interconnection Point when the other Unit is not operating as adjusted to Reference Conditions using the Correction Curves.

"**Emissions Guarantee**" is defined in Article 2.2 hereof.

"**Facility 60% Load Electrical Output Guarantee**" is defined in Exhibit CC.

"**Facility 60% Load Heat Rate Guarantee**" is defined in Exhibit CC.

"**Facility 60% Load Thermal Output Guarantee**" is defined in Exhibit CC.

["**Facility 60 percent Load Electrical Output Guarantee**" is defined in [Exhibit 1I.] [Article 11.1.]]

["**Facility 60 percent Load Heat Rate Guarantee**" is defined in [Exhibit 1.I.] [Article 11.1.]]

["**Facility 60 percent Load Thermal Output Guarantee**" is defined in [Exhibit 1I.][Article 11.1.]]

["**Facility Full Load Electrical Output Guarantee**" is defined in [Exhibit 1I.] [Article 11.1.]]

["**Facility Full Load Heat Rate Guarantee**" is defined in [Exhibit 1I.] [Article 11.1.]]

["**Facility Full Load Thermal Output Guarantee**" is defined in [Exhibit 1I.] [Article 11.1.]]

["**Facility Net Electrical Output**" shall mean the electrical output of the Facility when operating on Fuel (net of Facility electrical consumption) at <<110>> kV, measured in MW$_e$ at the Site Electrical Interconnection Point.][23]

["**HRSG**" shall mean the heat recovery steam generator and associated equipment described in the HRSG Purchase Agreement.]

["**HRSG Purchase Agreement**" shall mean the agreement between Sponsor and <<insert *supplier name*>> providing for the purchase of the HRSG.]

"**New Facility Buy Down Amounts**" are liquidated damages which attempt to compensate Sponsor for impaired performance of the New Facility and are set forth in this Exhibit 9.2 and are to be calculated in accordance herewith and therewith.

"**New Facility Buy Down Amounts**" are liquidated damages which attempt to compensate Project Company for impaired performance of the New Facility and are set forth in this Exhibit U and are to be calculated in accordance with Section 2.1 hereof.

"**New Facility Full Load Gas Electrical Output Guarantee**" means [_____].
"**New Facility Full Load Gas Btu Consumption Guarantee**" means [_____].
"**New Facility Full Load Gas Thermal Output Guarantee**" means [_____].
"**New Facility Full Load Oil Electrical Output Guarantee**" means [_____].
"**New Facility Full Load Oil Btu Consumption Guarantee**" means [_____].
"**New Facility Full Load Oil Thermal Output Guarantee**" means [_____].
"**New Facility 60 percent Load Fuel Consumption Guarantee**" means [_____].
"**New Facility 60% Load Fuel Consumption Guarantee**" means [___].
"**New Facility Full Load Gas Electrical Output Guarantee**" means [___].
"**New Facility Full Load Gas Btu Consumption Guarantee**" means [___].
"**New Facility Full Load Gas Thermal Output Guarantee**" means [___].
"**New Facility Full Load Oil Electrical Output Guarantee**" means [___].
"**New Facility Full Load Oil Btu Consumption Guarantee**" means [___].
"**New Facility Full Load Oil Thermal Output Guarantee**" means [___].

(a) simultaneously generating a Unit Net Electrical Output of <<*insert amount*>> and a Unit Net Thermal Output of <<*insert amount*>> at a Unit Heat Rate of no more than <<*insert amount*>> kJ/kWh;

(b) simultaneously generating a Unit Net Electrical Output of <<*insert amount*>> and a Unit Net Thermal Output of <<*insert amount*>> at a Unit Heat Rate of no more than <<*insert amount*>> kJ/kWh; generating a Unit Net Electrical Output of<<*insert amount*>> at a Unit Heat Rate of no more than <<*insert amount*>> kJ/kWh when operating in full condensing mode

(c) simultaneously generating a Facility Net Electrical Output of <<*insert number*>> and a Facility Net Thermal Output of <<*insert number*>> at a Facility Heat Rate of no more than <<*insert number*>> kJ/kWh;

(d) simultaneously generating a Facility Net Electrical Output of <<*insert number*>> and a Facility Net Thermal Output of <<*insert number*>> at a Facility Heat Rate of no more than <<*insert number*>> kJ/kWh;

(e) generating a Facility Net Electrical Output of <<*insert number*>> MWe at a Facility Heat Rate of no more than <<*insert number*>> kJ/kWh when operating in full condensing mode; and

(f) achieving each of the guaranteed emissions levels set forth in Article <<__>> and the Functional Specification.

["**STG**" means the steam turbine generator with duct burners and associated equipment described in the Vendor Purchase Agreements.]

["**Steam Turbine Purchase Agreement**" means <<*insert title of agreement/all parties*>>.]

["**Thermal <Customer> <Supplier>**" means <<insert name>>.]

["**Thermal Energy Supply Agreement**" shall mean the agreement between the Sponsor and <<*insert name*>>, regarding the sale and purchase of thermal energy <from> <for> the Facility.]

["**Thermal Interconnection Facilities**" shall mean the interconnection facilities required by the Thermal Customer in order to establish an interconnection between the Facility and the thermal distribution system of the Thermal Customer in accordance with the Thermal Energy Supply Agreement.]

["**Thermal Interconnection Point**" shall mean the physical points where the Thermal Interconnection Facilities and the Thermal Customer's distribution system are connected as further described in the Functional Specifications.]

"**Unit<<60>> Percent Load Gas Btu Consumption Guarantee**" means [____].

"**Unit Full Load Gas Btu Consumption Guarantee**" means [____].

"**Unit Full Load Gas Electrical Output Guarantee**" means [____].

"**Unit Full Load Gas Thermal Output Guarantee**" means [____].

"**Unit Full Load Oil Btu Consumption Guarantee**" means [____].

"**Unit Full Load Oil Electrical Output Guarantee**" means [____].

"**Unit Full Load Oil Thermal Output Guarantee**" means [____].

"**Unit Minimum Performance Level**" shall be met when a Unit has achieved all of the following:

(a) simultaneously achieving both an Adjusted Unit Net Electrical Output of at least <<ninety-seven percent (97 percent)>> of the Unit Full Load Gas Electrical Output Guarantee and an Adjusted Unit Thermal Output of at least <<ninety-seven percent (97 percent)>> of the Unit Full Load Gas Thermal Output Guarantee when operating on Gas during the Performance Tests at an Adjusted Unit Fuel Consumption of no more than <<one hundred and three percent (103 percent)>> of the Unit Full Load Gas Btu Consumption Guarantee;

(b) simultaneously achieving an Adjusted Unit Net Electrical Output of at least <<ninety-seven percent (97 percent)>> of the Unit Full Load Oil Electrical Output Guarantee when operating on Oil during the Performance Tests at an Adjusted Btu Consumption of no more than <<one hundred and three percent (103 percent)>> of the Unit Full Load Oil Btu Consumption Guarantee; and

(c) achieving the Emissions Guarantee.

["**Unit Net Thermal Output**" for a Unit shall mean the Unit's thermal energy output measured in Btu's (when the other Unit is not operating) at the Thermal Interconnection Point when operated on Fuel.][24]

"**Unit Net Thermal Output**" for a Unit shall mean the Unit's thermal energy output measured in MW_{th} (when the other Unit is not operating) at the Thermal Interconnection Point when operated on Fuel.[25]

["**Unit One <<60>> Percent Load Electrical Output Guarantee**" is defined in Exhibit 1I.]

["**Unit One <<60>> Percent Load Heat Rate Guarantee**" is defined in Exhibit 1I.]

["**Unit One <<60>> Percent Thermal Output Guarantee**" is defined in Exhibit 1I.]

["**Unit One Full Load Electrical Output Guarantee**" is defined in Exhibit 1I.]

["**Unit One Full Load Heat Rate Guarantee**" is defined in Exhibit 1I.]

["**Unit One Full Thermal Output Guarantee**" is defined in Exhibit 1I.]

"**Unit One 60% Load Electrical Output Guarantee**" is defined in Exhibit CC.

"**Unit One 60% Load Heat Rate Guarantee**" is defined in Exhibit CC.

"**Unit One 60% Thermal Output Guarantee**" is defined in Exhibit CC.

"**Unit Two 60% Load Electrical Output Guarantee**" is defined in Exhibit CC.

"**Unit Two 60% Load Heat Rate Guarantee**" is defined in Exhibit CC.

"**Unit Two 60% Thermal Output Guarantee**" is defined in Exhibit CC.

"**Unit Two Availability Guarantee**" shall have the meaning ascribed thereto in Article 7.2.10.

["**Unit Two <<60>> Percent Load Electrical Output Guarantee**" is defined in Exhibit 1I.]

["**Unit Two <<60>> Percent Load Heat Rate Guarantee**" is defined in Exhibit 1I.]

["**Unit Two <<60>> Percent Thermal Output Guarantee**" is defined in Exhibit 1I.]

["**Unit Two Full Load Electrical Output Guarantee**" is defined in Exhibit 1I.]

["**Unit Two Full Load Heat Rate Guarantee**" is defined in Exhibit 1I.]

["**Unit Two Full Thermal Output Guarantee**" is defined in Exhibit 1I.]

"**Warranty Notification Period**" shall mean a period commencing on the Facility Provisional Acceptance Date and ending <<one (1)>> year thereafter, provided that with respect to the HPB System only, the Warranty Notification Period (a) for the HPB System excluding the equipment related to the HPB System's Oil-firing gear shall commence on the HPB Gas-Firing Substantial Completion Date and end <<one (1)>> year thereafter, and (b) for the equipment related to the HPB System's Oil-firing gear only shall commence on the HPB Oil-Firing Substantial Completion Date and end <<one (1)>> year thereafter.

2. PERFORMANCE DAMAGES

2.0   The Contractor guarantees that at 90°F and forty percent (40%) relative humidity and also at 32°F and forty percent (40%) relative humidity the New Facility shall be capable of simultaneously achieving:

  (i)   the New Facility Full Load Gas Electrical Output Guarantee;
 (ii)   the New Facility Full Load Gas Thermal Output Guarantee;
(iii)   the New Facility Full Load Gas Btu Consumption Guarantee; and
 (iv)   the Emissions Guarantee.

Once the New Facility has achieved the New Facility Minimum Performance Level but failed to achieve the performance guaranteed in the previous sentence, the Contractor must, in full satisfaction of its obligations under the previous sentence, make a payment to Project Company calculated pursuant to the next three sentences.

The New Facility Buy Down Amount to be paid by the Contractor for the New Facility failing to achieve the New Facility Full Load Gas Btu Consumption Guarantee shall be equal to the product of:

(a) the difference (if positive) between the Adjusted New Facility Btu Consumption achieved during the Performance Tests when determining if the guarantee set forth in this Exhibit U has been met, minus the New Facility Full Load Gas Btu Consumption Guarantee; multiplied by

(b) $_____/Btu/Wh.

The New Facility Buy Down Amount to be paid by the Contractor for the New Facility failing to achieve the New Facility Full Load Gas Electrical Output Guarantee shall be equal to the product of:

(a) the difference (if positive) between the New Facility Full Load Gas Electrical Output Guarantee minus the Adjusted New Facility Net Electrical Output achieved during the Performance Tests when determining if the guarantee set forth in this Exhibit U has been met; multiplied by

(b) _____kW.

The New Facility Buy Down Amount to be paid by the Contractor for the New Facility failing to achieve the New Facility Full Load Gas Thermal Output Guarantee shall be equal to the product of:

(a) the difference (if positive) between the New Facility Full Load Gas Thermal Output Guarantee minus the Adjusted New Facility Net Thermal Output achieved during the Performance Tests when determining if the guarantee set forth in this Exhibit 9.2 has been met; multiplied by(b)<<insert number>> kW.

2.1   Sponsor may direct the Contractor to commence work on the HPB System before the Work Commencement Date, in which case the Contractor will commence all work on the HPB System and only the HPB System on the HPB Work Commencement Date.

2.2   The Contractor guarantees that the New Facility will be capable of simultaneously achieving all of the following (the "Emissions Guarantee"):

(a) the emissions from the New Facility shall not exceed the emissions delineated in Appendix 1 of this Exhibit 9.2 when either the New Facility or a Unit is operated at any load point above <<40>> percent (whether on Gas or Oil);
(b) not exceeding noise levels as set forth in Appendix 1 of this Exhibit 9.2;
(c) wastewater quality other than sanitary waste shall not be worse than that set forth in Appendix 1 of this Exhibit 9.2;

(d) the New Facility shall be designed such that clean rain water runoff will meet all limitations set forth in applicable Laws for effluent emissions; and

(e) all emissions rates from the New Facility shall not exceed the emission standards of the relevant Governmental Authorities and shall fully comply with all applicable Laws.

(f) If the HPB Gas-Firing Substantial Completion Date does not occur on or before the Guaranteed HPB Gas-Firing Substantial Completion Date, then the Contractor shall, within <<*fifteen (15)*>> Days of the Guaranteed HPB Gas-Firing Substantial Completion Date, submit to [_____] a detailed technical and narrative analysis of the root causes that prevented the achievement of the HPB Gas-Firing Substantial Completion Date and a detailed plan (the "HPB Remedial Plan") designating the new expected date for HPB Gas-Firing Substantial Completion and setting out the technical steps necessary for such and the HPB Remedial Plan shall include a reasonable schedule demonstrating that HPB Gas-Firing Substantial Completion can be achieved.

(g) If the HPB Oil-Firing Substantial Completion Date does not occur on or before the Guaranteed HPB Oil-Firing Substantial Completion Date, then the Contractor shall, within <<*fifteen (15)*>> Days of the Guaranteed HPB Oil-Firing Substantial Completion Date, submit to [_____] a detailed technical and narrative analysis of the root causes that prevented the achievement of the HPB Oil-Firing Substantial Completion Date and either create or amend the HPB Remedial Plan designating the new expected date for HPB Oil-Firing Substantial Completion and setting out the technical steps necessary for such and the HPB Remedial Plan shall include a reasonable schedule demonstrating that HPB Oil-Firing Substantial Completion can be achieved.

(h) If the CTG/HRSG Substantial Completion Date does not occur on or before the Guaranteed CTG/HRSG Substantial Completion Date, then the Contractor shall, within <<*fifteen (15)*>> Days of the Guaranteed CTG/HRSG Substantial Completion Date, submit to [_____] a detailed technical and narrative analysis of the root causes that prevented the achievement of the CTG/HRSG Substantial Completion Date and a detailed plan (the "CTG/HRSG Remedial Plan") designating the new expected date for CTG/HRSG Substantial Completion and setting out the technical steps necessary for such and the CTG/HRSG Remedial Plan shall include a reasonable schedule demonstrating that CTG/HRSG Substantial Completion can be achieved.

2.3   HPB Gas-Firing Commissioning Completion

2.3.1   "HPB Gas-Firing Commissioning Completion" with respect to the HPB System shall occur when all of the following conditions have been satisfied:

(a) all necessary Work with respect to the HPB System (except for the Oil-firing gear) has been mechanically completed in accordance with this Contract, including all Work that is necessary to allow the HPB System to safely and reliably handle Gas and generate and export thermal energy from Gas operation, provided that it is understood that the HPB System need not be connected to the distributed control system but may be connected only to its own local control system;

(b) all necessary Tie-Ins and Interconnection Facilities with respect to the HPB System have been completed by the Contractor in accordance herewith;

(c) all necessary Work with respect to the HPB System (except for the Oil-firing gear) has been inspected for completeness using the Specifications and Drawings and point-to-point checks to verify that they have been correctly installed so as to respond to simulated test signals which are equivalent to actual signals which would be received during operation as the basis of such inspection;

(d) a preliminary HPB Punch List with respect to the HPB System (except for the Oil-firing gear) has been provided in writing to the Sponsor;

(e) all HPB Systems have been cleared and are free of construction debris, satisfactorily cleaned and flushed where required, leak checked, hydrostatically, pneumatically, mechanically, electrically and functionally tested, and all systems, components, vessel internals, catalysts, chemicals, resins, desiccants, lubricants, and other similar materials thereof have been properly restored to operating condition;

(f) the following documents with respect to the HPB System have been transmitted to the Sponsor in a format consistent with the Functional Specifications, in semi-final draft form, without as-built drawings, but as reasonably complete as available information will allow, and at a minimum with sufficient information to permit the training of the Sponsor's operation and maintenance personnel and the normal operation and maintenance of the HPB System by persons generally familiar with facilities and plants similar to the New Facility:

   (i) Vendors' and other manufacturers' instructions and drawings relating to Equipment;

   (ii) QA/QC, and other test and inspection certificates and reports applicable to the Work, including those relating to relay settings and instrument calibration;

   (iii) Specifications and Drawings (current as of such time);

   (iv) construction turnover packages (as defined in the Functional Specification);

   (v) standard operating procedures and manuals, including all system operations and maintenance manuals; and

   (vi) HPB startup testing procedures for Gas firing approved in writing by [Sponsor];

(g) all material activities in the construction and commissioning phases for all HPB systems as more particularly described in the Functional Specifications, have been successfully completed;

(h) the Contractor's formal classroom training program for the Sponsor's operating personnel, with respect to the HPB System when operated on Gas, described in Article [__] has been offered;

(i) special tools, spare parts, and other items provided by the Contractor that are necessary for the startup of the HPB System when operated on Gas are available at the Site;

(j) all emission control and monitoring systems have been installed and calibrated as required by all applicable Laws;

(k) all system blows (by means of air and/or steam) have been completed; and

(l) all HPB operations manuals have been submitted to the Sponsor.

2.3.2 When, at any time, the Contractor believes that HPB Gas-Firing Commissioning Completion has occurred, the Contractor shall provide written notice thereof to the Sponsor, including such documentation and certifications necessary to verify the conditions for HPB Gas-Firing Commissioning Completion contained in Article [_____]. Thereafter, the Sponsor shall within <<*ten (10)*>> Days of receipt of same:

(a) notify the Contractor if the Sponsor believes HPB Gas-Firing Commissioning Completion has not occurred which notification shall specifically identify the conditions the Sponsor believes have not been satisfied; or
(b) issue to the Contractor a Certificate of HPB Gas-Firing Commissioning Completion acknowledging that HPB Gas-Firing Commissioning Completion has occurred on the first date on which all the conditions therefor were satisfied.

2.3.3 Upon being advised by the Sponsor that it believes HPB Gas-Firing Commissioning Completion has not occurred, the Contractor and the Sponsor shall meet promptly in order to resolve the disagreement and determine what, if any Work remains to be completed in order to achieve HPB Gas-Firing Commissioning Completion and once such occurs the Sponsor shall issue a Certificate of HPB Gas-Firing Commissioning Completion therefor.

2.4 HPB Gas-Firing Substantial Completion

2.4.1 "HPB Gas-Firing Substantial Completion" with respect to the HPB System shall occur when all of the following conditions have been satisfied:

(a) The Sponsor has issued a Certificate of HPB Gas-Firing Commissioning Completion;
(b) all material activities in the commissioning phase with respect to the HPB System when operated on Gas, except for items on the HPB Punch List, have been successfully completed, as more particularly described in the Functional Specifications;
(c) all HPB System tests with respect to HPB operation on Gas have been successfully completed as more particularly described in the Functional Specifications;
(d) the HPB System with respect to operation on Gas has passed the reliability and demonstration tests set forth in the Functional Specifications;
(e) the HPB Minimum Performance Level with respect to HPB operation on Gas has been achieved;
(f) the HPB System (except for the Oil-firing gear) shall have passed all HPB Performance Tests with respect to HPB operation on Gas, demonstrating full compliance with all HPB Performance Guarantees with respect to operation on Gas or the Contractor has paid all HPB Buy Down Amounts with respect to HPB operation on Gas;
(g) the Contractor shall have paid to the Sponsor any Delay Damages due with respect to the HPB System;
(h) no Adjustments related to the HPB System which are valued in excess of <<*insert currency and amount*>> in the aggregate are pending;

(i)   the Contractor has completed all Work with respect to the HPB System except as set forth on the HPB Punch List except with regard to Oil-firing gear only; and

(j)   the Parties have agreed in writing to a priced HPB Punch List and to a schedule of its completion.

2.4.2   When, at any time, the Contractor believes that HPB Gas-Firing Provisional Acceptance has occurred, the Contractor shall provide notice thereof to the Sponsor, including such operating data, documentation and certifications as necessary to verify the conditions for HPB Gas-Firing Provisional Acceptance contained in Section 6.4.1. Thereafter, the Sponsor shall within <<*fifteen (15)*>> Days:

(a)   notify the Contractor of any reason why HPB Gas-Firing Provisional Acceptance has not occurred which notification shall specifically identify the conditions the Sponsor believes have not been satisfied; or

(b)   issue to the Contractor a Certificate of HPB Gas-Firing Provisional Acceptance acknowledging HPB Gas-Firing Provisional Acceptance on the date such notice was submitted so long as all the conditions therefor were satisfied.

2.4.3   Upon being advised by the Sponsor that it believes HPB Gas-Firing Provisional Acceptance has not occurred, the Contractor and the Sponsor shall meet promptly in order to resolve the disagreement and determine what, if any Work remains to be completed with respect to the HPB System operated on Gas in order to achieve HPB Gas-Firing Provisional Acceptance and once such occurs the Sponsor shall issue a Certificate of HPB Gas-Firing Provisional Acceptance therefor.

2.4.4   So long as the Contractor is acting in good faith and reasonably in submitting notice that HPB Gas-Firing Provisional Acceptance has occurred, Delay Damages shall not accrue during the time in which the Sponsor is considering whether HPB Gas-Firing Provisional Acceptance has occurred.

2.4.5   "HPB Oil-Firing Provisional Acceptance" with respect to the HPB System shall occur when all of the following conditions have been satisfied or continue to be satisfied:

(a)   the Sponsor has issued a Certificate of HPB Gas-Firing Commissioning Completion;

(b)   any additions to the HPB Punch list regarding the HPB oil-firing gear have been made;

(c)   all material activities in the commissioning phase with respect to the HPB System when operated on Oil, except for items on the HPB Punch List, have been successfully completed, as more particularly described in the Functional Specifications;

(d)   all HPB System tests with respect to HPB operation on Oil have been successfully completed as more particularly described in the Functional Specifications;

(e)   the HPB System with respect to operation on Oil has passed the reliability and demonstration tests set forth in the Functional Specifications;

(f)   the HPB Minimum Performance Level with respect to HPB operation on Oil has been achieved;

(g) the HPB System has passed all HPB Performance Tests with respect to HPB operation on Oil, demonstrating full compliance with all HPB Performance Guarantees with respect to operation on Oil or the Contractor has paid all HPB Buy Down Amounts;

(h) no Adjustments related to the HPB System which are valued in excess of <<*insert amount and currency*>> in the aggregate are pending;

(i) the Contractor has completed all Work with respect to the HPB System when operated on Oil except as set forth on the HPB Punch List; and

(j) the Parties have agreed in writing to a priced HPB Punch List and to a schedule of its completion.

2.4.6   When, at any time, the Contractor believes that HPB Oil-Firing Provisional Acceptance has occurred, the Contractor shall provide notice thereof to the Sponsor, including such operating data, documentation and certifications as necessary to verify the conditions for HPB Oil-Firing Provisional Acceptance contained in Section 6.4.5. Thereafter, the Sponsor shall within <<*fifteen (15)*>> Days:

(a) notify the Contractor of any reason why HPB Oil-Firing Provisional Acceptance has not occurred which notification shall specifically identify the conditions the Sponsor believes have not been satisfied; or

(b) issue to the Contractor a Certificate of HPB Oil-Firing Provisional Acceptance acknowledging HPB Oil-Firing Provisional Acceptance on the date such notice was submitted so long as all the conditions therefor were satisfied.

2.4.7   Upon being advised by the Sponsor that it believes HPB Oil-Firing Provisional Acceptance has not occurred, the Contractor and the Sponsor shall meet promptly in order to resolve the disagreement and determine what, if any Work remains to be completed with respect to the HPB System operated on Oil in order to achieve HPB Oil-Firing Provisional Acceptance and once such occurs the Sponsor shall issue a Certificate of HPB Oil-Firing Provisional Acceptance therefor.

2.4.8   So long as the Contractor is acting in good faith and reasonably in submitting notice that HPB Oil-Firing Provisional Acceptance has occurred, Delay Damages shall not accrue during the time in which the Sponsor is considering whether HPB Oil-Firing Provisional Acceptance has occurred.

2.5   HPB Final Acceptance.

2.5.1   "HPB Final Acceptance" shall occur when all of the following conditions have been satisfied:

(a) the Sponsor has issued a Certificate of HPB Gas-Firing Provisional Acceptance and a Certificate of HPB Oil-Firing Provisional Acceptance;

(b) all items on the HPB Punch List have been completed;

(c) the Contractor shall have submitted HPB As-Built Drawings in a format consistent with the Functional Specifications;

(d) the Sponsor has received from the Contractor an Application for Payment in substantially the form set forth in Exhibit 10.3.1 for the Milestone Payment for HPB Final Acceptance, as set forth in Exhibit M;

(e) the Sponsor has received releases and waivers of any Liens established in connection with work on the HPB System against the HPB System and the Site and the Sponsor and its property, employees and agents from the Contractor and each Subcontractor performing Work at the Site and such other documentation as the Sponsor [or any Financing Party] may reasonably request to establish proof thereof; and

(f) the Contractor has completed all work on the HPB System.

2.5.2   When the Contractor believes that HPB Final Acceptance has occurred, the Contractor shall provide written notice thereof to the Sponsor, including such operating data, documentation and certifications as necessary to verify the conditions for HPB Final Acceptance contained in Article [___]. Thereafter, the Sponsor shall within <<*ten (10)*>> Days:

(a) notify the Contractor of any defects in the HPB System or of any other reason why HPB Final Acceptance has not occurred which notification shall specifically identify the conditions the Sponsor believes have not been satisfied; or

(b) issue to the Contractor a Certificate of HPB Final Acceptance acknowledging that HPB Final Acceptance has occurred on the first date on which all conditions therefor were satisfied; provided that such date shall be no earlier than the date of the Contractor's last written notice to the Sponsor under the previous sentence.

   Upon being advised by the Sponsor that it believes HPB Final Acceptance has not occurred, the Contractor and the Sponsor shall meet promptly in order to resolve the disagreement and determine what, if any Work remains to be completed in order to achieve HPB Final Acceptance and once such occurs the Sponsor shall issue a Certificate of HPB Final Acceptance therefor.

2.6   The Contractor agrees that:

(a) HPB Gas-Firing Provisional Acceptance shall occur (the "Guaranteed HPB Gas-Firing Provisional Acceptance Date"):
  (i) on or before *<insert date>* if the HPB Work Commencement Date or the Work Commencement Date occurs by *<insert date>*;
  (ii) on or before the date mutually agreed to by the Parties in a Change Order, if the HPB Work Commencement Date occurs after *<insert date>*;

(b) at the Sponsor's request, the Contractor will meet with the Sponsor to determine an expedited schedule for HPB Gas-Firing Provisional Acceptance and, if agreement on such an expedited schedule is reached, memorialize such agreement in a Change Order;

(c) HPB Oil-Firing Provisional Acceptance shall occur on or before a date which is the next Day following the Day on which [_____] has provided the Contractor with <<*ninety (90)*>> cumulative Days of access to the HPB System following the HPB Gas-Firing Provisional Acceptance Date (the "<u>Guaranteed HPB Oil-Firing Provisional Acceptance Date</u>");

(d) HPB Final Acceptance shall occur on or before the date which is <<*sixty (60)*>> Days after the HPB Oil-Firing Provisional Acceptance Date (the "<u>Guaranteed HPB Final Acceptance Date</u>");

(e) CTG/HRSG Provisional Acceptance shall occur (the "<u>Guaranteed CTG/HRSG Provisional Acceptance Date</u>"):
   (i) on or before <insert date>, if the Work Commencement Date occurs by <insert date>;
   (ii) on or before the date mutually agreed to by the Parties in a Change Order, if the Work Commencement Date occurs after <insert date>;

(f) New Facility Provisional Acceptance shall occur (the "<u>Guaranteed New Facility Provisional Acceptance Date</u>"):
   (i) on or before <insert date>, if the Work Commencement Date occurs by <insert date>;
   (ii) on or before the date mutually agreed to by the Parties in a Change Order, if the Work Commencement Date occurs after <insert date>; and

(g) New Facility Final Acceptance shall occur on or before the date which is sixty (60) Days after the New Facility Provisional Acceptance Date (the "<u>Guaranteed New Facility Final Acceptance Date</u>").

2.7   <u>Delay Damages</u>

2.7.1   If the HPB Gas-Firing Provisional Acceptance Date does not occur on or before the Guaranteed HPB Gas-Firing Provisional Acceptance Date, then the Contractor shall pay the Sponsor liquidated damages in the amount of:

(a) <<*insert amount and currency*>> for each of the first (1st) through the tenth (10th) Days following the Guaranteed HPB Gas-Firing Provisional Acceptance Date;

(b) <<*insert amount and currency*>> for each of the eleventh (11th) through the twentieth (20th) Days following the Guaranteed HPB Gas-Firing Provisional Acceptance Date; and

(c) <<*insert amount and currency*>> for each Day after the twentieth (20th) Day following the Guaranteed HPB Gas-Firing Provisional Acceptance Date.

   Those payment amounts shall cease to accrue on any Day occurring after the earlier of (i) the Guaranteed New Facility Provisional Acceptance date and (ii) the HPB Gas-Firing Provisional Acceptance Date.

2.7.2   If the HPB Final Acceptance Date does not occur on or before the Guaranteed HPB Final Acceptance Date, then the Sponsor shall be entitled (but shall not be required) to complete any or all HPB Punch List items remaining uncompleted at such time with respect to such Punch List (either itself or through other contractors) and to apply to the reasonable cost thereof any or all of the unpaid Adjusted Lump Sum Contract Price or Retainage (including draw upon any Retainage LC), and, if such is insufficient to cover the cost of completing all HPB Punch List items, the Contractor shall pay the excess cost thereof to the Sponsor promptly following demand therefor.

2.7.3   If the CTG/HRSG Provisional Acceptance Date does not occur on or before the Guaranteed CTG/HRSG Provisional Acceptance Date, then the Contractor shall pay the Sponsor liquidated damages in the amount of:

(a) <<*insert amount and currency*>> for each of the first (1st) through the tenth (10th) Days following the Guaranteed CTG/HRSG Provisional Acceptance Date;
(b) <<*insert amount and currency*>> for each of the eleventh (11th) through the twentieth (20th) Days following the Guaranteed CTG/HRSG Provisional Acceptance Date; and
(c) <<*insert amount and currency*>> for each Day after the twentieth (20th) Day following the Guaranteed CTG/HRSG Provisional Acceptance Date.

Those payment amounts shall cease to accrue on any Day occurring after the earlier of (i) the Guaranteed New Facility Provisional Acceptance date and (ii) the CTG/HRSG Provisional Acceptance Date.

2.7.4   If the New Facility Provisional Acceptance Date does not occur on or before the Guaranteed New Facility Provisional Acceptance Date, then the Contractor shall pay the Sponsor liquidated damages in the amount of:

(a) <<*insert amount and currency*>> for each of the first (1st) through the tenth (10th) Days following the Guaranteed New Facility Provisional Acceptance Date;
(b) <<*insert amount and currency*>> for each of the eleventh (11th) through the twentieth (20th) Days following the Guaranteed New Facility Provisional Acceptance Date; and
(c) <<*insert amount and currency*>> for each Day after the twentieth (20th) Day following the Guaranteed New Facility Provisional Acceptance Date.

Those payment amounts shall cease to accrue on any Day occurring after the New Facility Provisional Acceptance Date.

2.7.5   If the New Facility Final Acceptance Date does not occur on or before its specified date, then the Sponsor shall be entitled (but shall not be required) to complete any or all Punch List items remaining uncompleted at such time with respect to such Punch List (either itself or through other contractors) and to apply to the reasonable cost thereof any or all of the unpaid Adjusted Lump Sum Contract Price or Retainage (including draw upon any Retainage LC), and, if such is insufficient to cover the cost of completing all Punch List items, the Contractor shall pay the excess cost thereof to the Sponsor promptly following demand therefor.

2.7.6   If the Contractor (i) requires more than <<seventy two (72)>> Existing Plant Shutdown Hours, the Contractor shall pay to the Sponsor an amount equal to <<*insert amount and currency*>> for each Existing Plant Shutdown Hour in excess of <<seventy two (72)>> Existing Plant Shutdown Hours or (ii) commences any Existing Plant Outage before <<10:00 p.m.>> or concludes any Existing Plant Outage after <<6:00 a.m.>>, provided that no alternate commencement or conclusion time has been specified by the Sponsor's Project Manager Contractor shall pay to the Sponsor an amount equal to <<*insert amount and currency*>> for each Existing Plant Shutdown

Hour that occurs either before <<10:00 p.m.>> or after <<6:00 a.m.>> ("Shutdown Liquidated Damages").

2.7.7   The Sponsor may issue an invoice to the Contractor for any Shutdown Liquidated Damages accrued hereunder at any time and from time to time.

With Desalinization

["**Net Dependable Water Capacity**" means, in respect of each Hour, the total potable water desalination capacity of the desalination facility (expressed in cubic meters per hour) which is available at the <<Water Delivery Point>>, on an continuous and reliable basis, in conformity with the <<Operating Parameters>> and by reference to the <<Reference Conditions>>, as demonstrated by the <<Acceptance Test>>. The Net Dependable Water Capacity shall be established separately for operation on natural gas and fuel oil in accordance with Exhibits 1G, 1H, 1K and 1Q.]

["**Provisional Net Dependable Water Capacity**" shall mean the capacity of the [Desalination Facility] as notified by the Contractor to the Sponsor upon the Facility Final Acceptance.]

Coal-Fired Power Plants

["**Ash Disposal Agreement**" shall mean <<*insert name*>>.]

["**Ash Hauler**" shall mean <<*insert name*>>.]

["**Limestone Disposal Agreement**" shall mean <<*insert title of agreement*>>.]

["**Limestone Hauler**" shall mean <<*insert name*>>.]

["**Limestone Ratio**" shall mean the kilograms of limestone necessary to flow through the furnace of a Unit during one hour so that the SO_2 emissions from the Facility meet those set forth in Exhibits 1G, 1H and 1K divided by <<*insert number*>> kilograms.]

["**Limestone Supply Agreement**" shall mean <<*insert title of agreement*>>.]

"**Unit Heat Rate**" shall mean:

(a) [3.6 multiplied by (the heat added to the steam and water of a Unit's boiler (kW)) multiplied by 100 divided by the boiler efficiency measured as a percentage based on the net calorific value of the Fuel consumed according to DIN 1942; all divided by

(b) the sum of the electrical power (MW_e) measured at the high voltage side of the generator transformer and delivered by one Unit to the Site Electrical Interconnection Point and the thermal power (MW_{th}) from one Unit at the Thermal Interconnection Point.][26]

Hydroelectric and Geothermal Power Plants

"**Priced Bill of Quantities**" is set forth in Exhibit <<*insert number*>>.

"**Ready for Water Testing Date**" is <<*insert date*>> as may be adjusted as expressly provided under the Contract.

Certain Underground Risks. The Sponsor will be responsible only for the unit costs of the items below (as such unit rates are set out in the Priced Bill of Quantities) if the underground conditions below vary from baseline estimates set forth in the Sponsor's Information as follows:

(a) injection or support or injection or the water inflow rate from the rock into the tunnel in excess of that set forth in [Exhibit <<*insert name*>>];

(b) calculated on the basis of the unit rates set forth in [Exhibit <<*insert name*>>];

(c) dam and tunnel at agreed unit rates set forth in [Exhibit <<*insert name*>>];

(d) during tunnel excavation if rock type in any segment varies from the expected baseline rock type for such segment then the Sponsor will pay linear meter unit rates for the rock type actually encountered;

(e) in connection with the foregoing excavation, if the tunnel support necessary to be installed as a result of rock type varying from the rock type baseline estimates set forth in the Sponsor Information is different from the tunnel support in the agreed baseline tunnel support estimate set forth in the Sponsor Information, the Sponsor will pay the additional or be rebated the reduced cost (as the case may be) for costs for such additional or reduced structural support measures such as additional shotcrete, rock bolts, etc.;

(f) after hydro fracture testing to be carried out following tunnel excavation, the exact amount of high pressure pipe will be established and the difference between the estimated amount set forth in the Sponsor Information and the actual amount required will be calculated and the variance will be paid to the Contractor or rebated to the Sponsor; and

(g) if in relation to the construction and design of the intake Work the soil conditions vary from those set forth in the baselines estimate set forth in the Sponsor Information then the Sponsor will pay for additional lining, or other acceptable impermeability measure, etc., required.

Solar Power Plants

"**Array Block**" the as-designed quantity of Modules and Inverters connected to a single medium voltage step-up transformers.

"**Balance of System**" or "**BOS**" shall mean all aspects and components of the Facility, including but not limited to the Equipment, other than the Modules.

"**BOS Buydown Amount**" has the meaning set forth in Article <<*insert number*>>.

"**BOS Criteria**" means the Net Energy of the Balance of System are less than or equal to the design losses of the Balance of System, calculated in accordance with Exhibit 9.2.

"**BOS Performance Test**" means a test of the performance of the Balance of System in accordance with the testing protocol set out in Exhibit 1Q and the Functional Specifications.

"**Design Output**" means the designed electricity production of the Facility as designed by Contractor expressed in kWh for the first year of electricity production following the Facility Provisional Acceptance Date, determined by Contractor using the Output Criteria Model, Fixed Parameters and Variable Parameters values resulting from Contractor's latest drawing for the Facility, as certified by Contractor.

"**Equipment**" means all of the equipment, materials, apparati, structures, components, instruments, appliances, supplies and other goods required to complete the

Work and constituting the Facility, including but not limited to any trackers, inverters, transformers, SKIDS, wiring devices, substation equipment, DAS instrumentation, DAS control, combiner/re-combiner boxes, mounting system, revenue meters, plane-of-array pyranometers, horizontal pyranometers, back-of-module thermocouples, wind-speed sensors, wind-direction sensors, and ambient air temperature sensors. Equipment shall not include any (i) Modules or (ii) materials, apparati or tools owned by Contractor Person that are used to complete the Work but are not contemplated under this Contract to become part of the Work.

"**Module Delivery Schedule**" means the schedule of delivery of the Modules set forth in Exhibit <<*insert name*>> reflecting: (1) quantity of Modules; and (2) dates available for delivery to the Site.

"**Module Pick Up Date**" shall have the meaning set forth in the Module Warranty Terms and Conditions.

"**Module Product Specifications**" has the meaning given to such term in the Module Supply Agreement.

"**Module Supplier**" means the module supplier under the Module Supply Agreement.

"**Module Supply Agreement**" means is set forth in Exhibit <<*insert name*>>.

"**Module Warranty Terms and Conditions**" means those terms and conditions of the Module Supply Agreement providing warranties for the Modules.

"**Modules**" means multi-crystalline silicon photovoltaic modules to be procured by Owner as more particularly described in the Module Supply Agreement.

"**Net Energy Losses**" has the meaning given to the value of "*Net percentE$_{loss}$*" in the BOS performance test protocol set forth in Exhibit 1Q.

"**Output Criteria**" means the designed electricity production of the Facility for the first year of electricity production following the Facility Provisional Acceptance Date shall be no less than <<*insert name*>> kWh.

"**Output Criteria Model**" means the Base Case Model using the values set forth in the Functional Specifications for the Fixed Parameters of Variable Parameters.

"**Performance Criteria**" shall mean that the actual AC power rating of the Facility (P_{ACTUAL}) is equal to or greater than the contract AC power rating ($P_{CONTRACT}$) for the Facility at the time the Final Acceptance Tests are performed, where the value of (P_{ACTUAL}) and ($P_{CONTRACT}$) are expressed in terms of kWac and determined in accordance with Exhibit 1Q.

"**Solar Module Logistics Provider**" means Module Supplier or any other Person identified in writing by Sponsor to Contractor from time to time.

"**STC**" means standard test conditions defined at 1000W/m^2, 25 degrees Celsius, wind speed of 1m/s, and air mass of 1.5.

"**Technical Assessment**" means a thorough visual, mechanical and technical investigation of the Facility, including but not limited to Modules and BOS, and operating data related thereto resulting in a written diagnostic report prepared by Contractor in accordance with the Functional Specification, containing detail sufficient, as reasonably determined by Sponsor, to indicate the root cause(s) of the Facility's failure to achieve the Performance Criteria in any Final Acceptance Tests, and specifically indicating whether or not such root cause(s) is attributable to the BOS, Module workmanship and installation for which Contractor is responsible under this Contract, Module Defects covered by the Module Warranty Terms and Conditions, or any combination thereof. No later than <<*insert number*>> Days prior to the Facility Provisional Acceptance Date, Contractor shall present a written description of Contractor's proposed visual, mechanical and technical investigation approach for any Technical Assessment.

"**Testing Model**" has the meaning set forth in Exhibits 1K and 1Q.

"**Variable Parameters**" means the following parameters of the Base Case Model: (a) ground coverage ratio; (b) number of Modules; (c) Module nameplate (Wdc at STC); (d) PV losses due to temperature and irradiance level; (e) DC wiring losses; inverter losses; and (f) AC side losses. For greater certainty, AC side losses include transformation losses, AC ohmic losses, and AC power consumption of station service/parasitic loads.

Supply and Procurement of Equipment. Contractor shall at its own expense procure or supply and pay for all of the Equipment, shall arrange and pay for the delivery of all Equipment to the Site, shall arrange and pay for the receipt, off-loading and storage of all Equipment and Modules at the Site, and, shall arrange for the off-loading of all Modules at the Site (and, if applicable, the return of contains for Modules to the appropriate person). Within <<*insert number*>> Days of the Notice to Proceed, Contractor shall identify to Owner in writing an appropriate location on the Site for delivery of the Modules.

Module Responsibilities. Contractor's responsibilities with respect to Modules procured by Owner under the Module Supply Agreement shall be as follows. Contractor shall request the delivery of Modules to the Site by the Solar Module Logistics Provider, at the times and in the quantity required to enable Contractor to achieve the Completion Guarantees, using the Logistics Monitoring System. Upon their delivery to the Site by the Solar Module Logistics Provider, Contractor shall be responsible for;

(a) safe off-loading of the Modules from the Solar Module Logistics Provider's vehicle in accordance with the Installation and User Manual and <Good> <Prudent> <Utility> Practices, and in such a manner so as to ensure no damage occurs or exclusions to the Module Warranty Terms and Conditions shall apply;

(b) completing safe off-loading from the Solar Module Logistics Provider's vehicle on the day any Modules are delivered to the Site (and, for all Module deliveries made to the Site after <<*insert time*>> a.m. and prior to <<*insert time*>> p.m. local time, if Contractor fails to so safely off-load the Modules within a <<*insert time*>> hour period, for reimbursing Sponsor for any overtime payments payable by Sponsor to the Solar Module Logistics Provider pursuant to the Solar Module Logistics Agreement as a result of such delay);

(c) conducting a visual inspection of the shipping containers and packaging for the Modules;

(d) providing written certification in a form satisfactory to the Sponsor indicating that Contractor has conducted such visual inspection and that the shipping and/or storage containers for the Modules exhibit no visible damage from prior shipping or storage, with any exceptions (including, without limitation, damage or volume shortfalls) noted and documented;

(e) scanning the barcodes of each palette of Modules using the On-Site Logistics Portal and, by the end of the day on which the Modules are delivered, transmitting to Solar Module Logistics Provider through the Logistics Monitoring System an <<*insert computer program name*>> file or batch file in the pre-agreed format that identifies the delivered Modules by serial number; and

(f) risk of loss for the Modules. Following Contractor's inspection pursuant to this Article [____], Contractor shall be responsible for the Modules in the same manner and to the same extent that Contractor is responsible for all Work, and Contractor shall fulfill those Warranty obligations set forth at Article 13.

(iii) Contractor shall install and incorporate the Modules into the Work in accordance with this Contract, and shall interface with the Module Supplier and Sponsor as necessary and as requested by Sponsor to ensure the proper installation, incorporation and operation of the Modules. Upon installation in full of each Array Block, Contractor shall record the location of such Array Block and the serial number of each Module incorporated into such Array Block and, by the end of the day on which the Modules are installed, transmit to Solar Module Logistics Provider through the Logistics Monitoring System an <<*insert computer program name*>> file or batch file in the pre-agreed format that identifies the Modules installed into such Array Block by serial number.

IT Monitoring System. Contractor shall (i) install at the Site a Module barcode scanning system and information portal (the "On-Site Logistics Portal") and (ii) cooperate with Sponsor and Solar Module Logistics Provider in the development of electronic linkages between the On-Site Logistics Portal and the integrated electronic database being developed by Owner and Solar Module Logistics Provider (together with the On-Site Logistics Portal, the "Logistics Monitoring System"), to enable Contractor to request delivery of Modules pursuant to Article [____], record the delivery of Modules to Site pursuant to Article [____], and record the location of the Array Block in which each Module is installed pursuant to Article [____].

Modules

Sponsor has entered into (i) the Module Supply Agreement whereby the Module Supplier has agreed to make the Modules available to [_____] according to the terms and conditions of the Module Supply Agreement and (ii) the Solar Module Logistics Agreement whereby the Solar Module Logistics Provider has agreed to warehouse (if required) and deliver Modules to the Site. Sponsor shall procure that (A) Modules are available for delivery to the Site by Solar Module Logistics Provider in accordance with Exhibit 1C and (B) Modules are delivered to the Site by the Solar Module Logistics Provider within <<*insert number*>> Business Days after the date requested by the Contractor pursuant to Article [____]. Once the Modules are delivered to the Site, Contractor shall be responsible for the off-loading, inspecting the Modules in accordance with Article [____], storing the Modules at the Site, and installation and incorporation of the Modules into the Facility, risk of loss and any other obligations expressly assigned to Contractor with respect to the Modules under the terms of this Contract.

Contractor acknowledges that Sponsor has provided to Contractor the Installation and User Manual for the Modules, certified by the Module Supplier and the Module Warranty Terms and Conditions. Contractor shall promptly inform Sponsor if

Contractor believes such installation and user manual requires amendment. In the event any amendment to such installation and user manual is agreed between Sponsor and Module Supplier, Sponsor shall inform Contractor of the amended terms as soon as reasonably practicable, subject to Module Supplier's approval.

In the event Sponsor and Module Supplier agree to an amendment to the Module Supply Agreement that affects the scope or extent of the Module Warranty Terms and Conditions during the Facility Warranty Notification Period, Sponsor shall inform Contractor of the amended terms as soon as reasonably practicable, subject to Module Supplier's approval.

Output Criteria. With each drawing submittal provided by Contractor pursuant Exhibits 1K and 5.1(z), Contractor shall provide a certification setting forth the value of the Variable Parameters implied by such drawing for the Facility, the Design Output provided by such drawing and confirmation that the drawing is in accordance with Contractor's obligation to perform the Work to achieve the Output Criteria. In the event Contractor is not able to make the certification required by this Article [__] at any time, Contractor shall immediately: (i) provide written notice to Sponsor setting forth the reasons therefor; (ii) within <<*five (5)*>> Business Days of such notice, provide a revised drawing to Sponsor that achieves the Output Criteria, and (iii) conform the Work to such revised drawing.

Module Warranty. Sponsor appoints Contractor to administer and enforce the Module Warranty Terms and Conditions until the Facility Provisional Acceptance Date. Contractor shall be responsible for enforcing the Module Warranty terms and Conditions with respect to the Modules throughout performance of the Work until the Facility Provisional Acceptance Date. Until the Facility Provisional Acceptance Date, if Contractor discovers any Defective Module or any breach of the Module Warranty Terms and Conditions, it shall promptly (and in any event within <<*five (5)*>> Days of discovery thereof) notify Sponsor in writing. Upon knowledge or notice of any breach of the Module Warranty Terms and Conditions, Contractor shall enforce any rights of Sponsor pursuant to the Module Warranty Terms and Conditions with respect to such Module, and (a) if already installed, disconnect and extract any such Module from the Facility, (b) inspect any repaired or replacement Module that is provided by Module Supplier, (c) install and incorporate any repaired or replacement Modules into the Work, and (d) otherwise remedy any breach of the Module Warranty Terms and Conditions. Upon completion of any Work relating to the Modules, Contractor shall update the active inventory list to reflect the serial numbers and all other relevant information with respect to any new Modules installed as part of such Work. In the event Contractor has been unable to enforce Sponsor's rights despite compliance with the Module Warranty Terms and Conditions, Contractor shall provide notice to Sponsor. Contractor shall not engage in any activity or use Modules in any way that will void the Module Warranty and Terms and Conditions. Upon receipt of such notice from Contractor, Sponsor shall make commercially reasonable efforts to support Contractor in the enforcement of such rights; provided, however, Contractor shall not be relieved of its obligations under this Article <<*insert number*>>. In the event that more than twenty percent (20 percent) of any Modules submitted by Contractor pursuant to a single return merchandise authorization for claims under the Module Warranty Terms and Conditions are found to not be defective, Contractor shall indemnify Sponsor for any "no defect found charge" payable to the Module Supplier pursuant to the "Return Policy" set forth at Schedule [___] of the Module Warranty Terms and Conditions. For any outstanding administration, enforcement and remedy obligations under this Article <<*insert number*>> outstanding as of the Facility Provisional Acceptance Date, Contractor shall provide a detailed written notice to Sponsor regarding such outstanding obligations in sufficient detail to enable Sponsor's separate operations and maintenance provider to complete such outstanding obligations following the Facility Provisional Acceptance Date pursuant to a separate operations and maintenance agreement. {For clarity, notwithstanding any other provision of this Contract, Sponsor agrees and acknowledges that Contractor is only responsible for the performance of the Modules, including but not limited to the failure of the Modules to meet the required performance specifications, to the extent arising from or relating to the acts or omission of Contractor.}

Wind Power Plants

["**Balance of Plant Work**" means the Work associated with provision of the Facility other than the supply (to the point of import in <<*insert country name*>>) of the <<*insert equipment list*>>.]

"**Power Curve**" means <<_____>>.

"Warranted Performance" occurs when:

(a) [the Turbine in question can safely and reliably generate and export electrical power at not less than <<*ninety percent (90 percent)*>> of the Warranted Power Curve.][27]

(b) the Work has been substantially completed in accordance with this Contract, including all work that is necessary to allow the Turbine in question to safely and reliably generate and export electrical power at not less than <<*ninety percent (90 percent)*>> of the Warranted Power Curve. [28]

(c) the sound emitted from any Turbine under normal operating conditions and following completion of any repairs, modifications or adjustments performed pursuant to Article 11.3 shall not, as measured in accordance with the IEC standard for acoustic noise measurement techniques "Wind Turbine Generator Systems, IEC 61400-11 Acoustic Measurement Techniques, International Electrotechnical Commission, Netherlands, 1998" (such sound level measurements to be corrected back to the Turbine using American National Standards Institute Standards or other generally recognized acoustical engineering calculation methods), exceed 104 dBA measured at the hub of such Turbine.

(d) Verification of Warranty Compliance. If at any time during the Warranty Notification Period Arkadia reasonably suspects that the Power Curve of any Turbine is less than that warranted in accordance with Article 9.2 above, the Sponsor may require that the Contractor perform a power curve verification test pursuant to the test procedures set forth in the [IEC publication. "Wind Turbine Generator Systems, Part 12—Wind Turbine Power Performance Testing, Final Draft International Standard IEC-61400-12 (latest edition)]. If the results of such additional testing demonstrate that the Power Curve of any Turbines so tested is less than that warranted, the costs related to the testing of such Turbine shall be at the Contractor'ssole expense, and the Contractor shall remedy the condition in accordance with Article 11.3. If the results of such additional testing demonstrate that the Power Curve of any Turbines so tested is greater than or equal to that warranted, the costs related to the testing of such Turbine shall be borne by the Sponsor.

(e) Serial Defects. A "Serial Defect" in the Work shall be deemed to exist if during the Warranty Notification Period twenty percent (20%) or more of the same part or component of the Work contain the same defect or deficiency ("Threshold Percentage"). If a Serial Defect is discovered in any Turbine component. The Sponsor shall notify the Contractor and the Contractor shall perform a thorough investigation of all Turbines as soon as possible after such notice and shall provide the Sponsor with a detailed written report identifying the cause of the Serial Defect, whether the Serial Defect is one that is likely to affect other Turbines in the Facility and, if so, which Turbines may be affected. In the event of the discovery of a Serial Defect during the original Warranty Notification Period, the Warranty Notification Period for the part or component containing the Serial Defect shall be extended by an additional <<*insert number*>> Days. Any Serial Defects that occur during the Warranty Notification Period and are agreed by the Parties to be attributable to Turbine commissioning only and not anticipated by the Parties to affect normal operation of the Turbines or the Facility shall not be considered in calculating whether such defect or deficiency exceeds the Threshold Percentage and comprises a Serial Defect.

Electrical Transmission Facilities

"**Apparent Currents**" means the effective cable system current, from the three high voltage cables that contributes to the magnetic field resulting from the vector sum of the Load Currents and induced currents traveling on cable sheaths, shielding and other cable system components.

"**BIL**" means basic impulse insulation level, a reference impulse insulation strength expressed in terms of the crest value of the withstand voltage of a standard impulse voltage wave.

"**BSL**" means basic switching impulse insulation level, a reference switching impulse insulation strength expressed in terms of the crest value of the withstand voltage of a standard switching impulse voltage wave.

"**Cable Route**" means the route extending between the Injection Point and the Delivery Point. With the exception of the Cable Route inside the Substation, the Contractor must locate the cable in the permissible corridor for the Cable Route as shown on the Route Planning Maps (included as Attachment <<*insert number*>>). The Route Planning Maps illustrate the right-of-way limits and easement limits established by the Sponsor through negotiation with the interested public agencies and private landowners, as recorded in the Sponsor Land Use Agreements for installation and maintenance of the Generator Lead Facility.

"**Cable System**" is defined as any equipment or ancillary system required for the Generator Lead Facility between and including the Contractor's points of interconnection as described in the Functional Specification. This includes potheads, SCFF

Cable, FO cable, splices, grounding, vaults, duct banks, conduit, TPS system, GLF Control House, oil pressurization system, leak detection system, DTS, discrete temperature sensor system, auxiliary power systems, manholes, transition stations, cofferdams, foundation, support structures, corrosion protection, and fire detection and protection system.

"**dB**" means decibel.

"**DCS**" means distributed control system.

"**DTS**" means distributed temperature sensor system.

"**EMI**" means electromagnetic interference in the normal operation of adjacent systems as a result of the operation of the Cable System.

"**Generator Lead Facility 600 MVA Delivery Capability Guaranty**" is defined in Section [__].

"**Generator Lead Facility 670 MVA Delivery Capability Guaranty**" is defined in Section [__].

"**Generator Lead Facility Line Loss Guaranty**" is defined in Section [__].

"**FO**" means fiber optic.

"**Forced Outage Hours Guaranty**" is defined in Article [__].

"**Forced Outage Hours Liquidated Damages**" are defined in Article [__].

"**Generator Lead Facility 3—Phase Line Loss Guaranty**" is defined in Article [__].

"**Generator Lead Facility << >>MVA (<< >>KV, << >>A) Performance Guaranty**" is defined in Article [__].

"**Generator Lead Facility**" means the ___ Kilovolt alternating current generator lead and related facilities necessary to transmit ___MVA of electrical power from the Facility to the [_____] Station as more fully described in the Functional Specification.

"**GIS**" means gas insulated substation.

"**GSU**" means generator step up transformer.

"**HDD**" means horizontal directional drilling.

"**HDPE**" means high density polyethylene plastic material.

"**Line Loss**" is defined as the sum of all losses in the cable including the conductor, dielectric, sheath, and armor carried out to the third decimal place.

"**Load Current**" means the positive and negative currents carried by the high voltage conductors while transmitting electrical power from its source to its load.

["**Loss of Availability Damages**" are defined in Article 7.2.13.]

"**Magnetic Fields**" means the area influenced by the flow of current through the Cable System.

"**Maximum Line Loss Level**" is defined in Article [__]

"**Maximum Line Loss Guaranty**" is defined in Section [__].[29]

"**Minimum Delivery Capability Guaranty**" is defined in Section [__].

"**MVA**" means megavolts amperes.

"**Net Current**" means the vector sum of all of the currents flowing in all of the conductors of a three-phase cable system.

"**Performance Guarantees**" are the Generator Lead Facility [___] MVA (<<__>> kV, <<__>>A) Performance Guaranty and the Generator Lead Facility 3-Phase Line Loss Guaranty.

"**RTU**" means remote terminal unit.

"**SCADA**" means supervisory control and data acquisition system.

"**Stray Current**" means the sum of all currents that flow on the conductors of a three-phase cable system that are not caused by the transmission of power over the three-phase cable system. An example of Stray Current in this project is current that flows on the three-phase cable system of the Generator Lead Facility when no Load Current is present.

"**Submarine Cable**" means the armored SCFF Cable to be installed between the [___] Transition Station and the [___] Transition Station.

(a) Factor performance test certificates for the land and submarine cable have been accepted by the Company and all equipment has been installed in accordance with manufacturers' requirements and the requirements of this Contract.

(b) The DC high voltage test has been completed. The DC high voltage test procedures shall be submitted to the Sponsor <<*one hundred (100)*>> Days prior to testing.

(c) the Performance Tests have been completed and the results thereof meet the requirements set forth in this EPC Contract; and (i) the Generator Lead Facility [___] MVA (<<__>> kV, <<__>>A) Performance Guaranty and the Generator Lead Facility 3-Phase Line Loss Guaranty have been achieved or (ii) the Minimum Performance Level and the Maximum Line Loss Level have been achieved.

The Contractor shall have <<*thirty (30)*>> Days from the Final Acceptance Date to attempt to meet the Generator Lead Facility [___] MVA (<<__>> kV, <<__>>A) Performance Guaranty and the Generator Lead Facility 3-Phase Line Loss Guaranty (the "Cure Period") and failing such achievement by the end of the Cure Period all Performance Liquidated Damages pursuant to Article [__] must be paid within <<*five (5)*>> Days of the end of the Cure Period. If Final Acceptance has occurred but the Contractor has not achieved the Generator Lead Facility [___] MVA (<<__>> kV, <<__>>A) Performance Guaranty and the Generator Lead Facility 3-Phase Line Loss Guaranty, the Contractor and the Sponsor shall use commercially reasonable efforts to minimize the economic impact on the Sponsor during the Cure Period, if any, and the Contractor shall be responsible for any damage or degradation to the Generator Lead Facility it causes during the Cure Period.<here>

1. Based on the thermal resistivities and soil temperatures contained in Attachment 3-G, and in accordance with the Performance Testing requirements contained in the Scope, the Contractor warrants that the Generator Lead Facility shall be capable of continuously delivering <<__>> MVA (<<__>> kV, <<__>>A) to the [___] Delivery Point from the [__] Injection Point and while providing such service the cable conductor will remain at or below <<90>> degrees Celsius ("Generator Lead Facility [___] MVA (<<__>> kV, <<__>>A Performance Guaranty").

2. The Contractor warrants that the Generator Lead Facility shall be capable of delivering electrical power, as measured in MVA, from the Injection Point to the Delivery Point with Line Losses between the Injection Point and the Delivery Point not exceeding <<*insert number*>> kW per phase and <<*insert number*>> kW for the entire 3-phase system at [___] MVA (<<*insert number*>> kV, <<*insert number*>>A) ("Generator Lead Facility 3-Phase Line Loss Guaranty"). The

Generator Lead Facility 3-Phase Line Loss Guaranty shall be demonstrated by measuring the cable resistance during factory testing of each cable length and calculating the Line Loss in accordance with the Functional Specification.

3. The Minimum Performance Level shall be met if the Generator Lead Facility can deliver [_____] MVA to the Delivery Point from the Injection Point and while providing such service the cable conductor temperature will not exceed <<90>> degrees Celsius.

4. The Maximum Line Loss Level shall be met if, during the testing to determine whether the Generator Lead Facility 3-Phase Line Loss Guaranty has been achieved, Line Losses do not exceed <<insert number>> kW per phase and <<insert number>> kW for the entire 3-phase system at [___] MVA (<<insert number>> kV, <<insert number>>A).

5. In the event the EPC Contractor fails to achieve the Generator Lead Facility [__] MVA (<<insert number>> kV, <<insert number>>A) Performance Guaranty during the Performance Tests but achieves the Minimum Performance Guaranty, the Contractor must, within <<five (5)>> Days of the end of the Cure Period, pay the Company liquidated damages as follows:**Formula**: ([___] MVA – Actual Tested MVA) * [_____] dollars <<insert currency amount>> = Payment <<insert currency amount>>.

6. In the event the Contractor fails to achieve Generator Lead Facility 3-Phase Line Loss Guaranty during factor testing but achieves the Maximum Line Loss Level, the EPC Contractor must, within <<five (5)>> Days of the end of the Cure Period, pay the Sponsor liquidated damages (together with the liquidated damages for failure to achieve the Generator Lead Facility [___] MVA (<<insert number>> kV, <<insert number>>A) Performance Guaranty "Performance Liquidated Damages") as follows:**Formula:** (Actual 3-Phase Line Loss – [___])/[___] * [_____]dollars <<insert currency amount>> = Payment <<insert currency amount>>.

7. The Contractor warrants that, during each year of the Warranty Notification Period, the Generator Lead Facility shall have Forced Outage Hours of no more than [___] hours (the "Forced Outage Hours Guaranty"). If during the Warranty Notification Period the Generator Lead Facility has Forced Outage Hours in excess of [___] hours in any year, the Contractor must immediately pay to the Sponsor liquidated damages in the amount of [_____] (<<insert currency amount>>) per hour for each hour in excess of [___] Forced Outage Hours in each such year ("Forced Outage Hours Liquidated Damages"), subject to the following conditions:

(a) Forced Outage Hours shall not include any hour or portion thereof not reported to (the "Grid Operator") as Forced Outage Hours;

(b) the cause of the Forced Outage is the Generator Lead Facility;

(c) the Forced Outage is attributable to the action or omission of the Contractor;

(d) the Sponsor maintains an inventory of spare parts as listed in Exhibit 4.29.1; and

(e) Forced Outage Hours Liquidated Damages will not accrue during any period that the Sponsor Permits required for the operation of the Generator Lead Facility are not in effect or the Contractor is not permitted by the Sponsor to carry out warranty work.

8. "Forced Outage" shall mean an interruption or reduction of the transmission capacity in MVA of the Generator Lead Facility that is not due to (a) a Forced Majeure Excused Event; (b) planned maintenance outage or ISO approved unplanned maintenance outage; or (c) the act or omission of any non-Contractor Person which is not in accordance with the Contractor's instruction manuals or prudent industry practices.

9. Forced Outage Hours for any year in which Forced Outage Hours are measured shall equal:

$$\sum_{i}^{n} FOH_i$$

where:

n = the total number of periods of Forced Outage during the year in which the Forced Outage Hours are measured:<here>

$$FOH_i = H_i \left(1 - [R_i \div F]\right)$$

where:

H_i =the number of hours in the period of Forced Outage in question during the year in which Forced Outage Hours are being measured;

R_i =the transmission capacity in MVA of the Generator Lead Facility during the Forced Outage I;

F =<<*insert number*>> MVA.

The Contractor warrants that the Generator Lead Facility can be used and operated in a manner that complies with all Permits, which were in effect during the execution of the Work, for the design life of the Generator Lead Facility.

10. The Contractor's liability for Liquidated Damages is limited as follows:
    (a) Delay Liquidated Damages – <<*insert number*>> percent of Contract Price;
    (b) Performance Liquidated Damages – Generator Lead Facility [___] MVA (<<*insert number*>> kV, <<*insert number*>>A) Performance Guaranty [___] percent of Contract Price;
    (c) Line Loss Liquidated Damages – Generator Lead Facility 3-Phase Line Loss Guaranty (<<*insert number*>> percent of Contract Price);
    (d) Forced Outage Hours Liquidated Damages – Forced Outage Hours Guaranty (<<*insert number*>> percent of Contract Price); and
    (e) The Contractor's aggregate liability for all Liquidated Damages shall not exceed [*insert number*] percent of the Contract Price.
        1. At ISO conditions, the EPC Contractor warrants that the Generator Lead Facility shall be capable of delivering [___] MVA to the Delivery Point from the Injection Point at a temperature of no more than [ ] ("Generator Lead Facility [___] MVA Delivery Capability Guaranty").

2. At ISO conditions, the Contractor warrants that the Generator Lead Facility shall be capable of delivering <<*insert number*>> MVA to the Delivery Point from the Injection Point at a temperature of no more than [____] ("Generator Lead Facility [____] MVA Delivery Capability Guaranty").

3. The Contractor warrants that the Generator Lead Facility shall be capable of delivering electrical power, as measured in MVA, from the Injection Point to the Delivery Point with line losses at the Delivery Point not exceeding [____] percent of the MVA injected at the Injection Point ("Generator Lead Facility Line Loss Guaranty").

4. The Generator Lead Facility [____] MVA Delivery Capability Guaranty, the Generator Lead Facility [____] MVA Delivery Capability Guaranty and the Maximum Line Loss Guaranty shall collectively be referred to as the "Performance Guarantees."

5. The Minimum Delivery Capability Guaranty shall be <<ninety five percent (95 percent)>> of the Generator Lead Facility [____] MVA Delivery Capability Guaranty at a temperature of no more than [____].

6. The Maximum Line Loss Guaranty shall be <<two hundred percent (200 percent)>> of the Generator Lead Facility Line Loss Guaranty.

7. The Minimum Delivery Capability Guaranty and Maximum Line Loss Guarantee shall be collectively referred to as the Minimum Performance Level.

8. In the event the Contractor fails to achieve the Generator Lead Facility [____] MVA Delivery Capability Guaranty during the Performance Tests but achieves the Minimum Performance Level, the Contractor must, within <<five (5)>> Days of the end of the Cure Period, pay the Sponsor liquidated damages as follows:
   Formula: ([____] MVA – Actual Tested MVA) * [_____] dollars (<<*insert currency amount*>>) per MVA = Payment <<*insert currency amount*>>.

9. In the event the Contractor fails to achieve the Generator Lead Facility Line Loss Guaranty during the Performance Tests but achieves the Minimum Performance Level, the Contractor must, within <<five (5)>> Days of the end of the Cure Period, pay the Sponsor liquidated damages as follows:
   Formula: (Actual Line Loss (percent) – Maximum Line Loss Guaranty (percent)) * [_____] dollars (<<*insert currency amount*>>) per percentage point = Payment (<<*insert currency amount*>>)

10. The Contractor warrants that, during each year of the warranty notification period, the Generator Lead Facility shall have: (a) an Annual Availability of no less than <<0.99>> and (b) Forced Outage Hours of no more than [____] hours. If at the end of the end of the warranty notification period the Generator Lead Facility has an Annual Availability of less <<0.99>> in any year of the warranty notification period, the EPC Contractor must immediately pay to the Company liquidated damages in the amount of <<*insert currency and amount*>> per 0.001 for each 0.001 that the Annual Availability is below <<0.99>> in each such year. If at the

end of the end of the warranty notification period the Generator Lead Facility has Forced Outage Hours in excess of <<*insert number*>> hours in any year, at the end of the warranty notification period, the Contractor must immediately pay to the Sponsor liquidated damages in the amount of [_____] dollars per hour for each hour in excess or [___] hours in each such year.

11. "Annual Availability" shall be determined according the following formula:

Annual Availability =

$$(8760 - AMH - \sum_i^n FOH_i \div 8760$$

$$FOH_i = H_i[1 - (R_i \div F)]$$

F = [___] MVA, or, if the Contractor has made payment in full of all required Performance Damages with respect to the Generator Lead Facility [___] MVA Delivery Capability Guaranty, the MVA which forms the basis for the Performance Damages.

FOHi = the number of hours during which the reduction or elimination of the transmission in MVA is in effect during the particular Forced Outage i;

AMH = the actual number of hours the Generator Lead Facility is out of service as the result of the annual scheduled maintenance during the year which Annual Availability Period is being measured but in no event greater than <<*insert number*>> hours;

Ri= the transmission capacity in MVA of the Generator Lead Facility during the Forced Outage i;

Hi = the number of hours in the period of Forced Outage in question during the year in which Annual Availability is being measured.

N = total number of periods of Forced Outage during the year in which the Annual Availability is being measured.

12. Forced Outage Hours for any year in which Annual Availability is measured shall equal:

$$\sum_i^n FOH_i$$

["**Facility [___] MVA Delivery Capability Guaranty**" is defined in _____.]
["**Facility [___] MVA Delivery Capability Guaranty**" is defined in _____.].
["**Facility Line Loss Guaranty**" is defined in _____].
["**Maximum Line Loss Guaranty**" is defined in _____.]
["**Minimum Delivery Capability Guaranty**" is defined in _____.]
["**MVA**" means megavolts amperes.]

At ISO conditions, the EPC Contractor warrants that the Generator Lead Facility shall be capable of delivering 670 MVA to the [___] Delivery Point from the [___] Injection Point at a temperature of no more than [__] ("Generator Lead Facility 670 MVA Delivery Capability Guaranty."

At ISO conditions, the EPC Contractor warrants that the Generator Lead Facility shall be capable of delivering 600 MVA Delivery to the [___] Delivery Point from the [____] Injection Point at a temperature of no more than [__] ("Generator Lead Facility 600 MVA Delivery Capability Guaranty").

The EPC Contractor warrants that the Generator Lead Facility shall be capable of delivering electrical power, as measured in MVA, from the [____] Injection Point to the [____] Delivery Point with line losses at the [____] Delivery Point not exceeding [_____]% of the MVA injected at the [____] Injection Point ("Generator Lead Facility Line Loss Guaranty").

The Generator Lead Facility 670 MVA Delivery Capability Guaranty, the Generator Lead Facility 600 MVA Delivery Capability Guaranty, and the Maximum Line Loss Guaranty shall collectively be referred to as the "Performance Guarantees."

The Minimum Delivery Capability Guaranty shall be ninety five percent (95%) of the Generator Lead Facility 670 MVA Delivery Capability Guaranty at a temperature of no more than [_____].

The Maximum Line Loss Guaranty shall be two hundred percent (200%) of the Generator Lead Facility Line Loss Guaranty.

The Minimum Delivery Capability Guaranty and the Maximum Line Loss Guarantee shall be collectively referred to as the Minimum Performance.

In the event the EPC Contractor fails to achieve the Generator Lead Facility 670 MVA Delivery Capability Guaranty during the Performance Tests but achieves the Minimum Performance Level, the EPC Contractor must, within five (5) Days of the end of the Cure Period, pay the Company liquidated damages as follows:

Formula: (670 MVA – Actual Tested MVA) * [_____] dollars ($[_____])
per MVA = Payment ($)

In the event the EPC Contractor fails to achieve the Generator Lead Facility Line Loss Guaranty during the Performance Tests but achieves the Minimum Performance Level, the EPC Contractor must, within five (5) Days of the end of the Cure Period, pay the Company liquidated damages as follows:

Formula: (Actual Line Loss (%) – Maximum Line Loss Guaranty (%)* [_____]
dollars ($[_____]) per percentage point = Payment ($)

"Forced Outage" shall mean an interruption or reduction of the transmission capacity in MVA of the Generator Lead Facility that is not due to (a) a Forced Majeure Excused Event: (b) scheduled maintenance in excess of 88 hours; or (c) the act or omission of any non-Contractor Person.

Forced Outage Hours for any year in which Annual Availability is measured shall equal:

$$\sum_{i}^{n} FOH_i$$

The EPC Contractor warrants that the Generator Lead Facility can be used and operated in a manner that complies with all Permits for the design life of the Generator.

Liquid Natural Gas Terminals

Gasification Facilities and Regasification Facilities

1.0 DEFINITIONS

"**Commissioning LNG**" means such quantity of LNG reasonably necessary to carry out the Tests to achieve completion of the Work, which are estimated to be:

(a) in respect of completion of the First Gas Works, not less than 100,000 cubic meters of LNG;
(b) in respect of completion of the First Tank Export Works, not less than 30,000 cubic meters of LNG; and
(c) in respect of completion of the remainder of the Work, not less than 30,000 cubic meters of LNG.

"**Commissioning LNG Tanker**" means a vessel used for the transportation of Commissioning LNG.

"**Commissioning Procedure**" means the procedure for the Commissioning Tests to be developed pursuant to Clause [_____] (Commissioning Tests).

"**Commissioning Spare Parts**" means those spare parts which are necessary to enable the Contractor to carry out commissioning and testing of the relevant Phase.

"**Commissioning Tests**" means the tests and procedures described as such in the Scope of Work or reasonable to be inferred therefrom or as agreed between the parties in accordance with this Contract.

"**Company Delay Event**" means, in relation to the Performance Texts:

(a) a failure by the Sponsor to provide Commissioning LNG;
    (i) in respect of the First Gas Export Works, within the Fourth Window Period as defined in Clause [__] (Commissioning LNG);
    (ii) in respect of the First Tank Works, within the window period nominated under Clause [__] (Commissioning LNG);
    (iii) in respect of the remainder of the Work, within the window period nominated under Clause [__] (Commissioning LNG);
(b) subject to Clause [__] (Managing Connections) and [__] (Co-operation with Interface Contractors) and provided the Regas Facilities have achieve completion in respect of the First Gas Export Works, a failure by the Sponsor to have in place a gas pipeline that is capable of receiving Fuel; or
(c) an instruction from the Sponsor to the Contractor that the Contractor may not commence the Performance Tests because the Sponsor would otherwise be prevented from discharging its commercial contracts or where an instruction is otherwise given for the Sponsor's convenience; save where any failure to comply with such obligation in (a) or (b) above was caused or contributed to by the Contractor or where the instruction in (c) above was necessitated by the Contractor or where the instruction in (c) above was necessitated by the poor performance of the Contractor and providing always that such event is the sole cause preventing the Contractor from carrying out the Performance Tests in respect of such Phase.

"**Defects Correction Period**" means:

(a) In respect of each Phase, the period commencing on the earlier of:
  (i) the First Gas Export Date, the First Tank Export Date or the Remaining Works Completion Date (as the case may be); or
  (ii) if, after Facilities Completion of the relevant Phase, the Contractor has been delayed, for a continuous period of more than <<ninety (90) Days>>, in commencing the Performance Tests for that Phase solely by reason of one or more Company Delay Events, <<ninety-one (91)>> Days after Facilities Completion of the First Gas Export Works, the First Tank Export Works or the Remaining Works Completion Date (as the case may be); and ending on the date falling twenty-four (24) Months after the relevant date, each as extended in accordance with Clause [_____] (Correction of Defects); and
(b) in respect of any Discrete Asset, the period commencing on the date specified in the relevant Taking Over Certificate issue pursuant to Clause [_____] (Taking Over) and ending on the date falling twenty-four (24) Months after that date, as extended in accordance with Clause [_____] (Correction of Defects).

"**Demonstrated Capacity**" means the maximum aggregate quantity of Regasified LNG, expressed in MMBtus, that Facility is capable of delivering at the Delivery Point on a sustainable basis over any consecutive twenty-four (24) hour period, as determined from time to time in accordance with Clause [__] using a reference standard Gas having a Gross Heating Value of 1050 Btu per scf and operating with no more than N minus one (N − 1) vaporizing units.

"**Firm Send-Out Capacity**" means 1,103,000 MMBtus in any continuous twenty-four (24) hour period with N minus one (N − 1) vaporizing units in operation, using a reference standard Gas Btu content of 1050 Btu per scf.

"**First Gas Export**" means the achievement of completion of the First Gas Export Works.

"**First Gas Export Date**" means the date on which completion of the First Gas Export Works is achieved.

"**First Gas Export LNG Date**" has the meaning given to that term in Clause [__] (Commissioning LNG).

"**First Gas Export Delay Liquidated Damages**" means the liquidated damages for delay in achieving Completion of the First Gas Export Works by the Guaranteed First Gas Export Date, specified in Paragraph 1(A) (First Gas Export Delay Liquidated Damages) of Exhibit 9.2.

"**First Gas Export Works**" means those parts of the Work described as First Gas Export Works in the Functional Specification.

"**First Tank Export**" means the achievement of completion of the First Tank Export Works.

"**First Tank Export Date**" means the date on which the First Tank Export Works are complete.

"**First Tank Export LNG Date**" shall have the meaning set out in Clause [__] (Commissioning LNG)

"**First Tank Export Delay Liquidated Damages**" means the liquidated damages for delay in achieving First Tank Export by the Guaranteed First Tank Export Date, as specified in Exhibit 9.2.

"**First Tank Export Works**" means those parts of the Works described as First Tank Export Works in the Functional Specification.

"**Full Facilities Operation**" means when all of the following have occurred:

(a) the Company's Representative has issued the facilities Completion Certificate in respect of:
  (i) the First Gas Export Works;
  (ii) the First Tank Export Works; and
  (iii) the Works;
(b) the Sponsor has issued to the Contractor Performance Certificates in accordance with Clause [__] (Performance Certificate) for the First Gas Export Works, the First Tank Export Works and the Work;
(c) the Sponsor has issued to the Contractor a Reliability Certificate for the Work;
(d) the Contractor has provided the copies of:
  (i) the O&M Manual specified in Clause [__] (O&M Manual); and
  (ii) all designs, drawings;
(e) The Contractor has provided the Operational Spare Parts requested by the Company in accordance with Article 4.29.1.

"**Gross Heating Value**" means the quantity of heat, expressed in Btus, produced by the complete combustion in air of one (1) cubic foot of anhydrous Fuel, at a temperature of 60.0 degrees Fahrenheit and at a pressure of 14.696 psia, with the air at the same temperature and pressure as the Fuel, after cooling the products of the combustion to the initial temperature of the Fuel and air, and after condensation of the water formed by combustion.

"**Industry Standards**" shall mean those prudent standards, practices, methods and procedures of design, engineering, procurement, construction, manufacture, fabrication, assembly, erection, installation, workmanship, inspection, monitoring, testing, commissioning, startup and operation (and all other activities of the type included in the Works) (and all property, assets, resources, services, equipment, systems and components necessary and advisable therefor), and that degree of skill, diligence, prudence and foresight, in each case that is expected from a skilled, experienced, professional and prudent international contractor engaged in the same type of undertaking as the Facility and the Works under the same or similar circumstances; and in any event, shall include:

(i) such prudent standards, practices, methods and procedures that are not lower than those expected from a skilled, experienced, professional and prudent contractor holding all such characteristics in Canada and in the U.S.A.; and
(ii) any applicable international standards, practices, methods and procedures established by the International Maritime Organization (IMO), the Oil Companies International Marine Forum (OCIMF) and the Society of International Gas Tankers and Terminal Operators Limited (SIGTTO), and by any other internationally recognized Authority or Person with the standards and practices of

which it is customary for reasonable and prudent owners and/or operators of facilities similar to the Facility to comply;

"**International Standards**" means the international standards and practices applicable to the ownership, classification, design, equipment, operation or maintenance of LNG Tankers (in the case of Customer) or LNG unloading equipment (in the case of [_____] LNG) established by (a) IMO, OCIMF or SIGTTO, or (b) any other internationally recognized agency or organization with whose standards and practices it is customary for reasonable and prudent owners or operators of vessels similar to LNG Tankers, or equipment similar to Facility unloading equipment and mooring facilities (in the case of [_____] LNG), to comply.

"**LNG**" means Fuel in a liquid state at or below its boiling point at a pressure of approximately one (1) atmosphere.

"**LNG Tanker**" means an ocean-going vessel suitable for transporting LNG.

"**LNG Transporter**" means any Person that owns or operates an LNG Tanker.

"**LTBP**" means the London Tanker Brokers Panel or, if the London Tanker Brokers Panel ceases to exist, another mutually acceptable internationally recognized independent panel, body or other entity comprised of shipbrokers that provides demurrage awards and settlements to the international shipping community.

"**Major Subcontract**" shall mean any Subcontract relating to supply of Equipment that comprises submerged combustion vaporizers, LNG pumps, BOG compressors, unloading arms or the 9 percent nickel plates for the LNG tanks.

"**Maximum Regasified LNG Delivery Rate**" means the maximum quantity of Regasified LNG, expressed in MMBtus, that Facility is capable of delivering at the Delivery Point in any continuous twenty-four (24) hour period with N vaporizing units in operation, as determined from time to time in accordance with Clause [___] using a reference standard Fuel Btu content of 1050 Btu per scf.

"**Metering Data**" has the meaning set forth in Article 8.2.

"**Minimum Blending Inventory**" means the amount of LNG that when blended with a cargo of LNG (net of heel) from an LNG Tanker would result in an amount of commingled LNG with an average molar compositional analysis meeting the Pipeline Specification, provided that such average compositional analysis shall be determined with eighty percent (80 percent) of the quantity of the LNG (either the intended discharge quantity of Hot LNG or the Blending Inventory Account) with the lower Btu analysis.

"**Minimum Inventory**" means the minimum quantity of LNG, expressed in cubic meters, determined by the Sponsor in accordance with Prudent Operating Practices and with reference to design operating parameters provided by the Contractor and notified to Customer that is required to maintain [_____] LNG Terminal in a sufficiently cold state so as to be able at all times to provide the Services.

"**Minimum Regasified LNG Delivery Rate**" means the minimum quantity of Regasified LNG, expressed in MMBtus, that [_____] LNG Terminal is capable of delivering at the Delivery Point, in accordance with the design parameters specified in the relevant Contract, below which the operation of Facility would not be in accordance with Prudent Operating Practices.

"**MMscf**" means one million (1,000,000) scf.

"**MMscfd**" means one million (1,000,000) scf per Day.

"**NGL**" means liquid hydrocarbons capable of being extracted from the Facility, including, separately or collectively, ethane, propane, butane and natural gasoline.

"**Notice of Readiness**" means a notice stating that an LNG Tanker has received all necessary customs, port and security clearances and Permits necessary to berth and is in all other respects ready to be berthed and start the LNG discharge process.

"**OCIMF**" means the Oil Companies International Marine Forum.

"**Pilot**" means any Person who is certified by the Atlantic Pilotage Authority and is engaged by an LNG Transporter to come onboard an LNG Tanker to assist the master in pilotage, berthing and unberthing of such LNG Tanker at the Facility.

"**Pipeline**" means the downstream Gas pipeline system and related equipment that will be constructed and operated after the date hereof to transport Regasified LNG from the Pipeline Delivery Point to <<*insert name*>> existing pipeline system, including any modification or expansion of such pipeline system.

"**Pipeline Delivery Point**" means the point of interconnection between the Facility and the <<*insert name of pipeline*>>.

"**Pipeline Specification**" means the compositional specification for Gas capable of being accepted for transportation on the Pipeline, as identified in the Pipeline Tariff, provided that until there is a Gas Transportation Tariff of the Pipeline Owner filed with and approved by <<*insert Governmental Authority*>> and applicable to the Pipeline that identifies a compositional specification that is applicable to the Pipeline.

"**Port Charges**" means all charges of whatsoever nature (including rates, tolls, and dues of every description) in respect of an LNG Tanker entering or leaving the Port and approaching and leaving the Facility, including charges imposed by fire boats, tugs and escort or other support vessels, the coast guard, linesmen, a Pilot, and any other Person assisting an LNG Tanker to enter or leave the Port and approaching and leaving the Facility.

"**Point of Receipt**" shall mean the point at the Facility at which the flange coupling of the Facility's LNG unloading line joins the flange coupling of the LNG unloading manifold on board an LNG vessel.

"**Point of Redelivery**" shall mean (i) in the case of redeliveries of Gas (i.e., deliveries of Gas vaporized from LNG previously delivered to the Facility) to [_____], the point of interconnection between [_____]'s Complex and the Facility; (ii) in the case of redeliveries of Gas to the power facility for use as a fueled by [_____] to operate the power facility, the inlet of the gas turbines at the power facility, and (iii) in the case of redeliveries of Gas for the downstream market, or when neither of the delivery points in clauses (i) and (ii) is specified, the point of interconnection with the Downstream Pipeline.

"**Primary Vaporizers**" shall mean the vaporizers owned, installed and operated by [_____], which comprise part of the LNG facilities.

"**Prudent Operating Practices**" means any of the then current practices, methods and acts engaged in or approved by a significant portion of the LNG industry (including with regard to LNG shipping, LNG regasification and Gas transportation) for the operation of facilities similar to the Facility or of LNG Tankers, as the case may be, that, in the exercise of reasonable judgment in light of the facts known (or as would have reasonably expected to should have been known) at the time the decision was made, would have been expected to accomplish the desired results consistent with good business practices, taking into account such factors as reliability, safety, consideration of the environment, lawfulness, cost, efficiency, prudent operation, maintenance and use of the Facility or the LNG Tankers, as the case may be. Prudent Operating Practices is not intended to be limited to the optimum practice, method or act to the exclusion of all other currently applicable standards but, rather, to be a spectrum of standard industry operating practices, methods or acts.

"**Ready for Cool Down Date**" shall have the meaning stated in subsection [____].

"**Ready for Cool Down Date Tests**" shall have the meaning stated in subsection [_____].

"**Ready for Cool Down Date Certificate**" shall have the meaning stated in subsection [____].

"**Receipt Point**" means the point at the Facility at which the flange of the Facility's loading arms joins the flange of the Facilities unloading manifold on board an LNG Tanker.

"**Regas Facilities**" means a regasification terminal and associated facilities, comprised by the entirety of the Work, as further described in the Functional Specification.

"**Reliability Certificate**" means a certificate to be issued by the Company pursuant to Clause [_] (Reliability Test) in the form set out in Part E (Form of Reliability Certificate) of Annex 5 (Testing and Completion Certificates).

"**Reliability Tests**" means the reliability tests to be carried out in accordance with the procedures and to the standards referred to in the Functional Specification.

"**Remaining Works Completion**" means the completion of all Work, including the First Gas Export Works and the First Tank Export Works.

"**Remaining Works Completion Date**" means the date on which Remaining Works Completion is achieved.

"**Remaining Works Liquidated Damages**" means liquidated damages for delay in achieving Completion of the remainder of the Works by the Guaranteed Remaining Works Completion Date, as specified in Paragraph [_] Remaining Works Liquidated Damages of Exhibit 9.2.

"**SIGTTO**" means the Society of International Gas Tankers and Terminal Operators Limited.

"**Standard Cubic Foot**" or "**scf**" means the volume of Gas that occupies one actual cubic foot at a temperature of 60 degrees Fahrenheit and a pressure of 14.696 psia.

"**System Quantities**" shall mean the amount of Gas or LNG used for or lost during maintenance activities, normal operation or as a result of a Force Majeure Excused Event or otherwise unaccounted for, the ownership of which cannot be reasonably attributed or identified, determined in accordance with Section [_].

"**Tests**" means collectively:

(a) the Tests During Construction;
(b) the Pre-commissioning Tests;
(c) the Commissioning Tests;
(d) the Tests on Completion;
(e) the Performance Tests; and
(f) the Reliability Test,

as described in the Scope of Work or reasonably to be inferred therefrom or as agreed between the parties in accordance with this Agreement.

1.1 Cargo Handling Manual

The Parties shall work together to develop and agree on a manual (the "Cargo Handling Manual") setting forth the specific procedures for the receipt, storage and in-tank blending of LNG delivered to the Facility. On or before the date that is <<*insert number*>> Days before the expected in-service date of the Facility, the Contractor shall provide to the Sponsor a draft Cargo Handling Manual. Within <<*insert number*>> Days of receiving the draft Cargo Handling Manual, Customer shall provide to [_____] LNG written comments thereupon, failing which the Sponsor shall be deemed to have approved the draft Cargo Handling Manual such that it shall be the final Cargo Handling Manual. If the Sponsor provides written comments on the draft Cargo Handling Manual within the time period set forth above, the Parties shall meet to discuss those comments within <<*insert number*>> Days.

1.2 Onshore Operations Manual

The Parties shall work together to develop and agree on a manual (the "Onshore Operations Manual") setting forth the specific procedures for operating the Facility (other than marine and cargo handling operations). On or before the date that is <<*insert number*>> Days before the Expected In-Service Date, the Contractor shall provide to the Sponsor a draft Onshore Operations Manual. Within <<*insert number*>> Days of receiving the draft Onshore Operations Manual, the Sponsor shall provide to [_____] LNG written comments thereupon, failing which the Sponsor shall be deemed to have approved the draft Onshore Operations Manual such that it shall be the final Onshore Operations Manual. If the Sponsor provides written comments on the draft Onshore Operations Manual within the time period set forth above, the Parties shall meet to discuss those comments within <<*insert number*>> Days.

1.3 Marine Operations Manual

The Parties shall work together to develop and agree on a manual (the "Marine Operations Manual") that is in accordance with International Standards, governs marine activities at the Facility, is applicable to all LNG Tankers delivering LNG to the Facility and includes rules and procedures for the vetting of LNG Tankers (including a Facility questionnaire). On or before the date that is <<*insert number*>> Days before the Expected In-Service Date, the Contractor shall provide to Sponsor a draft Marine Operations Manual. Within <<*insert number*>> Days of receiving the draft Marine Operations Manual, the Sponsor shall provide to the Contractor written comments thereupon, failing which the Sponsor shall be deemed to have approved the draft Marine Operations Manual such that it shall be the final Marine Operations Manual. If the Sponsor provides written comments on the draft Marine Operations Manual within the time period set forth above, the Parties shall meet to discuss those comments within <<*insert number*>> Days.

1.4 Commissioning

During the period from the Start Date until the Facility Provisional Acceptance Date there shall be a period (the "Initial Supply Period") during which Customer shall deliver a cargo of LNG (the "Commissioning Cargo") to the Facility and the Contractor shall take delivery of such cargo to enable the testing and commissioning of the Facility. At the Sponsor's sole cost and expense, all Sponsor Persons shall be entitled to be present to observe all commissioning and testing activities.

1.5 Commissioning Cargo Delay Damages

If the Contractor confirms that it requires the Commissioning Cargo, but the Sponsor fails to tender delivery of such Commissioning Cargo before the end of the Allowed LNG Unloading Time for the delivery of the Commissioning Cargo, the Sponsor shall pay to the Contractor liquidated damages of <<*insert amount and currency*>> as the sole remedy therefor and Contractor shall not be entitled to an Adjustment pursuant to Clause [__].

1.6 Functional Test of [_____] LNG Terminal

Prior to the delivery of the Commissioning Cargo, the Sponsor may at its sole cost and expense conduct an inspection to confirm that the berth at the Facility and all safety, discharge and unloading systems are ready for the Commissioning Cargo. If the Sponsor notifies the Contractor of any issue or such safety or security requirements, and if the failure might compromise the ability of the Contractor, the Contractor shall promptly remedy such issue as soon as possible, and in such event, the Sponsor shall have the right to confirm that such remedies have been effected. The Sponsor's failure to notify the Contractor of any failure shall not relieve the Contractor of any obligation hereunder.

1.7 LNG Vessel Arrival Notices

Customer shall cause the operator of any vessel that delivers LNG for Customer's account hereunder to provide the following notices to [_____], except as otherwise agreed to in writing by the Parties:

(a) Promptly following the departure from the port of origin;
(b) The location of the vessel at noon each Day during the voyage;
(c) 48 hours prior to the estimated time of arrival ("ETA") at the LNG Facilities, if the notice given under clause (a) was given more than 48 hours prior to the ETA;
(d) 24 hours prior to the ETA; Five hours prior to the ET; and
(e) Notice of Readiness.

   In addition, the Parties shall agree on the timing and content of any other notices that are necessary for the safe and efficient delivery of LNG.

1.8 LNG Quality

The LNG received by [_____] at the Point of Receipt, whether for its own account or the account of Customer, shall be of a quality that, if gasified, would meet the requirements established for Gas in this Article [__].

1.9 Required Properties

(a) Moisture, impurities, helium, natural gasoline, butane, propane and any other hydrocarbons except methane may be removed from the Gas delivered to or for the account of Customer prior to such delivery or redelivery, provided that Customer is redelivered an equal number of Btus as Customer delivered into Storage. [_____] may subject, or permit the subjection of, the Gas to compression, cooling, cleaning and other processes.

(b) LNG delivered by or on behalf of Customer to the Point of Receipt for storage and vaporization:

 (i) shall be commercially free from objectionable odors, dust, gum, gum-forming constituents, gasoline, PCB's or other solid or liquid matter that might interfere with its merchantability or cause injury to or interference with proper operation of the lines, regulators, meters or other appliances through which it flows;

 (ii) shall have a Heating Value of not less than 1000 Btu per Standard Cubic Foot, and not more than 1150 Btu per Standard Cubic Foot; the Heating Value shall be measured by methods in accordance with accepted industry practice, such as, but not limited to, recording-calorimeter(s) or gas chromatograph(s) located at appropriate points; Customer shall use reasonable endeavors to advise [_____] of the estimated Heating Value and density of the LNG scheduled for delivery to the Point of Receipt for Customer's account prior to unloading the LNG cargo at the LNG Facilities;

 (iii) shall contain no more than two-tenths grain of hydrogen sulfide per hundred Standard Cubic Feet of Gas volume, as measured by methods in accordance with accepted industry practice;

 (iv) shall not contain more than two grains of total sulfur per hundred Standard Cubic Feet of Gas Volume, as measured by methods in accordance with accepted industry practice;

 (v) shall not contain in excess of one percent of nitrogen, as measured by methods in accordance with accepted industry practice; and

 (vi) shall otherwise comply with all applicable Permits to which the LNG Facilities or the Power Facility is subject.

(c) Gas redelivered by [_____] to or for the account of Customer at Point of Redelivery:

 (i) shall not contain sand, dust, gums, crude oil, impurities or other objectionable substances which may be injurious to pipelines or may interfere with the transmission of the Gas;

 (ii) shall not contain more than three-tenths grams of hydrogen sulfide per hundred Standard Cubic Feet of Gas volume, as measured by methods in accordance with accepted industry practice;

 (iii) shall not contain more than two grains of total sulfur per hundred Standard Cubic Feet of Gas volume, as measured by methods in accordance with accepted industry practice;

 (iv) shall not contain more than 0.25 grams of mercaptan sulfur per hundred Standard Cubic Feet of Gas volume, as measured by methods in accordance with accepted industry practice, or such higher content as, in Customer's judgment, will not result in deliveries by Customer to its customers of gas containing more than three-tenths grams of mercaptan sulfur per hundred Standard Cubic Feet of Gas volume, as measured by methods in accordance with accepted industry practice;

(v)   shall not contain more than 2 percent by volume of carbon dioxide, as measured by methods in accordance with acceptable industry practice;

(vi) shall not have a water vapor content in excess of seven pounds per million Standard Cubic Feet of Gas volume, such vapor content to be measured by methods in accordance with accepted industry practice;

(vii) shall be as free of oxygen as it can be kept through the exercise of all reasonable precautions and shall not in any event contain more than 0.4 percent by volume of oxygen, as measured by methods in accordance with acceptable industry practice;

(viii) shall have a Heating Value of not less than 950 Btu per Standard Cubic Foot and not more than 1165 Btu per Standard Cubic Foot. The Heating Value shall be measured by methods in accordance with acceptable industry practice, such as, but not limited to recording-calorimeter(s) or gas chromatograph(s) located at appropriate points;

(ix) shall be delivered to a Point of Redelivery at a temperature of more than 40°F and less than 100°F, and at the maximum pressure available when operating the Primary Vaporizers or Tolling Vaporizers at a pressure of 650 psig.

1.10 Commissioning LNG

1.10.1   Subject to Clause [__] (Commissioning LNG), the Company will supply and make available to the Contractor, at the Company's cost, for use in the performance of its obligations under this Agreement, the Commissioning LNG, having the characteristics set out in the Scope of Work, on:

(i)   the First Gas Export LNG Date;
(ii) the First Tank LNG Export Date; and
(iii) the Second Tank LNG Export Date.

1.10.2   The "First Gas Export LNG Date" shall be the date nominated by the Contractor in accordance with this Clause [__] as the date on which the Contractor requires Commissioning LNG in order to achieve Completion of the First Gas Export Works. For the purposes of this Clause [__]:

(i)   such date shall be no earlier than [_____] and no later than [_____] (the "First Window Period") and provided that such dates (and the date of any subsequent window period below) shall be extended:

(a)   for any period that an event of Force Majeure affects a party's ability to fulfill its obligations under this Agreement, as such period is determined in accordance with Clause [__] (Force Majeure); and

(b)   by the same amount as the Guaranteed Dates are extended pursuant to Clause [__] (extension of the Guaranteed Dates); and

(ii) such date shall be determined in accordance with the following:

   (a) no later than <<*one-hundred and seventy (170)*>> Days after the date of this Agreement, the parties will agree on a six (6) Month window (the "Second Window Period") falling within the First Window Period during which the First Gas Export LNG Date is expected to occur. In agreeing upon the Second Window Period, the parties will take into consideration the Program to achieve the First Gas Export pursuant to this Agreement. If the parties are unable to reach agreement by the end of the <<*one-hundred and seventy (170)*>> Day period referred to above, the Company shall decide and notify the Contractor as to when the Second Window Period is to occur;

   (b) no later than <<*one-hundred and ninety (190)*>> Days in advance of the first Business Day of the Second Window Period, the Contractor will advise the Company of a <<*forty-five (45)*>> Day window (the "Third Window Period") falling with in the Second Window Period during which the First Gas Export LNG Date is expected to occur and the Contractor's reasonable estimate of the actual date on which the First Gas Export LNG Date will occur. The Contractor shall update such reasonable estimate on a monthly basis prior to making its notification pursuant to Clause [__]

   (c) no later than <<*one-hundred (100)*>> Days in advance of the Third Window Period, the Contractor will advise the Company of a <<*fifteen (15)*>> Day window (the "Fourth Window Period") during which the First Gas Export LNG Date is expected to occur and the Contractor's reasonable estimate of the actual date on which the First Gas export LNG Date will occur. The Contractor shall update such reasonable estimate on a weekly basis prior to making its estimate pursuant to Clause [__]; and

   (d) no later than <<*forty-five (45)*>> Days in advance of the first Business Day of the Fourth Window Period, the Contractor will notify the Company of the First Gas Export LNG Date, which shall fall within the Fourth Window Period. The First Gas Export LNG Date shall be the date notified pursuant to this Clause [__]; provided that, failing notification by the Contractor pursuant to this Clause [__], the First Gas Export LNG Date shall be the last Business Day of the last notified window period, or such later date as may be agreed by the parties.

1.10.3   The "First Tank Export LNG Date" shall be the date nominated by the Contractor in accordance with this Clause [__] as the date it requires Commissioning LNG in order to achieve Completion of the First Tank Export Works. For the purposes of this Clause [__], such date shall be determined as follows:

  (i) the Contractor shall from time to time give the Company notice of the Contractor's best estimate of the expected First Tank Export LNG Date;

  (ii) no later than six (6) Months after Completion of the Early Gas Works, the Contractor will advise the Company of a one (1) Month window during which the First Tank Export LNG Date is expected to occur and the Contractor's reasonable estimate of the actual date on which the First Tank Export LNG Date will occur. The Contractor shall update such reasonable estimate on a monthly basis prior to making its notification pursuant to Clause [__];

(iii) no later than one (1) Month in advance of the first Business Day of the window period referred to in Clause [__], the Contractor will notify the Company of the First Tank Export LNG Date, which must fall within such window period. The First Tank Export LNG Date shall be the date notified pursuant to this Clause [__]; provided that, failing notification by the Contractor pursuant to this Clause [__], the First Tank Export LNG Date shall be the last Business Day of the window period notified pursuant to Clause [__], or such later date as may be agreed by the parties.

1.10.4   The "Second Tank Export LNG Date" shall be the date nominated by the Contractor in accordance with this Clause [__] as the date it requires Commissioning LNG in order to achieve Remaining Works Completion. For the purposes of this Clause [__] such date shall be determined as follows:

 (i) the Contractor shall from time to time give the Company notice of the Contractor's best estimate of the expected Second Tank Export LNG Date;
 (ii) no later than <<*fifteen (15)*>> Business Days after the Completion of the First Tank Export Works, the Contractor will advise the Company of a <<*ten (10)*>> Day window during which the Second Tank Export LNG Date is expected to occur and the Contractor's reasonable estimate of the actual date on which the Second Tank Export LNG Date will occur. The Contractor shall update such reasonable estimate on a monthly basis prior to making its notification pursuant to Clause [__];
 (iii) no later than <<*fifteen (15)*>> Business Days in advance of the first Business Day of the window period referred to in Clause [__]; the Contractor will notify the Company of the Second Tank Export LNG Date, which must fall within such window period. The Second Tank Export LNG Date shall be the date notified pursuant to this Clause [__]; provided that, failing notification by the Contractor pursuant to this Clause [__], the Second Tank Export LNG Date shall be the last Business Day of the window period notified pursuant to Clause [__], or such later date as may be agreed by the Parties.

1.10.5   The Company shall be under no obligation to supply any Commissioning LNG unless the Contractor has demonstrated to the Company's reasonable satisfaction, as the same time as the notice is given pursuant to Clause [__], Clause [__] or Clause [__], that all facilities (including tanker offloading and storage facilities) necessary to receive the supply of Commissioning LNG will be completed in accordance with this Agreement.

1.10.6   In the event that the Company obtains for the Contractor any Commissioning LNG in response to a request from the Contractor in accordance with Clause [__], Clause [__] or Clause [__], but the Contractor is unable to unload such delivered Commissioning LNG as a result of the failure by the Contractor to comply with its obligations under this Agreement, the Commissioning LNG Tanker may depart when required in order to satisfy its next scheduled obligation.

1.10.7   If the LNG Tanker does depart in the circumstances set out in Clause [__], the Company will use its reasonable efforts to reschedule an LNG tanker to meet the

Contractor's revised requirements but shall have no liability to the Contractor for any extension of time or delay and expense claims if the procedures in Clauses [__] to [__] (inclusive) has not been complied with or if all facilities (including tanker offloading and storage facilities) necessary to receive the supply of Commissioning LNG have not been completed in accordance with this Agreement.

1.10.8   The Company shall have no obligation to provide any Commissioning LNG other than in accordance with this Clause [__]

1.10.9   The Company shall own, be entitled to sell and may retain any proceeds of sale from any Gas derived from Commissioning LNG.

2.0 COMMISSIONING AND TESTING LNG

2.1 Quality and Quantity of LNG

The quality of the LNG that Owner shall make available shall satisfy the gas analysis contained in the Scope of Works and Technical Requirements, but is otherwise made available without warranty or representation of any kind. Owner shall provide LNG for commissioning and testing (the "Commissioning LNG") in one full cargo of not less than 120,000 cubic meters.

2.2 LNG Delivery

2.2.1 Contractor shall provide written notice to Owner at least 150 days prior to the date that Contractor then estimates will be the Ready for Cool Down Date.

2.2.2 Contractor shall further provide written notice to Owner of its best estimate of the Ready for Cool Down Date at least ten Business Days prior to the first day of the month preceding the month in which such estimated Ready for Cool Down Date falls.

2.2.3 At the same time Contractor notifies Owner pursuant to subsection 7.2.2, Contractor shall specify and confirm a seven day period beginning on or after the Ready for Cool Down Date estimated in subsection 7.2.2 during which Contractor requests Owner to deliver the Commissioning LNG.

2.2.4 Not later than 24 days prior to the Ready for Cool Down Dale estimated in subsection 7.2.2, Contractor shall provide written notice to Owner of a three day period, within the seven day period referenced in subsection 7.2.3 above, within which Contractor requests Owner to deliver the Commissioning LNG.

2.2.5 Notwithstanding the steps to be followed by Contractor pursuant to subsections 7.2.1, 7.2.2, 7.2.3 and 7.2.4, Contractor shall notify Owner in writing as soon as is practical of any change in the estimated Ready for Cool Down Date, if Contractor reasonably determines that it will not be able to meet the requirements necessary for the issuance of the Ready for Cool Down Date Certificate on the estimated Ready for Cool Down Date pursuant to subsection 13.2.1. Upon receipt of such notice, Owner shall be entitled to require Contractor to repeat the steps in subsections 7.2.1, 7.2.2, 7.2.3 and 7.2.4, provided, however, that Owner may waive any of such steps in its absolute and unfettered discretion.

2.2.6 Any and all revenues generated from the sale of sent-out or other Gas during commissioning, testing or at any other time as a result of the operation of the Facility shall not be for Contractor's benefit.

2.3 Commissioning Cargo Delay Damages

If Owner, on behalf of Contractor and in accordance with subsection 7.2 above, schedules the delivery of the Commissioning LNG and Contractor fails to be ready to unload the delivered LNG within the three day period specified in subsection 7.2.4 such that Owner incurs demurrage charges to a Customer or ship owner, Contractor shall pay to Owner the amount of the actual demurrage charges incurred by Owner (not to exceed<<*insert currency and amount*>>per day). Owner shall use reasonable commercial efforts to minimize demurrage charges if schedule changes occur. In addition, the LNG tanker will be free to depart when required to satisfy its next schedule obligation, in which case Owner will use its reasonable efforts to reschedule an LNG tanker to meet Contractor's revised requirements once established.

2.4 Expected Commissioning Loss

Owner accepts that up to 10,000 m^3 of LNG (the "Expected Commissioning Loss") provided by Owner pursuant to subsection 6.4 may be lost in the commissioning and testing of the Facility.

2.5 Excess Commissioning Loss

In the event more than the Expected Commissioning Loss is lost during commissioning and testing of the Facility (the "Excess Commissioning Loss"), there shall be deducted from the Contract Price an amount equal to the Excess Commissioning Loss (expressed in MMBtus) multiplied by the Henry Hub Price(s) applicable to those quantities. Such amounts in respect of the Excess Commissioning Loss shall be deducted from the next Milestone payment(s).

2.6 No Reimbursement

For the avoidance of doubt, in the event that Contractor loses less LNG than the Expected Commissioning Loss, Contractor shall not be entitled to any reimbursement and shall not be entitled to retain or have any claim to any quantity of Gas produced from such LNG.

2.7 Reliability Run LNG Costs

Subject to subsection 7.5, the cost of any LNG used or lost during or after the Reliability Run or any repetitions of the Reliability Run shall be for the account of Owner.

3.0 READY FOR COOL DOWN DATE REQUIREMENTS, START DATE REQUIREMENTS; PROVISIONAL ACCEPTANCE

3.1 Requirements for Ready for Cool Down Date

3.1.1   From the date on which each and all of the following circumstances and activities have been satisfactorily completed (the "<u>Ready for Cool Down Date</u>"), Contractor shall be entitled to request that a Ready for Cool Down Date Certificate be issued in accordance with subsection 13.5:

(a) Contractor has completed all the Works in respect of the Start Date System in accordance with the Contract, and the Start Date System is free from all Defects except solely for items that Owner determines might not have an adverse effect on the safety, security, operability or integrity of the Start Date System and use of the Start Date System for its intended purpose (the "<u>SD Punch List Items</u>");

(b) Contractor has provided to Owner final and complete versions of all the Contractor Data as specified in the Scope of Works and Technical Requirements and in subsections 24.3, 24.4 and 24.7 to the extent relating to the Start Date System;

(c) Contractor has provided training (in accordance with the Scope of Works and Technical Requirements) to Owner and/or its nominated operations and maintenance Personnel pursuant to Section 26.0 and to Owner's satisfaction;

(d) Contractor has received test certificates signed by Owner's Representative confirming Owner's acceptance of all of the Ready for Cool Down Date Tests, pursuant to Section 12.0; and

(e) all Authorizations necessary to operate and maintain the Start Date System as contemplated under the Contract, and which are the responsibility of Contractor pursuant to subsection 5.4.2, have been obtained and are in full force and effect.

3.2 Requirements for Start Date

3.2.1   From the date on which each and all of the following circumstances and activities have been satisfactorily completed (the "<u>Start Date</u>"), Contractor shall be entitled to request that a Start Date Certificate be issued in accordance with subsection 13.5:

(a) Contractor has received the Ready for Cool Down Date Certificate;

(b) Contractor and Owner have prepared an update of the SD Punch List Items, adding any items that that have become known and deleting any items that have been completed; and

(c) Contractor has provided to Owner, and Owner has accepted, a certificate confirming that in Contractor's reasonable opinion the Start Date System is ready for operation and is capable of sending out regasified LNG at a rate of at least 425 MMscfd on a reliable basis (with all components of the Start Date System operable). Owner shall not begin send-out of commercial quantities of regasified LNG unless it has accepted such certificate or waived Contractor's obligation to pay Liquidated Damages for Delay in respect of the Start Date.

DESIGN PARAMETERS

The design parameters for [_____] LNG Terminal are as follows:

(a) systems for communications and shore/vessel link between [_____] LNG and LNG Tankers;

(b) berthing and discharging facilities that are capable of receiving the LNG Tankers listed on Schedule [__]; and

(c) LNG Tankers can safely reach, fully laden, and safely depart, and at which LNG Tankers can lie safely berthed and unload safely always afloat;

(d) lighting sufficient to permit customary unloading operations (other than berthing or departing berth) by day or by night;

(e) unloading arms, pipes and appurtenant facilities permitting the discharging of LNG at the following rates 14,000 cubic meters per hour;

(f) tie-ins for installation of infrastructure for stripping and storage for re-injection and/or removal from [_____] LNG Terminal of NGLs;

(g) tie-ins for installation of infrastructure for the loading of LNG on trucks;

(h) a vapor return line system of sufficient capacity to transfer to an LNG Tanker quantities of Regasified LNG necessary for the safe unloading of LNG at required rates, pressures and temperatures;

(i) facilities allowing ingress and egress between [_____] LNG Terminal and an LNG Tanker by an independent surveyor for purposes of conducting tests and measurements of LNG on board an LNG Tanker;

(j) at least two (2) full-containment LNG Storage Tanks, each with a total gross capacity of at least one hundred sixty thousand (160,000) cubic meters of LNG;

(k) LNG regasification facilities comprising N vaporizing units with a combined design send-out capacity of approximately 1,103,000 MMBtus utilizing N minus one (N – 1) vaporizing units and at least 1,260,000 MMBtus utilizing N vaporizing units, using a reference standard Gas Btu content of 1,050 Btu per scf;

(l) pipeline facilities interconnecting [_____] LNG Terminal and the Pipeline;

(m) a gangway on the Jetty to provide safe access for personnel to and from an LNG Tanker;

(n) mooring facilities in compliance with the recommendations of OCIMF, SIGTTO and IMO, allowing a safe berthing operation to facilitate the discharge of LNG in accordance with Prudent Operating Practices; and

(o) adequate spacing and design requirements on the Jetty and the piperack thereon to accommodate one (1) additional twenty-four inch (24″) diameter loading arm and one (1) thirty inch (30″) diameter pipeline to shore, for future installation.

LNG TANKER SPECIFICATIONS

Ship Characteristics	Limits	Units
Length Overall (MAX)		Meters
Beam (MAX)		
Draft (MAX)		Meters
Manifold Height Above Water	MIN:	Meters
(MIN/MAX)	MAX:	

Parallel Midbody Forward of Manifold Center (MIN) See Note 1		Meters
Parallel Midbody Aft of Manifold Center (MIN) See Note 1		Meters
Maximum Deadweight		Tonnes
Cargo Capacity	MIN:	Cubic
(MIN/MAX)	MAX:	Meters
Height of Centerpoint of manifold above any obstructions (MIN/MAX) See Note 2	MIN: MAX:	Meters

Note 1:   The design of the facilities includes ____ (__)–breasting dolphins. All LNG tankers shall be designed so as to touch a minimum of ____ (__) of these breasting dolphins while moored securely alongside the berth.

Note 2:   Cargo manifold and drip tray dimensions are to be guided by OCIMF standards as closely as possible. Vessels, which do not meet standard, will need to be reviewed on a case by case basis to ensure compatibility with unloading facilities.

Cargo Systems

Vapor Return Capacity (MAX)		
Vapor Return Manifold Connections (on both Port and Starboard sides)		Centered Between Cargo Line Connections Must comply with ANSI 150
Cargo Manifold Connections (on both Port and Starboard sides)	3 × 12" (2 forward of the Vapor return line and 1 aft of the vapor return line)	Must comply with ANSI 150
Unloading Rate		

Mooring Lines:	_____ lines for vessels up to _____ m² windage	_____ lines. For larger vessels
Mooring lines to be arranged as follows:		
Headlines		
Forward Springs		
Breast Lines (Fore/Aft)		
After Springs		
Stern Lines		

Note 3: The LNG berth will require a minimum of ____ mooring lines, mandatory mooring wires on drums. No mixed mooring lines are acceptable.

Additionally, all vessels shall be equipped with suitable emergency shutdown systems, including communications systems, for use during cargo operations, which are compatible with the Facility.

Notes

1 Part D of this volume contains a checklist of issues from power purchase agreements, which may be helpful in reviewing these provisions.

2 Articles 7.2.1, 7.2.2, and 7.2.3 test three separate load points for the Facility, which is better than just testing the Facility at maximum load because unless the Facility is to be "baseloaded," which means that the grid will keep it on all the time as one of its base plants (as opposed to a peaking facility that just comes on at peak hours when electricity demand increases, such as in the morning when people wake up and turn on coffee makers, etc.), the Facility may very well spend most of its time running at 75 percent load (i.e., there is no point in buying a racecar designed to run at 180 mph if you will only be driving it in the city and at those speeds it gets four blocks per gallon). This is how project companies go bankrupt, by miscalculating fuel costs.

3 These Articles test different load points for each Unit because each Unit must be able to run efficiently while the other Unit is under repair or maintenance, which could be months in cases of major overhauls which are usually done every 4–6 years on gas turbines.

4 Have engineering professionals review.

5 For gas power plants only.

6 For gas power plants only.

7 For gas power plants only.

8 Simple cycle guarantees after stack installation completed could be added.

9 Always have technical people provide this. This example is from a coal-fired heat and power plant.

10 Have technical people check this measurement point.

11 Have technical people check this point.

12 Have engineering professionals check this measurement point.

13 You can include Article 7.2.8 also if it is applicable.

14 This form does not contain a heat rate guarantee for Start-up Fuel, but the Sponsor may want one.

15 This Contract does not include a Buy Down Amount for MW_{th} so one must be added if the Sponsor wants one.

16 This Article makes sure that the Facility meets the permit emissions levels no matter at what load it is running.

17 This is only applicable for fluid bed boilers.

18 The Contractor will usually ask for these periods.

19 This Article eliminates the Contractor's "double jeopardy" for Buy Down Amounts if the Facility performance tests are for more than one load point because if the Contractor cannot make the 100 percent load point guarantee, it will also miss the 75 percent load point guarantee and it should not pay more than once for the same problem; otherwise, the liquidated damages start to look like penalties and might be argued to be unenforceable.

20 This Article prevents "double jeopardy" if more than one Unit and also Facility are tested separately because if one Unit is not meeting its guarantee point the facility will not either unless another Unit is exceeding its own guarantee point. Generally this only applies in coal plants with common handling facilities so a Unit could meet its output when the other Unit is off but the Facility running with both Units cannot meet its guarantee.

21 Always have engineering professionals provide this definition. This example is from a coal-fired heat and power plant. The heat rate is the amount of energy (usually measured in Btu or calories required to produce 1 kWh of electricity).

22 Note that combustion turbine vendors generally quote heat rate in higher heating value, which can be misleading to a Sponsor that is accustomed to dealing with vendors directly.

23 Have engineering professionals check this measurement point.

24 Have engineering professionals review this.

25 Always have technical people check this.

26 Always have technical people check this.

27 For wind turbines only.

28 Used only for wind power projects.

29 Only for transmission-lines.

Power Purchase Agreement Checklist

Risk Assessment Matrix

POWER PURCHASE AGREEMENT

Item/Issue	Result/ Consequence under PPA	Typical result in other PPAs	Equity Concern	Lender Concern	Section Reference
1. Delay in obtaining permits/ approvals leading to delay in guaranteed in-service date		Schedule relief as required if not result of generator's fault			
2. Delay in financial closing leading to delay in guaranteed in-service date		Daily liquidated damages up to capped amount and termination right of power purchaser after 3-6 months			
3. Power Purchaser force majeure leading to delay in guaranteed in-service date (including customs clearances/visas)		Schedule and monetary relief as required			
4. Natural force majeure leading to delay in guaranteed in-service date		Schedule relief as required if in country of site or major equipment supply			

Item/Issue	Result/ Consequence under PPA	Typical result in other PPAs	Equity Concern	Lender Concern	Section Reference
5. Labor difficulties leading to delay in guaranteed in-service date		Sometimes schedule relief if national in nature			
6. Tardy performance of EPC contractor, equipment transporters, fuel supplier, fuel transporter, operator, transmission interconnection company for interconnection or back feed for testing/start-up leading to delay in guaranteed in-service date		Daily liquidated damages up to capped amount and termination right of power purchaser after 3-6 months			
7. Differing site conditions leading to delay in guaranteed in-service date		No monetary or schedule relief unless site chosen by power purchaser			
8. Failure to commence construction by date imposed in power purchase agreement		Daily liquidated damages up to cap then termination right of power purchaser			
9. Successful force majeure claims of EPC contractor, fuel supplier, limestone supplier/ transporter, fuel transporter, interconnection company leading to delay in guaranteed in-service date or interruption of commercial operations		Usually no relief unless force majeure under power purchase agreement provisions as well			

Item/Issue	Result/ Consequence under PPA	Typical result in other PPAs	Equity Concern	Lender Concern	Section Reference
10. Express scope of force majeure:					
(a) beyond party's (reasonable) control		(a) usually included			
(b) party used (commercially) reasonable efforts to avoid		(b) usually included			
(c) party did not provoke		(c) usually included			
(d) events worldwide		(d) usually silent			
(e) change in law/ permits		(e) sometimes			
(f) lack of fuel/ water/limestone		(f) only if due to force majeure			
(g) lack of transportation availability, shipping congestion		(g) usually silent			
(h) failure to perform of contractors, subcontractors, transporters, fuel suppliers		(h) only if due to force majeure			
(i) equipment failure		(i) sometimes for major equipment if has been properly maintained			
(j) war		(j) included			
(k) hostilities, riot, protests		(k) included			
(l) terrorism or threats		(l) included			
(m) civil strike		(m) included			
(n) blockade, embargo, sanctions, boycotts		(n) included			
(o) strikes/labor disputes/ slowdowns		(o) included			
(p) famine/plague/ epidemic		(p) included			
(q) explosion		(q) yes			
(r) fire		(r) yes			

Item/Issue	Result/ Consequence under PPA	Typical result in other PPAs	Equity Concern	Lender Concern	Section Reference
(s) flood, storm		(s) yes			
(t) landslide		(t) yes			
(u) volcano		(u) yes			
(v) earthquake		(v) yes			
(w) tsunami, typhoon, hurricane, tidal wave		(w) yes			
(x) other natural disaster catchall		(x) yes			
(y) any other event beyond control		(y) usually			
11. Exclusions from scope of force majeure:					
(a) labor shortages		(a) usually silent			
(b) financial difficulties/ bankruptcies of suppliers		(b) usually silent			
(c) financial market conditions		(c) usually			
(d) must delay critical parts of construction work		(d) sometimes mentioned			
(e) existence of hazardous materials on site		(e) sometimes mentioned			
12. Lenders stop lending		No relief			
13. Failure to attain guaranteed MW output at in-service date		95% minimum threshold and liquidated damages for shortfall			
14. Failure to maintain guaranteed MW capacity annually after degradation applied. Annual/ seasonal capacity test?		Proportional reduction in capacity payment			

Item/Issue	Result/ Consequence under PPA	Typical result in other PPAs	Equity Concern	Lender Concern	Section Reference
15. Failure to attain guaranteed heat rate at in-service date or thereafter after degradation applied. Annual heat rate test?		105% maximum rarely penalties			
16. Failure to deliver power when dispatched/ false declaration of availability other than in cases of force majeure/ scheduled maintenance		Annual or monthly availability guaranty and liquidated damages or reduction in capacity payment up to cap or reimburse or supply replacement power and with termination right of power purchaser after 3 years of poor performance			
17. Minimum load dispatch requirement, minimum run time of 24 hrs or penalty on power purchaser		Usually at least 70% of nameplate rating for solid fuel			
18. Ramp up/down limits		Always			
19. Maximum starts per day/yr for hot/warm cold starts		Always			
20. Ability to drop off system for frequency interruptions exceeding 0.5 seconds or drop in power factor		Usually			

Item/Issue	Result/ Consequence under PPA	Typical result in other PPAs	Equity Concern	Lender Concern	Section Reference
21. Automatic generation control by dispatcher		Usually not for coal plants			
22. Power purchaser's right to terminate for extended force majeure		Permitted if power purchaser buys plant at pre-determined price which at least covers debt>			
23. Power purchaser pays tariff during force majeure of either party or agreement term extended		Sometimes for pre-determined period			
24. Ability to set annual outage schedule/ ability to shift year of major overhaul/ additional outage hours for major overhaul years		Varies but often maintenance prohibited during peak season, "carryover" hour credits usually not permitted except for major overhauls which must be accelerated or are delayed at power purchaser's request			
25. Changes in law (i.e., tax and environmental) passed on to power purchaser as incurred		Usually passed on to power purchaser			
26. Power purchaser has right to renegotiate tariff for fundamental regulatory changes		Usually not			

Item/Issue	Result/ Consequence under PPA	Typical result in other PPAs	Equity Concern	Lender Concern	Section Reference
27. Power purchaser has right to terminate for material breach, failure to maintain insurance, bankruptcy of generator		Usually			
28. Power purchaser can assign power purchase agreement		Usually not			
29. Power purchaser has security interest in plant and step in right		Often			
30. Local law governs		Usually			
31. International arbitration		Usually			
32. Situs of arbitration outside country of site		Usually			
33. Direct agreement with lenders for cure rights		Usually anticipated by power purchase agreement provisions with 90-day additional cure periods			
34. Limit on number of times performance bond must be replaced		Often			
35. Power purchaser agrees to permit and cooperate in refinancing by sponsor		Sometimes			

Item/Issue	Result/ Consequence under PPA	Typical result in other PPAs	Equity Concern	Lender Concern	Section Reference
36. Overall liability cap and waiver of consequential damages and liquidated damages sole remedy if payable		Usually			
37. Limitation on sales of excess energy		Usually none			
38. Ability of power purchaser to lower firm capacity		Not typical			

Part E

Concession Agreement Checklist

CHECKLIST FOR ISSUES IN CONCESSIONS

Number	Topic	Government Perspective	Concessionaire Perspective	Financier Perspective	Comment
I	Approval of concessionaire's project contracts and financing agreements and subsequent amendments thereto	Desirable for material changes and notice of other changes	Not desirable—can create delays and additional costs	Not desirable	Often only applies to construction, operation and financing agreements
2	Refinancing by concessionaire in future	Desirable to prohibit or share in benefit in some manner	Not desirable—prefers unfettered ability without consent or sharing and government should cooperate to assist in refinancing	Not concerned—will be prepaid	

No.		Col A	Col B	Col C	Notes
3	Government remains owner of assets so lenders cannot foreclose and sell assets	Desirable	Indifferent	Highly undesirable	Often government will agree to continue to repay debt or service debt from tariff revenues (less operations costs) to address lender concerns if concession terminates for any reason
4	Government signs ''direct agreement'' with lenders regarding ability to cure concessionaire defaults and ''step in'' cure ability	Indifferent (though resistance can be expected if Government has not previously signed direct agreements)	Indifferent	Highly desirable and generally a requirement	The forms of the direct agreements (and the rights granted to Lenders vary in different jurisdictions
5	Minimum equity investment amount and cap on debt/equity ratio initially and during concession term	Desirable	Highly undesirable	Indifferent	Not so typical
6	Minimum ownership requirement and change in control restrictions on sponsors	Desirable	Not desirable	Indifferent	Change in control often permitted after certain successful operating period but subject to acceptable transferee test based on experience and resources

Number	Topic	Government Perspective	Concessionaire Perspective	Financier Perspective	Comment
7	Contractor performance bond to concessionaire pledged by concessionaire to government	Desirable	Not desirable	Not desirable	Not typical
8	Concessionaire to disclose financial model to government and provide updates and annual budget	Desirable	Not desirable	Indifferent	Not typical
9	Tariff/revenue subsidies included in national/state budget	Not desirable	Desirable	Highly desirable—sometimes a requirement	Not typical
10	Government can terminate for force majeure	Desirable	Not desirable	Indifferent if debt repaid	
11	Government can terminate for convenience	Desirable	Not desirable	Indifferent if debt repaid	
12	Government can terminate for concessionaire default	Desirable	Not desirable	Not desirable–often require debt to be repaid	
13	Concessionaire can terminate for force majeure	Not desirable	Desirable	Indifferent provided debt will be repaid—financiers may likely to require control over terminating	
14	Concessionaire can terminate for government default	Not desirable	Desirable	Desirable	
15	Insurance proceeds for catastrophic event used to rebuild concession property	Desirable	Desirable	Not desirable–often requirement that in the case of total loss debt must be repaid from proceeds	
16	Liquidated damages for delay in service commencement subject to cap	Not desirable	Desirable	Desirable	

#		Col1	Col2	Col3	Col4
17	Liquidated damages for failure to meet interim milestones on project schedule	Desirable	Not desirable	Not desirable	Often any liquidated damages collected will be returned if service commencement date is subsequently achieved on schedule
18	Service and capacity improvements required throughout term as technology advances	Desirable	Not desirable	Not desirable unless expected cash flow will cover capital costs	
19	Change in law protection for concessionaire	Not desirable	Desirable	Often a requirement	
20	Change in economic circumstances protection for both parties	Not desirable	Desirable	Desirable	
21	Concession exclusive and competing projects prohibited	Not desirable	Desirable	Desirable	
22	Exploitation of ancillary commercial opportunities	Not desirable unless sharing revenue	Desirable	Not desirable	Lenders often will not want concessionaire exposed to other business risks
23	Income tax holiday	Not desirable	Desirable	Indifferent	
24	Import duty/VAT exemption for all contractors/subcontractors	Not desirable	Desirable	Indifferent	
25	Subsurface risk protection for concessionaire	Not desirable	Desirable	Requirement	
26	Pre-existing contamination protection for concessionaire	Not desirable	Requirement	Requirement	

Number	Topic	Government Perspective	Concessionaire Perspective	Financier Perspective	Comment
27	Cost relief for force majeure for concessionaire	Not desirable	Desirable	Desirable	
28	Dispute resolution board formed to mediate disputes	Desirable	Desirable	Desirable	
29	International arbitration	Indifferent (may be resistant)	Requirement	Requirement	
30	Utility relocation cost protection for concessionaire for unknown utilities	Not desirable	Desirable	Desirable	
31	Concession not effective until financial closing	Not desirable	Desirable	Indifferent	

Printed and bound by CPI Group (UK) Ltd, Croydon, CR0 4YY

18/10/2024

01776204-0008